CLIMATE CHANGE
SCIENCE, STRATEGIES, & SOLUTIONS
Pew Center on Global Climate Change

About the Pew Center

The Pew Center on Global Climate Change is an independent, non-profit, non-partisan organization dedicated to providing credible information, straight answers, and innovative solutions in the effort to address global climate change. Formed by the Pew Charitable Trusts in 1998, the Center publishes reports on domestic and international policy, economics, environmental impacts, and practical solutions relating to climate change. The Center also hosts conferences and workshops to facilitate dialogue among business, government, and non-governmental organizations, and participates in international meetings, including the ongoing negotiations of the United Nations Framework Convention on Climate Change. The Pew Center is led by Eileen Claussen, the former U.S. Assistant Secretary of State for Oceans and International Environmental and Scientific Affairs.

CLIMATE CHANGE
SCIENCE, STRATEGIES, & SOLUTIONS
Pew Center on Global Climate Change

Eileen Claussen, Executive Editor

Vicki Arroyo Cochran, Managing Editor

Debra P. Davis, Editor

BRILL

Leiden • Boston • Köln

This book is printed on acid-free paper.

Library of Congress Cataloging-in-Publication Data

Climate change: science, strategies, & solutions / Eileen Claussen, executive editor,
Vicki Arroyo Cochran, managing editor, Debra P. Davis, editor.
   p. cm
   Includes bibliographical references and index.
   ISBN 9004120246 (alk. paper) – ISBN 9004122761 (pbk.)
   1. Climatic changes. I. Claussen, Eileen. II. Cochran, Vicki Arroyo. III. Davis, Debra P.

QC981.8.C5.C611 2001
363.738'747 – dc21   2001035130

Die Deutsche Bibliothek-CIP-Einheitsaufnahme
Climate change: science, strategies, and solutions / Eileen Claussen
...ed.; Leiden, Boston, Köln: Brill, 2001

ISBN 90 04 12024-6 (hardcover)
ISBN 90 04 12276-1 (paperback)

CONTENTS

+

Contributors

Richard M. Adams, Oregon State University, Corvallis, Oregon

Matthew Brown, National Conference of State Legislators, Denver, Colorado

Anthony J. Brunello, PA Consulting, Arlington, Virginia

William Chandler, Battelle Pacific Northwest National Laboratory, Washington, D.C.

Eileen Claussen, Pew Center on Global Climate Change, Arlington, Virginia

Vicki Arroyo Cochran, Pew Center on Global Climate Change, Arlington, Virginia

Jae Edmonds, Battelle Pacific Northwest Division, Washington, D.C.

Everett M. Ehrlich, ESC Company, Washington, D.C.

Kenneth D. Frederick, Resources for the Future, Washington, D.C.

Peter H. Gleick, Pacific Institute for Studies in Development, Environment, and Security, Oakland, California

Geraldina Grünbaum, Pew Center on Global Climate Change, Arlington, Virginia

John Gummer, Sancroft International Ltd., London, United Kingdom

Timothy G. Higgs, Intel Corp., Chandler, Arizona

Brian H. Hurd, Stratus Consulting Inc., Boulder, Colorado

Michael Isenberg, Arthur D. Little, Inc., Cambridge, Massachusetts

Henry Lee, John F. Kennedy School of Government, Harvard University, Cambridge, Massachusetts

Christopher P. Loreti, Arthur D. Little, Inc., Cambridge, Massachusetts

Chris MacCracken, Battelle Pacific Northwest Division, Washington, D.C.

Michelle Manion, Industrial Economics, Inc., Cambridge, Massachusetts

Naoki Matsuo, Institute for Global Environmental Strategies, Tokyo, Japan

Robert Moreland, Sancroft International Ltd., London, United Kingdom

James E. Neumann, Industrial Economics, Inc., Cambridge, Massachusetts

Robert Nicholls, Middlesex University, London, United Kingdom

Kimberly O'Neill Packard, McKinsey & Company, Boston, Massachusetts

Naomi Peña, Pew Center on Global Climate Change, Arlington, Virginia

John Reilly, Massachusetts Institute of Technology, Cambridge, Massachusetts

Forest Reinhardt, Harvard Business School, Cambridge, Massachusetts

Joseph M. Roop, Battelle Pacific Northwest Division, Richland, Washington

Manik Roy, Pew Center on Global Climate Change, Arlington, Virginia

Maria Sanders, International Council for Local Environmental Initiatives, Berkeley, California

Michael Scott, Battelle Pacific Northwest Division, Richland, Washington

Jeanne Shaheen, Governor, State of New Hampshire

Joel Smith, Stratus Consulting Inc., Boulder, Colorado

Robert N. Stavins, John F. Kennedy School of Government, Harvard University, Cambridge, Massachusetts

William Wescott, Arthur D. Little, Inc., Cambridge, Massachusetts

John P. Weyant, Stanford University, Stanford, California

Tom M.L. Wigley, National Center for Atmospheric Research, Boulder, Colorado

Kenichiro Yamaguchi, Mitsubishi Research Institute, Tokyo, Japan

Gary Yohe, Wesleyan University, Middletown, Connecticut

Abby Young, International Council for Local Environmental Initiatives, Berkeley, California

Foreword

What are we to make of the current debate on climate change? The vast majority of the world's best scientists now accept as fact that human influences, primarily from energy use and land-use practices, are affecting the climate. But the speed as well as the consequences of climate change over the next century remain somewhat uncertain, although they are likely to be very serious. It is therefore in our interest to obtain and synthesize the best possible information about the magnitude and rate of future climate change, the impacts of this change on human and environmental systems, and the policies that should be used to mitigate climate change and its impacts. Without good information to guide our actions and decisions in the months and years ahead, we run the risk of never rising above polarized debate and policy paralysis, which would be a terrible legacy to leave to future generations.

This volume provides straightforward information and analysis on a broad range of climate change issues — from the science to the economics and politics of mitigation, both nationally and internationally. Our goal is to inform the debate, to raise issues that may require uncomfortable solutions, to describe and celebrate successes in reducing greenhouse gas emissions, and to honestly describe what we know and do not know about future climate change. That achieving progress on this issue is a challenge is perhaps a gross understatement. The climate change debate has pitted country against country, scientist against scientist, and business leader against business leader. What we hope to do with this volume is to steer current discussions and actions toward the positive and practical, and to assist those with the power to define our future in making decisions that will sustain both our environment and our economy.

If there is anything to conclude from the essays in this volume, it is that this is a highly complex problem with no "off-the-shelf," short-term solutions. Despite uncertainties in the science, there is little serious questioning of a number of basic points: that the climate is changing, that human activity is contributing to these changes, that the effects of the changes are likely to be profound, and that we will have to adapt to these changes even as we move forward to reduce greenhouse gas emissions. Joel Smith's introduction to the chapter on Science and Impacts makes these points clearly, followed by essays discussing the basic science of climate change and its potential impacts on sea level, agriculture, and water resources.

Moving beyond the science — and in light of the profound environmental implications of a changing climate — one must ask whether and how governments are moving forward to address the problem. The picture is neither easy to articulate nor inherently positive or negative. From the chapter on Global Strategies, we learn that many strategies and opportunities exist for limiting and reducing emissions of greenhouse gases, both within developed and developing countries. But this observation must be tempered with the realization that some governments have been slow to seize these opportunities, and others have moved forward only in very tentative ways. More specifically, it is clear that the governments of Europe and Japan have initiated many programs to reduce their emissions, but few have been fully implemented, and even fewer are likely to result in the kinds of emissions reductions called for under the Kyoto Protocol. We can also note with some frustration that political obstacles in the United States have thus far prevented the adoption of comprehensive strategies for addressing global climate change.

One of the most obvious stumbling blocks in the United States has been the assertion that the costs of mitigating climate change will be too prohibitive. In his introduction to the chapter on Economics, Robert Stavins illustrates the bounds of what we can determine about these costs by explaining why and how economic analyses of climate change policy are conducted. Other reports in this chapter discuss: how large-scale economic models work and what inputs most influence their results, how technological change — especially changes driven by the Internet economy — can influence climate change policy, how the trading of emissions rights can lower the costs of mitigation, and how business leaders can steer their companies on a path that is healthy for business as well as the global climate.

Despite the roadblocks to national and international action, and despite the uncertainties in our economic analyses, it is refreshing to see action being taken by states, cities, and businesses. Governor Jean Shaheen reviews the progress we have seen to date. And while there is no clear model for how each entity should address climate change, the variety of approaches that we illustrate and the seriousness of purpose that they exhibit should be cause for some celebration. We hope that the descriptions we provide will both enliven the debate and stimulate other action as more states, cities, and businesses set out to develop and implement programs for reducing greenhouse gas emissions.

No book of this type could stand without a clear look at the status of international climate negotiations and actions the U.S. Congress has taken on climate change. These current developments are summarized in two reports: one authored by me, and another co-authored by Vicki Arroyo Cochran and Manik Roy. It is our hope that these discussions, together with the reports from our other authors, provide the grounding necessary to bring new urgency to this issue and to aid governments, businesses, communities, and others in developing a strategic plan of action for meeting the challenge of climate change.

The chapter of Facts & Figures provides key data, figures, and charts that should be useful to anyone engaged in the work or study of climate science, economics, and policy. Naomi Peña and Geraldina Grünbaum compiled this information and reference their primary sources (including websites) in order to provide our readers with a valuable and lasting guide to climate change.

Any project of this magnitude requires the assistance of numerous individuals, and this volume is no exception. Foremost among the institutions and people who made this book possible is the Pew Charitable Trusts in Philadelphia. The Pew Center's core funding comes from the Trusts, and we are, of course, grateful for the support of Rebecca Rimel and Josh Reichert — both financial and otherwise. Members of our Board of Directors have been generous in providing their time, resources, and advice to the Center on this book project and all of the Center's endeavors.

The Pew Center wishes to thank all of the contributing authors for sharing their expertise, and for working with our staff and editors to "tell the story" of what we know about climate change science, strategies, and solutions. Several of the papers appearing in this volume were first published in a longer form as Pew Center reports. Current and former Pew Center staff members involved in oversight and editing of reports (in their original and abbreviated form) include Vicki Arroyo Cochran, Tony Brunello, Sophie Chou, Debra Davis, Elliot Diringer, Judi Greenwald, Geraldina Grünbaum, Gerald Hapka, Christie Jorge, Rebecca Livermore, Lisa McNeilly, and Naomi Peña. The staff and I wish to extend our thanks to the many reviewers acknowledged by the authors in each chapter, and to our outside consultants, including Ev Ehrlich, who advises us on our economics series, Joel Smith and Brian Hurd, who assist us in managing our science and impacts work; and Bill Woodwell, who tirelessly writes and rewrites many of our work products.

The Center also benefits from our relationship with the companies on our Business Environmental Leadership Council. We appreciate the BELC members' insightful comments on our reports and on many of the pieces in this volume.

Managing Editor (and Pew Center Director of Policy Analysis) Vicki Arroyo Cochran has skillfully guided this project through its many stages (and occasional crises), and Debra Davis has brought her meticulous editing and project development experience to her job as the book's editor. Both have worked closely with the authors, designers, and publisher to prepare the final manuscript. Laura Keelan Miles assisted with proofreading, as did Shannon Heyck-Williams, Marrissa Hunt, and Taryn Fransen. Jan Bigelow and Jerry Hapka managed contracts, Sally Ericsson coordinated BELC input, and Katie Mandes assisted with marketing. Jan Mt. Pleasant, Jean Pierce, and Anne Adundo provided administrative support at key times.

Gibby Waitzkin and her staff at Gibson Creative, including John Brown, Kelly Bush, Mike Dustrude, Wendy Faxon, Christina Jones, Dina Lyons, and Dawn Sword worked with our team and with cover illustrator Brad Yeo to design a volume that is as beautiful as it is substantive.

In Brill Academic Publishers, we found an experienced and supportive publisher (thanks to Joe Wisnewski who introduced us). We appreciate the cooperation and guidance we received at the early stages of this project from Janjaap Blom, and during the final publication and marketing phases from Jan Reijer Groesbeek, Ria Appelman, Hans van der Meyden, and Els van Egmond. And, finally, we're thankful to Enid Zafran for preparing the index.

This very talented team has worked with me to produce a volume that we hope you'll find both enlightening and a pleasure to read.

+

+

+

SCIENCE & IMPACTS

Contents

Understanding the Science & Impacts of Changes in Global & Regional Climate

Joel Smith

Climate change will alter natural systems in the United States and will affect many parts of the U.S. economy. The four reports presented in this chapter[1] review what is known about the science of climate change and the potential impacts of climate change on some key environmental resources and economic sectors in the United States. The first report, on the science of climate change, addresses how increased concentrations of greenhouse gases in the atmosphere are likely to change global and regional (U.S.) climate. The other three reports review how changes in climate may affect agriculture, water resources, and coastal resources in the United States. The effect of climate change on these sectors has been the subject of a substantial amount of research.[2]

An objective assessment of these effects can help us craft appropriate responses to climate change. Such an assessment requires awareness of a number of cross-cutting factors concerning what is known and not known about how climate change will affect these sectors. Specifically, it is critical to consider the following:

The climate is changing and will continue to change as a result of increased greenhouse gas concentrations in the atmosphere. Although there are many uncertainties about how regional climate will change, the effects of climate change are likely to be profound. The global climate is changing, largely as a result of human activities, as Wigley discusses in the section that follows on climate. Average global temperatures have increased about 0.7°C since the late 1800s. This increase in temperature is largely responsible for average global sea levels rising by 10 to 25 cm (4 to 10 inches) since 1900. Current levels of greenhouse gases in the atmosphere commit us to further increases in temperature and sea level. Moreover, even the most optimistic greenhouse gas control scenarios estimate that atmospheric concentrations will be higher in the future than they are today. Even with an aggressive approach to controlling emissions, some future climate change is inevitable.

It is likely that climate change in the 21st century will be greater than in the 20th century. Based on a recently published set of greenhouse gas emissions scenarios from the Intergovernmental Panel on Climate Change (Nakićenović and Swart, 2000), Wigley estimates that average global temperatures are likely to rise 1.1 to 6.4°C (2 to

The author and the Pew Center are grateful for the input of Rich Adams, Ken Frederick, Peter Gleick, Brian Hurd, Jim Neumann, and Tom Wigley, who reviewed this paper and provided helpful comments.

11.5°F) by 2100. This warming will cause a rise in sea level of 16 to 120 cm (6 to 47 inches) by 2100. Precipitation patterns will change, and the frequency and intensity of extreme climate events such as storms, hurricanes, droughts, and floods could increase. However, there is much uncertainty about the character of regional climate change. For example, while northern states are likely to see increased precipitation (particularly in the winter), it is uncertain how average precipitation will change in most of the country. This uncertainty has led some to state that we cannot draw any conclusions about climate change impacts. Yet, knowing that temperatures and sea levels will rise, we can draw conclusions about effects such as inundation of unprotected low-lying areas and northward shifts in crops and natural vegetation. Even in sectors such as water resources, where uncertainty about change in precipitation patterns introduces substantial uncertainty about impacts, changes are expected in river flow, lake levels, water quality, and the potential for floods and droughts.

Different regions of the country are likely to experience different impacts. The impacts of climate change will not be the same across the country, given differences in local circumstances and regional climate. For example, in their discussion on coastal resources in this chapter, Neumann et al. point out that the low-lying East and Gulf coasts are, on the whole, more vulnerable to sea-level rise than the West Coast. Frederick and Gleick note that water resources across the country will be affected differently, depending on how climate changes, but also depending on the variability of current climate, the contribution of snowpack to water supplies,[3] and how water is managed and used. Likewise, as Adams et al. discuss, national food production may not change substantially, but regional changes could be large. Northern areas may have relative increases in yields of important grain crops, while southern areas may have relative decreases.

Sectors will not be affected in isolation. In many areas of the country, impacts in one sector will affect other sectors. For example, agriculture will be affected by changes in water supplies for irrigation. Where water supplies decrease as a result of a drier climate, irrigated agriculture is likely to have its water supplies reduced. At the same time, demand for water for irrigation is likely to increase. Increased demand and decreased supply could exacerbate already stressed water resources. In some cases, the effects in one sector could offset another. For example, increased runoff could partially repulse higher salinity levels in bays and estuaries caused by sea-level rise. Thus, when linkages across related sectors are accounted for, the changes can be different than when sectors are examined in isolation.

The impacts of climate change could be made worse by other stresses. Population growth, pollution, and other stresses can make the impacts of climate change worse than they would otherwise be. For example, some forms of water pollution reduce dissolved oxygen levels in the water that fish and other aquatic species need for survival. Higher water temperatures, which will result from climate change, will also reduce dissolved oxygen levels.

Other future changes will also significantly affect these sectors' vulnerability to climate change. Climate change will happen over many decades, and during that time societal conditions will change considerably. The population of the United States will most likely grow, as will the economy, and technology is highly likely to improve. These changes could substantially alter how climate change will affect the sectors addressed in this chapter. Economic growth results in more financial resources that can be used to adapt to climate change. In addition, economic growth means that a shrinking share of the economy will be devoted to climate-sensitive activities such as agriculture, reducing the sensitivity of the economy to climate change. Improved technology, such as better forecasts of floods and coastal storms, could reduce vulnerability to climate change, yet population and economic growth could increase exposure by leading to more people and infrastructure in floodplains and low-lying coastal areas. It should also be noted that there is tremendous uncertainty about how population, the economy, and technology will change.

Managed systems tend to be less vulnerable to climate change than unmanaged systems because managed systems have a much greater capacity to adapt. The three sectors discussed in this chapter all have significant human intervention. For example, agriculture is heavily managed: farmers can change crop varieties, planting dates, application of fertilizer, pesticides, and irrigation, and other factors. In contrast, the potential for ecosystems to adapt to climate change is much more limited, and their vulnerability to climate change is much greater because many species cannot migrate rapidly in response to climate change. In addition, the ability of ecosystems to adapt is further limited by human development that has led to habitat fragmentation and reduction in species populations.

Climate change impact studies have tended to assume the "easiest" climate change to adapt to – that is, a gradual and monotonic change in climate, with no change in variability. To date, most climate change impact studies have compared current climate with only a mean change in temperature and precipitation. These scenarios assume that only average conditions change and that the change in climate is gradual. The problem is that climate variability, that is, the frequency and intensity of hot days, cold days, dry periods, wet periods, storms, hurricanes, El Niños, etc., is likely to also change. It could be more difficult to adapt to changes in climate variability than to changes in mean conditions. For example, a dramatic increase in flood intensity or frequency could overwhelm current flood control systems.

The possibility of surprises cannot be ruled out. The climate may change in unexpected ways, and natural and societal systems may respond to climate changes in ways we are not considering. Likewise, climate change could come more rapidly than we anticipate. In addition, physical, biological, and societal systems may have thresholds that we are not aware of or are not accurately factoring into assessments. Thus, the results of impact studies should always be interpreted with a degree of caution.

Given the importance of adaptation, assumptions made about adaptation in impact studies have a major influence on results of these studies. The estimated differences in magnitude of impacts can be quite large when different assumptions about adaptation are used. For example, as Adams et al. note in their discussion on agriculture, assuming efficient adaptation by farmers can result in substantially more optimistic estimates of climate change impacts. It is fair to say that in adapting to climate change humans will be neither brilliant nor stupid (e.g., Smit et al., 1996) — i.e., people will try to change their behavior to adapt to new situations, but they often may not make the optimal adaptations for the changed climate. For example, since climate change could affect climate variability, a farmer who adjusts to a wetter climate may be more vulnerable should the climate suddenly get drier. In addition, the literature on climate change impacts has generally not assessed either rapid changes in climate or increases in variability. These changes would make adaptation more challenging.

Taking steps that anticipate the effects of climate change can often be justified because they also are beneficial under current climate conditions. The discussion on water resources by Frederick and Gleick notes that while water conservation will reduce vulnerability of water resources to future climate change, it will also have benefits today. This is an example of an adaptation that makes sense, even without considering climate change, and is even more important if climate change is considered. In addition, other measures such as coastal construction setbacks from the high-tide line to anticipate sea-level rise, or enhanced flood control or drought response capability in anticipation of changes in floods and droughts can reduce risk to climate change at minimal cost.

Finally, while the focus of these reports is domestic (United States), developing countries may be at much greater risk from the impacts of climate change than developed countries for three reasons. First, developing country economies rely more on sectors that are much more sensitive to climate and climate variability than the economies of developed countries. For example, African economies typically derive 20 to 30 percent of their income from agriculture, which comprises only about 2 percent of the gross national product of the United States. Moreover, in many developing countries most of the population is engaged in agriculture; only a few percent of U.S.

workers are in the agriculture sector (World Bank, 2000). Second, some impacts of climate change may be worse in low latitudes than in high latitudes. For example, as Adams et al. note, many studies estimate that agricultural production of grain crops will shift from lower latitudes to higher latitudes. Most developing countries happen to be in low latitudes, while most developed countries happen to be in middle and high latitudes. Third, the capacity of developing countries to adapt to climate change is much more limited than the capacity of the United States and other developed countries, since many developing countries currently do not have the financial resources, access to technology, and institutional arrangements that would help them cope with climate change. Many are already burdened by high population and pollution levels and depleted resources such as fresh water.

In summary, there is still much to learn regarding climate change, and resolving uncertainties about regional climate change, future development, and adaptive responses will enable more accurate estimates of climate change impacts to be made. However, in spite of these uncertainties, there is an important body of literature — summarized here — that provides important insights into the sectors and regions that are likely to be affected.

This literature on climate change impacts can be useful to policy-makers and the public in two ways. First, it can be used to determine what magnitude of climate change will cause adverse consequences and what levels of greenhouse gas emissions controls we should strive to attain. Second, since as noted above some future climate change is inevitable, these studies identify the types of climate change that we will eventually have to adapt to. Such information can be useful for long-term planning in sectors affected by climate change.

Endnotes

1. The reports presented here are condensed versions of four reports previously published by the Pew Center.

2. Other sectors also likely to be significantly affected by climate change, including human health, terrestrial ecosystems, forestry, and aquatic ecosystems, are not addressed in this chapter. The Pew Center published reports on human health and terrestrial ecosystems in December 2000, and will publish other reports in this series through 2001.

3. Many areas of the country, particularly in the West, are dependent on snowpack for their annual water supplies. Warmer temperatures could result in a smaller snowpack and earlier runoff, which could cause water supply problems.

References

Nakićenović, N., and R. Swart, eds. 2000. *Special Report on Emissions Scenarios.* Cambridge University Press, Cambridge, UK.

Smit, B., D. McNabb, and J. Smithers. 1996. Agricultural Adaptation to Climatic Variation. *Climatic Change* 33: 7–29.

World Bank. 2000. World Bank Group Development Data. See http://www.worldbank.org/data/home.html (from "Country Data" and "Data by Topic" series).

The Science of Climate Change

Tom M.L. Wigley

Since the late 1800s, atmospheric concentrations of greenhouse gases have increased markedly, due almost entirely to human activities. At the same time, the average surface temperature of the earth has warmed by about 0.7°C. There is strong evidence that the two are related.

There are two leading climate change questions: How much additional warming will occur if we increase the atmospheric levels of greenhouse gases, and what will the consequences be for the climate system as a whole? These are particularly pressing questions because what we do to the atmosphere today will continue to affect the climate decades or even centuries into the future, and the efforts we make now to reduce the magnitude of future change will become apparent only very slowly.

This chapter describes what is currently known about factors — both human and natural — that influence the global climate. The chapter looks, in particular, at changes in temperature and precipitation over the past century, and, using the most up-to-date emissions data available, makes projections of global-mean temperature and sea level and regional climate changes likely in the next 100 years.

The chapter draws heavily on the work of the Intergovernmental Panel on Climate Change (IPCC), a group established by the United Nations to assess and disseminate credible scientific, technical, and socio-economic information relevant for understanding the risks of human-induced climate change.[1]

Some key findings, updated from Wigley (1999), are:

- Global-mean temperature is estimated to rise between 1.1°C and 6.4°C from 1990 to 2100 — two to nine times the rate of warming that has occurred over the past century. Estimates for sea-level rise over the same period range from 16 to 120 cm.

- For the United States, the rate of future warming is expected to be noticeably faster than the global-mean rate.

- Increased precipitation is likely over north-central and northeast regions of the United States in winter.

- High-temperature and high-precipitation extremes are likely to occur more frequently.

This paper is a condensed and updated version of a report published in June 1999 by the Pew Center: The Science of Climate Change: Global and U.S. Perspectives. The author and the Pew Center are indebted to many scientists for their constructive comments on early drafts of this paper, including E. Barron, B. Felzer, C. Hakkarinen, A. Henderson-Sellers, M. Hulme, M. MacCracken, M. McFarland, J. Mahlman, G. Meehl, N. Nakićenović, S.C.B. Raper, B.D. Santer, M.E. Schlesinger, K.P. Shine, and S.J. Smith.

Observed Changes: Human and Natural Influences

The IPCC's Second Assessment Report (SAR), released in 1996, found that "the balance of evidence suggests a discernible human influence on global climate." This statement broke new ground for the IPCC, which previously had stopped short of affirming a human role in climate change.

The IPCC was convinced by a range of evidence, most importantly that temperature changes observed over the past century are consistent, both in magnitude and patterns of change, with results produced by computer-driven models if those models take into account key human and natural factors believed to affect climate.

The primary cause of human-induced or "anthropogenic" climate change is the enhanced greenhouse effect. The word "enhanced" is important here because there is also a natural greenhouse effect, i.e., a warming of the earth's surface and lower atmosphere (troposphere) by the presence of certain gases. The so-called "greenhouse gases" include carbon dioxide (CO_2), methane (CH_4), and nitrous oxide (N_2O).

By adding more of these gases, and others that do not occur naturally — like halocarbons — to the atmosphere, humans have enhanced the natural greenhouse effect. This warming, in turn, increases evaporation rates and adds more water vapor to the atmosphere, causing further warming.

This greenhouse effect led scientists to predict substantial future global-mean temperature increases. But estimates of future warming must be tempered by consideration of the presence of sulfate aerosols. Sulfur dioxide (SO_2) emissions from the burning of fossil fuels have led to significantly higher concentrations of sulfate aerosols in the atmosphere. These aerosols, or small particles, partly offset greenhouse warming by reflecting solar radiation back into space.

Aerosols are important, not only in estimates of future climate change, but also in understanding past climate. It is only when scientists consider the combined effects of greenhouse gases and sulfate aerosols, along with the effects of changes in solar output, that observed climate changes are consistent with changes predicted by climate models.

Atmospheric Composition

The composition of the atmosphere has changed markedly since pre-industrial times: CO_2 concentration has risen from about 280 parts per million (ppm) to around 370 ppm today, CH_4 has risen from about 700 parts per billion (ppb) to over 1700 ppb, and N_2O has increased from about 270 ppb to over 310 ppb. Halocarbons, largely nonexistent prior to the 1950s, are now present in amounts that have a noticeable greenhouse effect.

Pre-industrial levels of greenhouse gases are known because the composition of ancient air trapped in bubbles in ice cores from Antarctica can be measured directly

(Etheridge et al., 1998; Güllük et al., 1998). These ice cores show that the concentrations of these gases are much higher than in pre-industrial times and far exceed levels of the preceding 10,000 years.

Human activity — fossil-fuel burning, land-use changes, production and use of halocarbons, etc. — is the dominant cause of these changes in atmospheric composition. Human activity is the undeniable source of atmospheric halocarbons (the most climatically important of which are the chlorofluorocarbons, CFC11 and CFC12) because the vast majority of these gases do not occur naturally. Today, many halocarbons are controlled under the Montreal Protocol,[2] and substitute chemicals, which do not cause ozone depletion and so are not controlled, are being introduced. These new gases, like all halocarbons, are strong greenhouse gases (although their net effects on future climate are expected to be small relative to CO_2).

For CO_2, CH_4, and N_2O, the human role is virtually certain too, partly because their changes since pre-industrial times have been so large and at such unprecedented rates, and also because computer simulations provide an unequivocal link between the emissions of these gases in recent decades and observed changes in atmospheric composition.

In addition to the gases mentioned above, anthropogenic emissions of the reactive gases, carbon monoxide (CO), nitrogen oxides (NO_x), and volatile organic compounds (VOCs) such as butane and propane have increased concentrations of tropospheric ozone. Tropospheric ozone (O_3) is a powerful greenhouse gas.

Most greenhouse gases also have natural sources. However, in pre-industrial times emissions were balanced by natural removal or "sink" processes. Human activities have disturbed this balance.

Human activities have also increased the aerosol loading of the atmosphere. Emissions of SO_2 from the burning of coal and other fossil fuels, and emissions of other substances from biomass burning, have increased atmospheric concentrations of aerosols, particularly sulfate aerosols. This increase is important because the presence of aerosols has a net cooling effect that may partly offset the warming effect of greenhouse gases.

Radiative Forcing

The changes in atmospheric composition described above have upset the balance between incoming and outgoing radiation. This imbalance is referred to as "radiative forcing." The climate system responds to positive radiative forcing by trying to restore the balance, which it does by warming the lower atmosphere.

The climate system has experienced more than just anthropogenic forcing since pre-industrial times. There is strong — but indirect — evidence that appreciable changes have occurred in the energy output of the sun. Reconstructions of changes in the sun's output prior to 1979, when satellite-based observational records first became available, are qualitatively similar to changes observed since 1979; but they also show

Climate Change: Science, Strategies, & Solutions

that larger magnitude changes have occurred over at least the past four centuries. These reconstructions, however, remain highly uncertain.

Figure 1 compares current estimates for the anthropogenic, solar, and total (anthropogenic plus solar) forcing histories. Until 1890, forcing changes were dominated by solar forcing. From 1890 to 1950, anthropogenic forcing increased by about 0.2 watts per square meter (W/m^2), while solar forcing showed a much larger upward trend (0.5 W/m^2).

Since 1950, the forcing record has been dominated by the anthropogenic component, particularly since 1970. Thus, anthropogenic forcing began to be appreciably larger than natural solar forcing only some 20 to 30 years ago.

Global-Mean Temperature

The simplest and most revealing index of climate change is the global-mean temperature near the earth's surface. Analysis of this record provides us with valuable insight into the causes of past climate change.

The latest temperature record, derived from painstaking analyses of records from meteorological stations on land and measurements taken by ships at sea over the past 140 years,[3] is

Figure 1

Solar & Anthropogenic Forcing

(1765–2000)

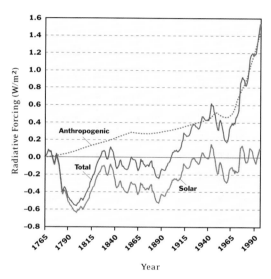

Note: Anthropogenic forcing (greenhouse gases plus aerosols) and solar forcing histories, together with their sum. Since it is change rather than the absolute level that is important, the anthropogenic and solar curves have both been zeroed to the start of the anthropogenic forcing record in 1765. The solar record is from Hoyt and Schatten (1993, updated).

shown in Figure 2. The most striking feature of this record is the overall warming trend, with the most recent years being the warmest. Overall, the warming amounts to about 0.7°C since the late 1800s (with a measurement uncertainty of about ±0.1°C).

Solar forcing and anthropogenic forcing together are enough to explain the overall warming trend (Santer et al., 1996; Wigley et al., 1997), although there could be additional influences from explosive volcanic eruptions (Robock, 2000) and factors internal to the climate system (Schlesinger and Ramankutty, 1996). Figure 3 shows how well observed temperatures correspond to model-predicted temperatures when the

Figure 2

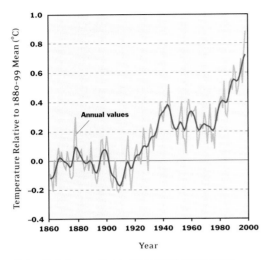

Global-Mean Temperatures

(1880-99 Reference Period)

Note: Global-mean (land plus marine) temperature changes relative to the 1880-1899 mean as a reference period. The last value shown is for 1998. The yellow line gives the annual values, while the green line gives a smoothed representation to show trends more clearly.

models include the effects of greenhouse gases, sulfate aerosols, and solar output.

The temperature record shows a number of other important features. First, there are large variations from year to year. Some of these variations are associated with El Niño, a periodic variation in marine temperatures in the tropical Pacific that has worldwide effects on climate. Others reflect short-term cooling due to volcanic eruptions, and the remainder are probably manifestations of other aspects of the climate system's own internally generated variability. Second, the record also shows large changes on the 10- to 30-year timescale. These probably reflect anthropogenic and solar forcing effects combined with internal variability.

Critics of the anthropogenic global warming hypothesis often point to the apparent discrepancy between the small greenhouse-gas forcing over 1910–1940 and the rapid global warming that occurred during this period. It is true that this warming was too rapid to be accounted for by anthropogenic forcing alone. However, when the possible effects of internally generated variability, solar forcing, and volcanic eruptions are accounted for, there is no serious discrepancy. Overall, the observed warming trend is consistent with what we know about the climate system and external forcing changes.

Scientists are also beginning to unravel the reasons for another apparent discrepancy: between temperature readings at the earth's surface and satellite-based temperature readings in the free atmosphere above the earth's surface (the troposphere). This discrepancy has caused some to discount the surface data and the warming trend.

In the troposphere, different records show different trends. Satellite data, collected since 1979, show no significant trend, while radiosonde (instruments carried aloft on weather balloons) data, which go back to the late 1950s, show a warming trend similar to that at the surface. Simple physics suggests that any warming at the earth's surface should extend throughout the troposphere, primarily because the convective activity associated with clouds keeps this part of the atmosphere well mixed.

The most obvious explanation for the difference is data uncertainties arising from factors such as changes and flaws in instruments, changes and deficiencies in coverage, and nonclimatic influences such as urban heat island effects. Wentz and Schabel (1998) have shown that an important correction associated with satellite orbital decay was originally neglected by satellite data producers. If a correction for orbital decay is applied, the satellite data show a noticeable warming trend and become more consistent with most other data sets (Santer et al., 1999).

More recently, other corrections have been applied to the satellite data, and additional analyses have been

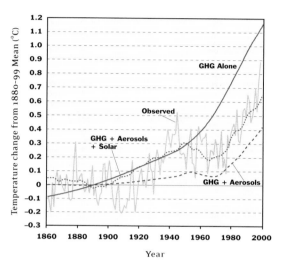

Figure 3

Observed Temperatures Compared with Model Predictions ($\Delta T_{2x} = 2.5^\circ C$)

Note: Comparison of observed and model-predicted temperature changes. The model results use best estimates of greenhouse-gas (GHG) forcing alone, GHG plus aerosol forcing, and GHG plus aerosol plus solar forcing. All three model results have been produced using a climate sensitivity of $\Delta T_{2x} = 2.5^\circ C$. All changes are shown relative to 1880–1899 as a reference period.

carried out to quantify factors that might affect the surface and troposphere (i.e., satellite data) differently. These factors include the effects of volcanoes and El Niño. Stratospheric ozone depletion, which could cause a cooling in the middle to upper troposphere, could also account for some of the difference (NRC, 2000). While the various candidate factors can, together, explain the surface/troposphere difference, the issue is still not fully resolved.

Precipitation

Precipitation is much more variable in both time and space than temperature, and reliable long-term records exist only over the earth's land areas (e.g., Hulme, 1992; Hulme et al., 1998); and, even here, the coverage is incomplete.

Figure 4 shows changes in annual total precipitation averaged over the land areas of the globe (excluding Antarctica) from the Hulme data set. The dominant characteristic of this record is its marked year-to-year variability, which is more pronounced on a regional scale. There is, however, no firm evidence of an overall trend in global-

Figure 4

Global-Mean Precipitation Changes

Over Land

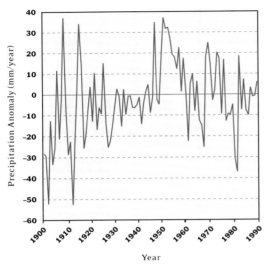

Year

Note: Annual-total precipitation changes averaged over the land areas of the earth between 55°S and 85°N. These data are from the Hulme (1992) gridded data set. An earlier version of these data was used in the IPCC SAR (Figure 3.11). The values represent anomalies from the 1961-1990 mean. Since the mean over this reference period was 989 mm, division of the mm/yr value by 10 gives the anomaly in percentage terms.

mean precipitation. The IPCC's Second Assessment Report noted a small (1 percent) increase per 100 years (Nicholls et al., 1996), but Figure 4 shows that this apparent trend arises solely because of the number of low-precipitation years prior to 1915, and records for that time frame are far less reliable.

Researchers have found some pronounced positive precipitation trends in specific regions (Groisman and Legates, 1995; Nicholls et al., 1996). In North America, for example, the frequency of extreme daily rainfall events (specifically, days with rainfall exceeding two inches) has increased in recent times, according to T.R. Karl and colleagues (Karl et al., 1995; Karl and Knight, 1998). They found that the changes are more than one would expect to have occurred by chance. Further, they note that similar changes might occur because of greenhouse-gas-induced global warming. These results are suggestive, but they do not prove a cause-effect relationship.

Predicting Future Global-Mean Climate

This section gives projections of global-mean temperature and sea-level changes over the next century. Projections are based on new emissions scenarios recently released by the IPCC (Nakićenović and Swart, 2000), referred to as the SRES (Special Report on Emissions Scenarios) scenarios. These supercede the six scenarios devised by the IPCC in 1992 (the IS92 scenarios). The SRES scenarios are the most up-to-date and comprehensive emissions scenarios available. They are grouped under four "storylines,"[4] for each of which there is an illustrative "marker" scenario (A1B, A2, B1 and B2). Two additional scenarios are given as illustrative cases for the A1 storyline,

A1FI (Fossil Intensive) and A1T (a high technology variant). Carbon dioxide emissions under A1FI are higher than under any of the marker scenarios.

A scenario — a plausible picture of the future that, in this case, defines future emissions for a range of gases — is the starting point for predicting future changes in climate, using computer-driven climate models (see Box 1). Such scenarios are, themselves, derived from scenarios for socioeconomic factors such as changes in population, economic growth, energy technology, etc. Emissions scenarios may incorporate the effects of policies to limit environmental change. The SRES scenarios considered here, however, are all no-climate-policy scenarios, i.e., they consider what might happen in the absence of new policies to limit climate change.

The SRES and earlier IS92 scenarios differ most markedly in their projections for SO_2 emissions, which are expected to decrease in the future as a result of current policies to combat air pollution and acid rain (Alcamo et al., 1995, pp. 281, 282). The IS92 scenarios did not fully consider the potential effects of these policies. Cuts in SO_2 emissions would reduce the atmosphere's sulfate aerosol loading and result in increased radiative forcing and warmer temperatures.

The original report (Wigley, 1999) used preliminary versions of the SRES marker scenarios, since this was all that was available at the time. A number of changes have been made from the preliminary scenarios, although in most cases these changes are only minor. The temperature and sea-level projections given here, however, differ noticeably from those for the preliminary scenarios given previously.

The differences arise because the new projections incorporate many new aspects of the science in order to be consistent, as far as possible, with the IPCC Third Assessment Report (TAR). The most important of these are: the inclusion of the effects of those changes in tropospheric ozone that are driven by changes in the emissions of the reactive gases CO, NO_x, and VOCs; and the incorporation of climate feedbacks on the carbon cycle in the model used to estimate future CO_2 concentrations. Tropospheric ozone generally gives a positive forcing contribution (except in the B1 scenario), while the carbon cycle feedbacks lead to larger CO_2 projections, so both of these factors lead to enhanced global warming relative to the earlier projections. In addition, the effects of reactive gas emissions are now included in the methane concentration projection calculations, and organic and black carbon aerosol effects are accounted for more comprehensively than previously. Note, however, that the climate and sea-level models used here are the same as those used in the original report (see Raper et al., 1996, for details) and in the SAR, with one exception: in the present analysis, a limit has been placed on how much the ocean's thermohaline circulation will change as the globe warms. (Further details are available at http://www.cgd.ucar.edu/cas/ACACIA/whatsnew.html.) For the TAR, a number of other changes have been made to these models, so the present results will differ from those in the TAR.

Box 1

Climate Models

Predicting future climate is a daunting task. To do so we must consider not just the atmosphere, but also the oceans, the cryosphere (i.e., land-based and marine snow and ice), the land surface, the stratosphere, and the sun, together with the interactions between these various components of the climate system. Each of these components involves processes that act on a vast range of spatial and temporal scales.

Because of these complexities, the only practical approach to climate prediction is to use mathematical models of the various processes and interactions and to run these models on computers. There is a hierarchy of models that may be used, depending on the degree of detail required in the prediction. For example, if we wish only to estimate how the global-mean temperature might change, we can use a relatively simple model that can be run on a personal computer.

If we wish, however, to estimate how temperature, rainfall, storminess, and other aspects of the weather might change at a particular place, then we need to use a much more sophisticated model (similar to, but even more complex than a weather forecasting model). These models, called general circulation models (GCMs), when used for extended (multi-decade) climate simulations, can be run only on the world's most powerful computers.

Simple models are best used for broad predictions since they cannot simulate all the complexities of climate. They have three advantages:

- They can be run quickly on microcomputers and so can be used to explore the implications of a wide range of future emissions scenarios.

- Users can specify model parameters (such as the value of the climate sensitivity, which could be specified as a low, intermediate, or high value) and thereby determine how the results might be affected by parameter uncertainties. In complex models, critical parameter values (such as the climate sensitivity) are determined internally by the model and are specific to that model.

- They produce more direct information about climate change by excluding changes generated by natural variability.

General circulation models are the best tools available for estimating the details of future climate. The most sophisticated GCMs simulate the key physical processes of both the atmosphere and the ocean in three dimensions — latitude, longitude, and altitude or depth.

These coupled ocean/atmosphere models (O/AGCMs) are necessary if one wishes to determine the spatial details of how climate will change over time in response to changes in atmospheric composition. Their spatial resolution, however, is still relatively coarse, typically a few hundred kilometers in the horizontal plane and a few kilometers in the vertical. The reason for this is that climate simulations extending over time intervals of decades or centuries are so computationally demanding that some level of detail has to be sacrificed.

Table 1

Radiative Forcing Estimates for the SRES Emissions Scenarios

Component	1765–1990	1990–2050						
		IS92a	A1B	A1FI	A1T	A2	B1	B2
CO_2	1.29	1.95	2.22	2.57	1.86	2.22	1.75	1.61
Aerosols	−1.30	−0.70	0.11	−0.15	0.47	−0.41	0.15	0.30
Other	1.07	0.88	0.84	1.43	0.97	1.09	0.40	0.78
Total	1.06	2.12	3.17	3.85	3.30	2.90	2.30	2.69

Component	1765–1990	1990–2100						
		IS92a	A1B	A1FI	A1T	A2	B1	B2
CO_2	1.29	3.69	3.71	5.41	2.51	4.71	2.22	2.93
Aerosols	−1.30	−0.65	0.61	0.59	0.80	0.16	0.82	0.46
Other	1.07	1.47	0.76	2.21	0.75	2.25	0.14	1.33
Total	1.06	4.51	5.08	8.21	4.06	7.12	3.18	4.72

Note: Radiative forcing breakdown for the SRES scenarios, compared with historical forcing over 1765–1990 and the central IS92 scenario (IS92a). Forcing values for 1765–1990 and for the SRES scenarios were calculated using relationships employed in the IPCC Third Assessment Report. "Aerosols" gives the total forcing for all aerosols. "Other" combines the influences of CH_4, N_2O, halo-carbons, tropospheric and stratospheric ozone, and stratospheric water vapor from CH_4 oxidation. Note that CO_2 forcing (and hence total forcing) depends slightly on the projected temperature changes because of climate feedbacks on the carbon cycle. The values given here correspond to projections with a climate sensitivity of 2.5°C equilibrium warming for a CO_2 doubling.

Table 1 compares radiative forcing values for the six SRES scenarios with values for the IS92a scenario, which lies in the middle of the range of IS92 possibilities. The table shows a dramatic difference in the forcing changes due to aerosols (primarily sulfate aerosols) over 1990 to 2100; the SRES scenarios show positive radiative forcing (0.16–0.82 W/m^2), compared with a strongly negative value (−0.65 W/m^2) for IS92a.

The table also shows the dominant role of CO_2 compared with the other greenhouse gases. All SRES scenarios have CO_2 as the dominant forcing agent, all show additional forcing due to the sum of other (non-CO_2) greenhouse gases, and all have a positive forcing contribution from aerosols from 1990 to 2100.

Model projections are shown for global-mean temperature in Figure 5 and for sea level in Figure 6. For best-estimate model parameters, 1990 to 2100 warming ranges from 1.7°C to 4.2°C. Sea-level rise estimates over the same period range from 45 to 72 cm. These results are noticeably higher than the best estimate results given in the SAR (Kattenberg et al., 1996; Warrick et al., 1996). The temperature results given here are similar to those given in the TAR (Cubasch and Meehl, 2001) in spite of the model differences noted above. Sea-level results obtained here, however, are substantially higher than those in the TAR (Church and Gregory, 2001), in part reflecting the large uncertainties that surround projections of future sea-level change.

Figure 5

Central **Temperature Estimates** Plus Extremes
for the SRES Illustrative Scenarios

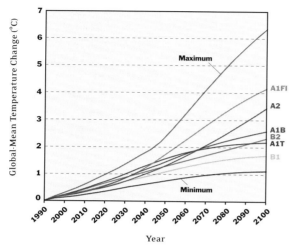

Note: The six central curves show best-estimate global-mean
temperature changes for the SRES illustrative scenarios (using
a climate sensitivity of $\Delta T2x=2.5°C$). The outer curves show
extreme values independent of scenario, using sensitivities of
1.5°C or 4.5°C. At different times, different scenarios may
represent the extremes. The results shown, therefore, span
both the range of scenarios and the range of model parameters.

Figures 5 and 6 also show the full range of results spanning the SRES illustrative scenarios and accounting for uncertainties in climate sensitivity and ice-melt parameters used in the models. For 1990 to 2100, the range of global-mean warming estimates is 1.1–6.4°C. Global-mean sea-level rise over the same period (using the models employed in the SAR) is between 16 cm and 120 cm. The corresponding ranges for the IS92 scenarios are 0.8–3.5°C and 13–94 cm. The maximum values are appreciably higher than in the earlier report (Wigley, 1999). This is partly because the A1FI scenario is now included in the analysis; this scenario has much greater radiative forcing than any of the marker scenarios (see Table 1). In addition, forcing values are generally higher than previously for reasons stated above.

The climate sensitivity, which indicates how strongly the climate system responds to a given change in radiative forcing, is the primary determinant of uncertainty in these projections. Higher sensitivity means greater warming for any given emissions scenario. The climate sensitivity is still subject to considerable uncertainty.[5] Ice melt is also uncertain because scientists cannot yet model accurately how glaciers and the Greenland and Antarctic ice sheets will respond to changing climate. The uncertainty ranges, shown as the lowest and highest curves in Figures 5 and 6, are derived by using the full ranges of climate sensitivity values, ice-melt parameter values, and illustrative SRES emissions scenarios.

Future U.S. Climate

The previous section gave a broad (global-mean) picture of the likely magnitude of future climate change. It is important also to determine future climate change for specific areas and regions in order to assess the potential impact of these changes and plan adaptive strategies. This section examines the results from 15 different climate models to gain insight into future changes in the United States.

Since each model has a different climate sensitivity (see Box 1) and produces different changes in global-mean temperature for a given emissions scenario, the models were first placed on an even playing field by dividing their regional precipitation and temperature changes by their particular global-mean temperature change. These so-called "normalized" results were then averaged to give what amounts to a "consensus" change per degree Celsius of global-mean warming. The advantage of this method is that the results can be applied to any future time, provided one has an estimate of the global-mean temperature change at that time; the normalized changes need only be scaled up by the global-mean change (e.g., the changes would be doubled if the warming were 2°C).

Figure 6

Central **Sea-Level Rise Estimates** Plus Extremes
for the SRES Illustrative Scenarios

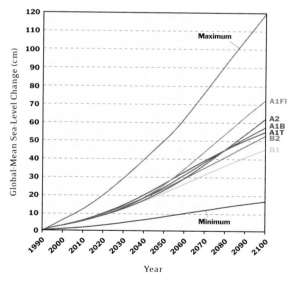

Note: The six central curves show best-estimate global-mean sea-level changes for the SRES illustrative scenarios (using a climate sensitivity of ΔT2x=2.5°C and central estimates of the various ice-melt model parameters). The outer curves show extreme values independent of scenario, using sensitivities of 1.5°C or 4.5°C and low and high ice-melt model parameters. At different times, different scenarios may represent the extremes. The results shown, therefore, span both the range of scenarios and the range of model parameters.

The result of this assessment is a set of flexible projections for average changes in temperature and precipitation over the United States during Northern Hemisphere winter, spring, summer, and fall. None of the models used up-to-date estimates of greenhouse gas and SO_2 emissions, so they do not include the effects of sulfate aerosols. Nonetheless, they provide useful information on future U.S. climate change possibilities.

The results of this analysis (Figure 7) showed a clear warming signal over the whole region and in all seasons. The best results (i.e., greatest consistency between models) were obtained

Figure 7

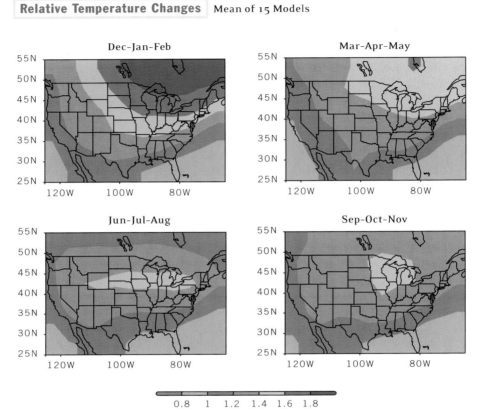

Relative Temperature Changes   Mean of 15 Models

Dec-Jan-Feb

Mar-Apr-May

Jun-Jul-Aug

Sep-Oct-Nov

0.8   1   1.2   1.4   1.6   1.8

Note: The maps show average temperature changes over the United States during each of the four seasons as a ratio of global-mean temperature change. Each value corresponds to a global-mean warming of 1°C. These model results may be used to estimate changes at any U.S. location and any future date simply by taking the global-mean temperature change for that year (e.g., from Figure 5) and multiplying it by the values shown in the maps.

for winter temperature patterns; many models showed an enhanced warming in higher latitudes during winter.

In almost all parts of the lower 48 states and in all seasons, the warming was projected to exceed the global-mean warming. The Southeast and Southwest were exceptions; warming in those regions was slightly below the global mean. At the other extreme, in winter, the northernmost states from North Dakota eastward to Maine showed enhanced warming by a factor of up to two relative to the global mean. The main reason for the enhanced warming over the United States is that land masses tend to respond more rapidly to, and are more sensitive to, external forcing. There is also a tendency for greater warming in higher latitudes as a result of melting sea ice.

Figure 8

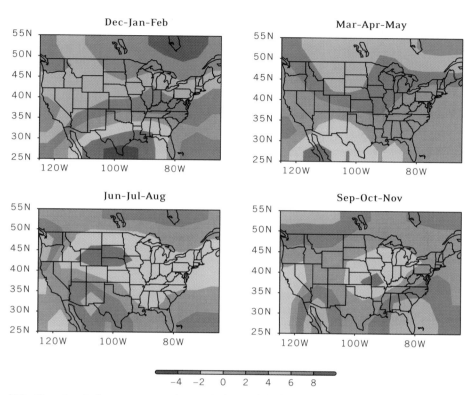

Relative Precipitation Changes Mean of 15 Models

Note: The values in the maps represent percent changes in U.S. precipitation during each of the four seasons. Each value corresponds to a global-mean warming of 1°C. These model results may be used to estimate changes at any U.S. location and any future date simply by taking the global-mean temperature change for that year (e.g., from Figure 5) and multiplying it by the values shown in the maps.

As presented in Figure 7, results are relative to changes in global-mean temperature, i.e., each value corresponds to a global-mean warming of 1°C. To make results applicable to a specific year, values in Figure 7 should be multiplied by the projected global-mean temperature change for that year (e.g., from Figure 5). For example, for a 2°C global-mean warming, the values should be doubled to find the corresponding regional and seasonal mean warming.

Results for precipitation change (Figure 8) are considered less reliable because of large differences among models. Again, these represent changes per 1°C global-mean warming. While the bulk of the study area shows precipitation increases, the changes (ranging between –4 percent and +8 percent per 1°C global-mean warming) are small everywhere relative to current levels of interannual variability, so clear patterns are not apparent.

Changes shown in Figure 8 — though small — could still have important consequences for sectors such as agriculture and water resources, as noted in other reports
in this chapter.

Sulfate aerosol effects will undoubtedly modify the results discussed above. At the
global-mean level, the forcing contribution from sulfate aerosols is small relative to the
total forcing (generally less than 15 percent; see Table 1). However, impacts on some
regions may be significant since emissions of SO_2 and the forcing from sulfate aerosols
vary greatly from region to region. In particular, SO_2 emissions have already begun to
decrease substantially over North America, and these decreases are expected to continue in the future. This will tend to cause additional (albeit relatively small) warming
over this region, over and above that shown in Figure 7.

Impacts will vary with emissions scenario and over time. At present, it is not possible to give any reliable indication of what they might be, partly because appropriate
climate modeling experiments have only recently been performed, but also because of
the very large uncertainties surrounding the quantification of the relationships between
SO_2 emissions and the resulting forcing and climate response effects.

Changes in Other Aspects of Climate

The impacts of climate change at any particular location will be determined not
only by changes in mean temperature and precipitation, but also by changes in factors
such as temperature and precipitation extremes and variability, and in the frequency
of hurricanes and tropical storms. A summary of what is known about these factors is
presented below.

Temperature Extremes

A general warming will shift the whole distribution of temperatures. Thus, relative
to any fixed threshold, the frequency of warm temperature extremes (on all timescales
— days, seasons, and years) will increase and the frequency of cold extremes (like frost
days) will decrease. This is a general result, applicable to any part of the globe. In the
absence of variability changes, the increase in the frequency of extreme warm events
will be disproportionately large (Wigley, 1985). For example, if a 1°C warming
increased the number of days over a particular threshold by 10 percent, then a 2°C
warming would cause an increase by substantially more than 20 percent.

Precipitation Extremes

The IPCC's Second Assessment Report notes that general circulation model results
suggest gradual increases in the frequency of intense precipitation events (heavy rainfall and snowfall) and, in some regions, increases in the probability of dry days and the
length of dry spells (see, e.g., Fowler and Hennessy, 1995; Gregory and Mitchell,

1995). Two more recent studies support this conclusion. Zwiers and Kharin (1998) found that heavy precipitation events over North America might occur twice as often in a world that was 3.5°C warmer than today. Frei et al. (1998) found a similar shift to more frequent heavy precipitation events in southern Europe.

While more frequent wet extremes and dry extremes are possible in the United States, there is no unequivocal evidence for either. Furthermore, the large inter-model differences in projections of mean precipitation change discussed elsewhere in this chapter imply that one should treat the predictions of single models cautiously, especially for changes in shorter timescale events.

Variability

Changes in variability could have a significant effect on agriculture and water resources (see, e.g., Mearns et al., 1996). The IPCC's Second Assessment Report (Houghton et al., 1996, p. 44) notes that "a small change in variability has a stronger effect [on the frequency of extremes] than a small change in the mean," as pointed out earlier by Wigley (1985).

There is, however, no consensus among models on changes in the interannual variability of climate elements like temperature and precipitation. Indeed, even the best models perform poorly in simulating such variability (e.g., Tett et al., 1998) — i.e., their simulations of current variability differ noticeably from observed variability. If any changes did occur, they would be regionally specific, so that some regions might experience increases in variability while nearby regions might experience changes in the other direction.

Hurricanes and Tropical Storms

Our knowledge of how climate change might affect the frequency, intensity, or tracks of tropical cyclones is highly uncertain. Current models used in climate studies do not adequately simulate individual tropical cyclones, and even the most sophisticated weather forecasting models are generally unable to predict the initiation of such storms.

Nevertheless, there is empirical evidence that there might be small increases in the frequency of Atlantic hurricanes (Raper, 1993), based on the positive correlation between sea-surface temperatures and hurricane frequencies in this region. There is also model evidence that minimum pressures may decrease and windspeeds may increase in tropical storms worldwide. Knutson et al. (1998), for example, project windspeed increases of 5 to 12 percent for a sea-surface temperature increase of 2.2°C (a rise that might occur by 2100). However, the projected changes are small relative to past interannual variability. Thus, even if these projections were reliable, it could be many decades before they are positively detected.

An associated possibility is that, along with a minor intensity increase, there could be substantially larger changes in the amount of precipitation associated with individual storms (Knutson and Tuleya, 1999). This may be a more robust result because, with increased ocean temperatures, it is almost certain that evaporation and the moisture-holding capacity of the atmosphere will increase. Along with this, one would expect increased precipitation at the global-mean level.

While the manifestation of this general increase over midlatitude land areas is highly uncertain, more confidence can be placed in the possibility of precipitation increases in areas currently frequented by tropical cyclones, and, as noted previously, in higher latitudes.

Changes in weather and climate extremes are certain to occur over the United States as the global climate changes. For all extreme events, however, it is unlikely that the projected changes will become evident in a statistically convincing way for many decades, with the exception of temperature extremes, which should become evident sooner.

Endnotes

1. In the United Nations Framework Convention on Climate Change, the term "climate change" is used to refer only to human-induced (or anthropogenic) change. The scientific usage of this term is more general, referring to all climate change, both natural and anthropogenic.

2. The Montreal Protocol on Substances that Deplete the Ozone Layer requires industry to phase out production and use of key chemicals that deplete the stratospheric ozone layer.

3. The standard record used by the IPCC combines land data developed in the Climatic Research Unit (Jones, 1994) and marine (sea surface) temperature data compiled by the U.K. Hadley Centre (Parker et al., 1995). Raw input data for these records come from many sources and must be carefully examined and adjusted to account for uncertainties due to factors such as changes in instruments and measuring techniques, and the exposure and location of instruments (Jones et al., 1999).

4. The SRES scenarios considered here are representative members of a wider set of emissions scenarios that are based on a set of four "storylines" (labeled A1, A2, B1 and B2) defining parameters such as future population levels, economic growth rates, energy technologies, levels of international cooperation, etc. The storylines and their background are described in the book edited by Nakićenović and Swart (see reference list), and a brief "Summary for Policymakers" can be downloaded from the Web site http://www.ipcc.ch/. In terms of future socioeconomic changes, the "A" versus "B" distinction primarily contrasts lower versus higher emphasis on sustainable development. The "1" versus "2" distinction corresponds to lower versus higher population growth, faster versus slower rates of technological change, and higher versus lower levels of international cooperation and integration. Anthropogenic emissions results for 2100 for CO_2 and SO_2 are shown below, compared with 1990 values.

	1990	A1B	A1FI	A1T	A2	B1	B2
Fossil CO_2 (GtC/yr)	6.0	13.1	30.3	4.3	28.9	5.2	13.8
Net deforestation (GtC/yr)	1.1	0.4	−2.1	0.0	0.2	−1.0	−0.5
SO_2 (TgS/yr)	71	28	40	20	60	25	48

Key:
GtC=gigatons carbon-equivalent (1 gigaton = 10^9 tons = 10^{15} gm)
TgS=teragrams sulfur-equivalent (1 teragram = 10^{12} gm)

5. The climate sensitivity is a measure of how strongly the climate system responds to external forcing (such as a change in the output of the sun, or an increase in greenhouse-gas concentrations). Strictly, it should be given in terms of the global-mean temperature increase that would eventually occur for a forcing of 1 W/m². In the present context, however, it is usually expressed by the global-mean temperature increase that would eventually occur if the amount of CO_2 in the atmosphere were doubled. (A doubling of CO_2, from any level, corresponds to a forcing of about 3.7 W/m².) The climate sensitivity is still highly uncertain: it lies in the range of 1.5–4.5°C warming for a CO_2 doubling with roughly 90 percent confidence (i.e., there is about a 10 percent probability that it might lie outside this range).

References

Alcamo, J., A. Bouwman, J. Edmonds, A. Grübler, T. Morita, and A. Sugandhy. 1995. An Evaluation of the IPCC IS92 Emissions Scenarios. In *Climate Change 1994: Radiative Forcing of Climate Change and An Evaluation of the IPCC IS92 Emission Scenarios.* J.T. Houghton, L.G. Meira Filho, J. Bruce, H. Lee, B.A. Callander, E. Haites, N. Harris, and K. Maskell, eds. Cambridge University Press, Cambridge, UK, pp. 247–304.

Church, J.A., and J.M. Gregory. 2001. Changes in Sea Level. In *Climate Change 2001: The Scientific Basis. Contribution of Working Group I to the Third Assessment Report of the Intergovernmental Panel on Climate Change.* J.T. Houghton, Y. Ding, D.J. Griggs, M. Noguer, P. Van der Linden, X. Dai, and K. Maskell, eds. Cambridge University Press, New York.

Cubasch, U., and G. Meehl. 2001. Projections for Future Climate Change. In *Climate Change 2001: The Scientific Basis. Contribution of Working Group I to the Third Assessment Report of the Intergovernmental Panel on Climate Change.* J.T. Houghton, Y. Ding, D.J. Griggs, M. Noguer, P. van der Linden, X. Dai, and K. Maskell, eds. Cambridge University Press, New York.

Etheridge, D.M., L.P. Steele, R.J. Francey, and R.L. Langenfelds. 1998. Atmospheric Methane Between 1000 A.D. and Present: Evidence of Anthropogenic Emissions and Climatic Variability. *Journal of Geophysical Research* 103: 15,979–15,993.

Fowler, A.M., and K.J. Hennessy. 1995. Potential Impacts of Global Warming on the Frequency and Magnitude of Heavy Precipitation. *Natural Hazards* 11: 282–303.

Frei, C., C. Schär, D. Lüthi, and H.W. Davies. 1998. Heavy Precipitation Processes in a Warmer Climate. *Geophysical Research Letters* 25: 1,431–1,434.

Gregory, J.M., and J.F.B. Mitchell. 1995. Simulation of Daily Variability of Surface Temperature and Precipitation in the Current and $2 \times CO_2$ Climates of the UKMO Climate Model. *Quarterly Journal of the Royal Meteorological Society* 121: 1,452–1,476.

Groisman, P. Ya., and D.R. Legates. 1995. Documenting and Detecting Long-Term Precipitation Trends: Where We Are and What Should be Done. *Climatic Change* 31: 601–622.

Güllük, T., F. Slemr, and B. Stauffer. 1998. Simultaneous Measurements of CO_2, CH_4, and N_2O in Air Extracted by Sublimation from Antarctica Ice Cores: Confirmation of the Data Obtained Using Other Extraction Techniques. *Journal of Geophysical Research* 103: 15,971–15,978.

Houghton, J.T., L.G. Meira Filho, B.A. Callander, N. Harris, A. Kattenberg, and K. Maskell, eds. 1996. *Climate Change 1995: The Science of Climate Change, Contribution of Working Group I to the Second Assessment Report of the Intergovernmental Panel on Climate Change.* Cambridge University Press, New York.

Hoyt, D.V., and K.H. Schatten. 1993. A Discussion of Plausible Solar Irradiance Variations, 1700–1992. *Journal of Geophysical Research* 98: 18,895–18,906.

Hulme, M. 1992. 1951-80 Global Land Precipitation Climatology for the Evaluation of General Circulation Models. *Climate Dynamics* 7: 57–72.

Hulme, M., T.J. Osborn, and T.C. Johns. 1998. Precipitation Sensitivity to Global Warming: Comparison of Observations With HadCM2 Simulations. *Geophysical Research Letters* 25: 3,379–3,382.

Jones, P.D. 1994. Hemispheric Surface Air Temperature Variations: Reanalysis and an Update to 1993. *Journal of Climate* 7: 1,794–1,802.

Jones, P.D., M. New, D.E. Parker, S. Martin, and I.G. Rigor. 1999. Surface Air Temperature and its Changes Over the Past 150 Years. *Reviews of Geophysics* 37: 173–199.

Karl, T.R., and R.W. Knight. 1998. Secular Trends of Precipitation Amount, Frequency, and Intensity in the United States. *Bulletin of the American Meteorological Society* 79: 231–241.

Karl, T.R., R.W. Knight, and N. Plummer. 1995. Trends in High-Frequency Climate Variability in the Twentieth Century. *Nature* 377: 217–220.

Kattenberg, A., F. Giorgi, H. Grassl, G.A. Meehl, J.F.B. Mitchell, R.J. Stouffer, T. Tokioka, A.J. Weaver, and T.M.L. Wigley. 1996. Climate Models — Projections of Future Climate. In *Climate Change 1995: The Science of Climate Change, Contribution of Working Group I to the Second Assessment Report of the Intergovernmental Panel on Climate Change.* J.T. Houghton, L.G. Meira Filho, B.A. Callander, N. Harris, A. Kattenberg, and K. Maskell, eds. Cambridge University Press, New York, pp. 285–357.

Knutson, T.R., and R.E. Tuleya. 1999. Increased Hurricane Intensities With CO_2-Induced Warming as Simulated Using the GFDL Hurricane Prediction System. *Climate Dynamics* 15: 503–519.

Knutson, T.R., R.E. Tuleya, and Y. Kurihara. 1998. Simulated Increase of Hurricane Intensities in a CO_2-Warmed Climate. *Science* 279: 1,018–1,020.

Mearns, L.O., C. Rosenzweig, and R. Goldberg. 1996. The Effect of Changes in Daily and Interannual Climatic Variability on CERES-Wheat: A Sensitivity Study. *Climatic Change* 32: 257–292.

Nakićenović, N., and R. Swart, eds. 2000. *Special Report on Emissions Scenarios, A Special Report of Working Group III of the Intergovernmental Panel on Climate Change.* Cambridge University Press, Cambridge, UK.

NRC (National Research Council). 2000. *Reconciling Observations of Global Temperature Change.* National Academy Press, Washington, DC.

Nicholls, N., G.V. Gruza, J. Jouzel, T.R. Karl, L.A. Ogallo, and D.E. Parker. 1996. Observed Climate Variability and Change. In *Climate Change 1995: The Science of Climate Change, Contribution of Working Group I to the Second Assessment Report of the Intergovernmental Panel on Climate Change.* J.T. Houghton, L.G. Meira Filho, B.A. Callander, N. Harris, A. Kattenberg, and K. Maskell, eds. Cambridge University Press, New York, pp. 133–192.

Parker, D.E., C.K. Folland, and M. Jackson. 1995. Marine Surface Temperature: Observed Variations and Data Requirements. *Climatic Change* 31: 559–600.

Raper, S.C.B. 1993. Observational Data on the Relationships Between Climatic Change and the Frequency and Magnitude of Severe Storms. In *Climate and Sea-Level Change: Observations, Projections and Implications.* R.A. Warrick, E.M. Barrow, and T.M.L. Wigley, eds. Cambridge University Press, Cambridge, UK, pp. 192–212.

Raper, S.C.B., T.M.L. Wigley, and R.A. Warrick. 1996. Global Sea-Level Rise: Past and Future. In *Sea-Level Rise and Coastal Subsidence: Causes, Consequences and Strategies.* J. Milliman and B.U. Haq, eds. Kluwer Academic Publishers, Dordrecht, The Netherlands, pp. 11–45.

Robock, A. 2000. Volcano Eruptions and Climate. *Reviews of Geophysics* 38: 191–219.

Santer, B.D., T.M.L. Wigley, T.P. Barnett, and E. Anyamba. 1996. Detection of Climate Change and Attribution of Causes. In *Climate Change 1995: The Science of Climate Change, Contribution of Working Group I to the Second Assessment Report of the Intergovernmental Panel on Climate Change.* J.T. Houghton, L.G. Meira Filho, B.A. Callander, N. Harris, A. Kattenberg, and K. Maskell, eds. Cambridge University Press, New York, pp. 407–443.

Santer, B.D., J.J. Hnilo, T.M.L. Wigley, J.S. Boyle, C. Doutriaux, M. Fiorino, D.E. Parker, and K.E. Taylor. 1999. Uncertainties in Observationally-Based Estimates of Temperature Change in the Free Atmosphere. *Journal of Geophysical Research* 104: 6,305–6,333.

Schlesinger, M.E., and N. Ramankutty. 1996. A 65–70 Year Oscillation in Observed Surface Temperatures. *Climate Sensitivity to Radiative Perturbations: Physical Mechanisms and Their Validation.* H. Le Treut, ed. NATO ASI Series I, Springer-Verlag, Berlin, Germany, pp. 305–316.

Tett, S.F.B., T.C. Johns, and J.F.B. Mitchell. 1998. Global and Regional Variability in a Coupled OAGCM. *Climate Dynamics* 13: 303–323.

Warrick, R.A., C. Le Provost, M. Meier, J. Oerlemans, and P. Woodworth. 1996. Changes in Sea Level. In *Climate Change 1995: The Science of Climate Change, Contribution of Working Group I to the Second Assessment Report of the Intergovernmental Panel on Climate Change.* J.T. Houghton, L.G. Meira Filho, B.A. Callander, N. Harris, A. Kattenberg, and K. Maskell, eds. Cambridge University Press, New York, pp. 358–405.

Wentz, F.J., and M. Schabel. 1998. Effects of Orbital Decay on Satellite-Derived Lower-Tropospheric Temperature Trends. *Nature* 394: 661–664.

Wigley, T.M.L. 1985. Impact of Extreme Events. *Nature* 316: 106–107.

Wigley, T.M.L. 1999. (Refers to original report for the Pew Center).

Wigley, T.M.L., P.D. Jones, and S.C.B. Raper. 1997. The Observed Global Warming Record: What Does it Tell Us? *Proceedings of the National Academy of Sciences* 94: 8,314–8,320.

Zwiers, F.W., and V.V. Kharin. 1998. Changes in the Extremes of the Climate Simulated by CCC GCM2 Under CO_2 Doubling. *Journal of Climate* 11: 2,200–2,222.

Impacts on the U.S. Agricultural Sector

Richard M. Adams, Brian H. Hurd, John Reilly

Food and fiber are essential for sustaining and enhancing human welfare; hence, agriculture has been a primary focus of debates about the effects of climate change. In fact, the United Nations Framework Convention on Climate Change views the sustainability of food production as paramount in its objectives for stabilizing greenhouse gas (GHG) emissions, stating that emissions should be stabilized at a level that "ensures that food production is not threatened."

This chapter analyzes the current state of knowledge about the effects of climate change on U.S. food production and agricultural resources. It considers likely regional and global changes in agricultural production, including how the effects are distributed across economic groups, such as producers and consumers, as well as across regions of the United States.

Overall, the consensus of economic assessments is that global climate change of the magnitude currently being discussed by the Intergovernmental Panel on Climate Change (IPCC) and other organizations (i.e., +0.8°C to +4.5°C) could result in some lowering of global production, but would have only a small, likely positive, overall effect on U.S. agriculture and its ability to provide sufficient food and fiber to both domestic and global customers over the next 100 years. However, the distribution of effects within the United States could be significant because producers, and local economies, will gain in some regions and lose in others, while consumers will see price changes.

Warming beyond that reflected in current studies is expected to impose greater costs, decreasing agricultural production in most areas of the United States and reducing global production. This reinforces the need to determine the magnitude and rate of warming that may accompany the greenhouse gas buildup currently underway in the atmosphere.

Since agricultural systems are managed, farmers can implement a number of strategies to mitigate potential losses from climate change. Adaptation strategies could include changing planting and harvest dates, rotating crops, selecting crops for cultivation, and altering fertilizer use rates. These strategies must be considered in assessments of climate change effects, because excluding them could overstate the potential negative impacts or understate potential gains associated with climate change (Rosenberg, 1992). However, the long time horizons involved in climate change assessments and uncertainties concerning the rate and magnitude of change make it difficult to predict how people will adapt. This is particularly challenging since

This paper is a condensed and updated version of a report published in February 1999 by the Pew Center: Agriculture & Global Climate Change: A Review of Impacts to U.S. Agricultural Resources. The authors and the Pew Center are grateful for the input of Harry Kaiser and Vernon Ruttan, who reviewed the report and provided comments.

adaptations are influenced by many factors, including government policy, access to capital, and development of new technology.

Numerical estimates presented here should be interpreted only as illustrative of the possible consequences of climate change. Although more may be known about the potential effects of climate change on agriculture than on any other sector, major uncertainties remain, primarily due to the lack of precise forecasts of climate change at geographic and time scales relevant to decision-makers.

This chapter focuses primarily on potential effects of climate change on agriculture. Agriculture can also affect climate. For example, it is a source of greenhouse gas emissions, and thus contributes to global warming, and it also could play an important role in mitigating global warming. The role of agriculture in climate change mitigation is discussed briefly here.

Dimensions and Trends in U.S. Agriculture

U.S. food security and sustainability are not likely to be threatened by changes currently estimated for global climate over the next several decades because of the extent and diversity of U.S. cropland resources and their resilience and capacity for adaptation. Additionally, if current trends in technological development continue, such as the development of new plant varieties and improvements in the use of fertilizers and other chemical inputs, new technology could limit the vulnerability of U.S. agricultural production to climate change.

Since the 1940s, U.S. agricultural output has increased by approximately 2 percent annually as a result of improved crop yields and production methods even though the agricultural land base has not increased (Huffman and Everson, 1992). Increases in productivity could continue into the future and contribute to agriculture's adaptation to a changing climate. For example, some studies suggest that crop yields could continue to rise nationally and globally, although most likely at a slower rate,[1] and could keep pace with growing U.S. and global populations (Reilly and Fuglie, 1998).[2]

The effects of these technological developments and their adoption can be seen in crop yield trends for major U.S. crops such as corn, soybeans, rice, barley, and cotton. For example, annual average U.S. wheat yields were about 18 bushels/acre from the first year of record, 1866, until about 1940. Since then, annual average wheat yields have risen in response to new wheat varieties, increased use of chemical inputs such as inorganic fertilizers and pesticides and larger equipment which improves the timeliness of planting and harvesting operations, reaching 44 bushels per acre in 1998.[3] It is important to note, however, that even as yields have increased, climate variability remains an important factor affecting crop production, leading to substantial harvest variations from year to year. Increases in climate variability and the incidence of extreme events — possibilities discussed by Rosenzweig et al. (2000) — could alter the trend of rising yields.

Box 1

Effects of World Hunger and Food Distribution

Research results indicate that global food production is most likely to be only modestly affected by climate change, although some countries could be more adversely affected than others. Consequently, global capacity to feed the world's population is not expected to be seriously threatened as a result of climate change in the foreseeable future.

However, global capacity to grow food is currently greater than that required to eliminate hunger, and yet hunger is endemic in a number of areas of the world because of poverty, scarce capital, civil strife, and droughts and famines. If climate change adversely affects agricultural markets in areas of the world where hunger is, or is expected to be, a significant problem, the added stress could pose a serious threat to the local or regional food supply. The result would be an increase in the risks of hunger in these regions. Unfortunately, these are precisely the areas that appear to be most prone to losses of agricultural production from climate change.

Population growth, economic pressures, land degradation, and political instability stress a nation's ability to satisfy food requirements and can diminish the ability to cope with climate change. Although these factors are difficult to estimate over the long run, Fischer et al. (1994) estimate that in the absence of climate change, the number of people at risk of hunger and malnourishment will increase from 500 million today to over 640 million by 2060 (though falling as a percentage of the world population).

Rosenzweig et al. (1995) found that all of the scenarios of future climate change used in their study (i.e., GISS, GFDL, and UKMO) increased esti-mates of the number of people at risk of hunger. Their analysis also showed that reduced population growth could do the most to minimize the impacts of climate change, followed by increased trade liberalization and higher economic growth rates.

Norse (1994) assessed the vulnerability of food security to threats from environmental degradation, economic growth, population growth, and climate change, and found that sub-Saharan Africa is the region most at risk in terms of food security. It is more vulnerable to reduced rainfall, change in rainfall variability, and greater evapotranspiration than any other region. About half of its arable land is already arid or semiarid, only 2 percent of its cropland is irrigated, and the high cost of irrigation development limits its use for low-cost staple foods. Much of the soil has low water holding capacity, and this could be reduced further by higher soil temperatures, leading to greater rates of soil organic matter breakdown. On the economic side, the anticipated low GDP growth rates imply that people and countries will be unable to overcome domestic food production problems through purchased imports.

In summary, overcoming the potential increased risk of hunger may require efforts to improve the food distribution system, to limit population growth, to raise the level of economic development, and to reduce trade barriers. Furthermore, continued agricultural research to improve crop varieties and production methods as well as to provide technical assistance to developing countries will be necessary to limit the vulnerability of at-risk countries.

Obtaining water to irrigate additional acreage also implies more and larger reservoirs, which in turn implies greater pressure to develop the relatively few remaining undammed rivers in the United States (Hurd et al., 1999). In addition, adaptation may have unintended environmental consequences; e.g., the drive to increase production increases pesticide use, irrigation, and use of marginal lands, all of which help degrade environmental quality (Adams et al., 1988; Crosson and Anderson, 1994; Adams et al., 2000; Rosenzweig et al., 2000).

Many climate change studies indicate that some regions are likely to get more rain, even as others get less. Increased intensity of rainfall is a threat to agriculture and the environment because heavy rainfall is primarily responsible for soil erosion, leaching of agricultural chemicals, and runoff that carries livestock waste and nutrients into water bodies. An example is the growth of a hypoxic zone (area of water depleted of oxygen and thus unable to support marine life) in the Gulf of Mexico following flooding of the Mississippi River, which carried heavy loads of nutrients into the Gulf. Normally, agricultural chemicals in surface water exist at levels that do not cause obvious harm. High concentrations during heavy rainfall — such as that associated with hurricanes or similar extreme events — can, however, result in fishkills.

Global warming could exacerbate air pollution through natural processes (increases in temperatures tend to increase formation of certain types of air pollutants, such as ozone and other photochemical oxidants) and through increased electricity use (e.g., associated with increased air conditioning). The adverse effects of air pollution on vegetation, including crops, are well documented (Heck et al., 1984; U.S. EPA, 1996). Tropospheric ozone, an important constituent of smog, is one of the major air pollutants in the United States. It has caused substantial damage to agricultural production. For example, Adams (1986) estimates that current levels of exposure of crops to ozone result in over $3 billion in damages in the United States. In addition, losses to forests and horticultural plants are estimated to be in excess of $2 billion in the United States (Callaway et al., 1985).

The degree to which concentrations of tropospheric ozone will increase due to rising global temperatures is uncertain, given the complex nature of the ozone formation process and the difficulty in forecasting future levels of precursor pollutants. However, there is evidence that ozone and its precursors will increase. For example, a 4°F warming (about 2°C) in the Midwest with no other change in weather or emissions could increase concentrations of ozone by as much as 8 percent in many regions where ozone precursors are an existing problem (U.S. EPA, 1996).

Human Response and Adaptation to Climate Change

Over time, agricultural systems and practices have adapted to changing economic and physical conditions. This has been accomplished by adopting new technologies and by changing crop mixes, cultivated acreage, and institutional arrangements. Such flexibility suggests significant human potential to adapt to climate change (CAST, 1992; Smit et al., 1996; Rosenberg, 1992; Easterling et al., 1993).

Farm-level adaptations can be made in planting and harvest dates, crop rotations, selection of crops and crop varieties for cultivation, water consumption for irrigation, use of fertilizers, and tillage practices. Price and other market-level changes can provide further opportunities to adapt. Each adaptation can lessen potential yield losses from climate change or even potentially improve yields in some regions.

In the longer term, adaptation might include the development and use of new crop varieties that offer advantages under possible future climate conditions, or investment in new irrigation infrastructure to insure against the possibility of less reliable rainfall. Inclusion of adaptations is thus a requisite feature of assessments of the effects of climate change on managed systems such as agriculture. Economic studies of climate change include varying degrees of adaptation. Procedures for including adaptation are discussed in Box 2.

Several studies (of both the structural and spatial analogue types discussed in Box 2) describe substantial opportunities for adaptation to offset the negative effects of climate change (e.g., Kaiser et al., 1993; Mendelsohn et al., 1999, 1994; Adams et al., 1999b). Most economic studies have demonstrated that inclusion of adaptation strategies will reduce economic losses or improve the gains from climate change. For example, Adams et al. (1999b) found that including adaptations in two climate change scenarios: GISS (described in Box 3) and a harsher climate forecast (a 5°C warming and 7 percent increase in precipitation) resulted in a 20 to 25 percent increase in the economic estimates ($10 billion to $12 billion gain in societal welfare, measured in 1990 dollars).

While adaptations can offset some of the impacts of climate change, they can be costly to implement. The strategy and timing of adaptation depend on many factors, including the rate, magnitude, and variability of climate change; access to capital; and changes in government policy.

Adaptations may involve significant time lags and long-term capital investment decisions that depend critically on the rate and variability of climate change. If climate changes at a rate that requires rapid adaptation, adaptation options would be limited and adjustment costs would be relatively high. A more gradual change would allow time for major infrastructure investments as systems depreciate (OTA, 1993). The magnitude of warming is also important. Studies to date examine changes in warming up to 5°C. Warming beyond this level would increase the adjustment costs (Hall, 1997).

Changes in climate variability and extreme events can also affect adaptation strategies. Rozenzweig et al. (2000) have considered the potential effects of changes in climate variability, extreme events, and crop pests, and have observed that should the frequency of drought, floods, or severe storms increase, farmers may find adapting more difficult. If climate uncertainty increases as the climate changes, adaptation

Box 2

Economic Approaches to Measuring Climate Change Effects

Two general approaches have been used to assess the potential economic consequences of climate change on agriculture: the structural approach and the spatial analogue approach (see Schimmelpfennig et al., 1996 or Adams et al., 1998). Each approach contains aspects of human response believed to be important in measuring economic effects. However, the approaches differ in terms of data requirements, assumptions, and the dimensions they measure.

Structural methods consider fundamental changes in crop yields and farmer response, and might also be called "decision duplication" methods because the analyst tries to duplicate the decisions of the farmer in choosing what crops to grow and how to grow them. These methods characterize the economic decision-making problem for farmers and consumers, identifying alternative ways of attaining objectives within existing resource and institutional constraints. Solutions to the decision problem are obtained by identifying the choices that result in the greatest economic welfare.

Structural methods are popular in climate change research because of their ability to: (1) assess the effects of as yet unrealized environmental changes such as additional warming, precipitation, or higher CO_2 levels, (2) include additional characteristics or changes in the structure of the decision problem, and (3) estimate changes in market prices and distributional effects on regional producers and consumers. One challenge to implementing the structural approach is to identify and incorporate adaptations that farmers and consumers might use to respond to climate changes. This becomes particularly difficult in light of the long time horizons associated with climate change. The main criticism of this approach is that if the analyst fails to anticipate correctly, the resulting estimates may be misleading.

The spatial analogue approach, in contrast, looks at how crop production currently varies across regions with different climates, and tries to infer the effects of climate change from these differences. This reduces the challenge of anticipating future adaptations by using information from past farm-level decisions collected from farmers operating across a range of climatic conditions. Using these data, it may be possible to estimate statistically how changes such as temperature might affect production and profits (Mendelsohn et al., 1994). The strength of the spatial analogue approach is that climate changes and farmer responses are implicit in the analysis (reflected in the data on farmer behavior across regions with different climates). An important weakness is that spatial analogue models abstract from the issues and costs of changes in infrastructure characteristics such as irrigation systems that may be necessary to mimic warmer climate practices. The approach also typically ignores likely changes in output and input prices that may result from global changes in production, and which in turn affect farm-level adaptation decisions. Another limitation is that the approach generally cannot include the effects of CO_2 changes.

Each provides useful, often distinct, information. Several recent studies have combined the approaches to gain the advantages of each. For example, the spatial analogue models have been used to improve the adaptation included in structural models (Darwin, 1995; Adams et al., 1998).

Box 3

General Circulation Models

General circulation models (GCMs) are sets of sophisticated computer programs that simulate the circulation patterns of the earth's atmosphere and oceans. The purpose of these climate models is to describe how major changes in the earth's atmosphere, such as changes in the concentrations of greenhouse gases, affect climatic patterns including temperature, precipitation, cloud cover, sea ice, snow cover, winds, and atmospheric and oceanic currents. The models are not intended to predict weather events, and their resolution is too coarse to account for the effects of local geographic features such as mountains that may influence regional climate. They are, however, useful tools for examining long-term climatic trends, patterns, and responses to significant changes.

GCMs remain simple, however, compared to the complexity of the real climate system. These models continue to evolve as better information on and understanding of physical relationships are developed, and as improvements in computing power are realized. Climate models differ with respect to their assumptions, detailed structure, spatial and temporal resolution, and complexity, and as a result there is significant variation in the projected results of different models. This variation illustrates the degree of uncertainty associated with climate projections but can also provide a sense of reliability to the extent that consistent patterns emerge across different models. Estimated changes in average global temperatures and precipitation of some of the climate models referenced in this paper are shown below.

Characteristics of Selected Climate Models Under a Doubling of CO_2

GCM	Change in Global Mean Temperature (°C)	Change in Global Precipitation (percent)*
GISS	+4.2	+11.0
GFDL-R30	+4.0	+8.3
UKMO	+5.2	+15.0
OSU	+2.8	+7.8

*Estimates at regional levels vary considerably across seasons and regions and are much less certain. In some cases, estimates show reduced regional precipitation.

Source: U.S. Country Studies Program, 1994.

responses will be affected. For example, if risk aversion is high among farmers in regions where water is limited, farmers may shift production to more drought-tolerant crops, even if expected returns are lower (Pope, 1982; Hurd, 1994).

Recent studies by Adams et al. (2000) and Chen et al. (in press) examine the consequences of increased climatic variability and extreme events (in this case El Niño-Southern Oscillation or ENSO events) on agriculture. The results suggest that increased variability and increased frequency of ENSO events will have adverse economic consequences because farmers cannot plan for such events as well as they can for changes in mean or average temperature and precipitation.

Implementing adaptation often requires local access to financial and physical capital, technical assistance, and other inputs such as water and fertilizer. Infrastructure costs (e.g., for irrigation, reservoirs, and distribution systems) are also important. To the extent that climate change results in significant geographic shifts in production, costs to move or add infrastructure capacity could be substantial.

Although recent trends toward a more market-driven agricultural economy are significant (and are generally assumed in climate impact assessments), government policies remain a driving force in U.S. agriculture. The impacts of climate change on U.S. agriculture will depend on how these policies evolve over time. For example, changes in the Bureau of Reclamation policies to develop and supply agriculture with irrigation water could greatly affect how western agriculture adapts.

In addition, government policies established to address global climate change could affect agriculture in significant ways, such as changing fuel and fertilizer costs, encouraging the planting of tree and biofuel crops, and encouraging agricultural methods that conserve and enhance soil. Over the long run, government policies can affect not only prices and farmer behavior but also the tools and strategies that farmers use in adapting to climate change.

Because explicit adaptation responses are difficult to project, an assessment of the agricultural effects of climate change cannot account for the full range of adaptation options likely to arise over the next century. Conversely, adaptation options incorporated into recent assessments may not be technically or economically feasible in some cases or in some regions. While U.S. agriculture may have the means to successfully adapt, the capacity for adaptation in developing countries is limited as a result of limited access to markets for crop inputs or outputs, and limited infrastructure development (Reilly and Hohmann, 1993).

Greater accuracy in climate forecasts is critical for making sound decisions regarding adaptation. One key area of uncertainty is the extent of CO_2 buildup in the future. Until recently, those who build and run GCMs (see Box 3) tended to estimate climate under a doubling of CO_2 in the atmosphere (typically referred to as $2xCO_2$). This presents a number of problems. One is that the climate associated with $2xCO_2$ will probably not be realized until late in this century. Thus, the models do not indicate how climate may change in coming decades. A second problem is that the $2xCO_2$ scenarios assume that atmospheric greenhouse gas concentrations have stabilized, resulting in stable climate. In fact, concentrations are likely to continue rising and climate is likely to continue changing. In examining impacts on agriculture, most studies have assumed that climate has stabilized and the agricultural system just needs to "catch up." It is far more likely that climate will continue to change and agriculture will need to continually adapt to it.

Climate Change: Science, Strategies, & Solutions

Role of Agriculture in Mitigating Climate Change

Agriculture is both a receptor of possible climate changes arising from greenhouse gas emissions and a source of greenhouse gases, including CO_2, methane (CH_4), and nitrous oxide (N_2O). The agricultural sector is an energy-intensive industry, but because it is a relatively small share (less than 3 percent) of the economy, it is a relatively minor user of fossil fuels and hence a minor contributor to U.S. CO_2 emissions. Nonetheless, agriculture constitutes 40 percent of anthropogenic sources of methane (primarily from rice and cattle production), and 68 percent of N_2O (mainly from nitrogen fertilizer).

The understanding of agriculture's contribution to these emissions has increased considerably over the past decade, leading to several potential strategies for reductions. Methane reduction strategies include changes in animal feed rations in the short run and genetic and dietary improvements in the long run, as well as changes in rice fertilization and other management practices. Reduced use of nitrogen fertilizer, particularly those easily volatilized forms such as anhydrous ammonia, could reduce N_2O emissions, as could the use of advanced fertilizer techniques (controlled release and better placement), better management of manure use, and better timing of applications.

In the short term, reduced use of nitrogen fertilizer and feeding systems that produce less methane from livestock are expected to reduce yields or increase costs. Such effects, in turn, suggest higher food costs and, hence, losses to consumers. In the long run, improved breeding programs for livestock, better management of nitrogen in rice and other crop production, and improved crop breeding to reduce fertilizer dependence are needed to reduce methane and nitrous oxide emissions.

Policies to reduce carbon emissions from fossil fuel combustion, such as a national or global tradeable permit scheme, are expected to result in energy price changes. Estimates of the carbon "price" in a permit trading system and its impact on energy prices vary significantly, depending on assumptions about how such a system is implemented. In the short to medium time frame, implementation of carbon emission control policies are more likely to adversely affect agriculture through, for example, higher fuel and fertilizer costs than climate changes over the same period.

Although policies to reduce greenhouse gas emissions may impose costs on agriculture, they also create substantial economic opportunities for agricultural producers. For example, afforestation (planting trees) to sequester carbon is a prominent strategy for mitigating greenhouse gas emissions, and is a potential opportunity for agriculture because enough marginal agricultural land exists in the United States to offset a considerable amount of carbon emissions (Adams et al., 1999a; Marland et al., 1999).

The potential benefits of this afforestation strategy are broad-based: planting trees creates a low-cost source of biomass, alternative fuels, and carbon-based materials. Some estimates suggest that tree planting on marginal agricultural lands can be a significant contributor to mitigation at a relatively low cost compared with reducing carbon emissions from fossil fuels (IPCC, 1996b).

Additional amounts of carbon can be sequestered in soils by relatively minor changes in agricultural practices. "Growing carbon" on agricultural lands would create a new crop for farmers. The use of tradable carbon permits that would allow firms that need to reduce carbon emissions to instead purchase reductions or sequestration, as allowed under provisions of the Kyoto accord, is one way that these opportunities for agriculture could be created. A tradeable permit system, in combination with government incentives to sequester carbon in soils, could create substantial income opportunities for farmers in some regions.

Endnotes

1. Other analyses find evidence of a plateau in aggregate yield trends and question the validity of assumed continued yield growth (Brown, 1994). Resource degradation (e.g., soil erosion) and exhaustion of yield enhancement potential are cited as limiting factors. Investigations of these factors in other studies show little evidence of a yield plateau, note remaining opportunities for yield enhancements, and see very limited effects of resource degradation on yield (Reilly and Fuglie, 1998).

2. Between 1950 and 1990, the world population grew at an annual rate of 2.25 percent. The United Nations projects that global population will grow at an annual rate of 1.13 percent through 2025 and slow to about 0.6 percent between 2025 and 2050. The U.S. population is projected to increase by nearly 60 million between 1998 and 2025 — an annual rate of 0.7 percent. The rate is then projected to fall to 0.017 percent between 2025 and 2050.

3. The average yield for winter, durum, and spring wheat, weighted by volume of production, is based on National Agricultural Statistical Service data. The weather in wheat growing areas of the United States in 1998 was unusually favorable. A trend analysis estimates that yields in 1998 would have been 39 bushels/acre had it been a normal weather year (NASS, 1998).

4. Most GCM-based forecasts of climate change assume a doubling of CO_2 from either current or pre-industrial levels. Some forecasts also assume an effective doubling of CO_2, which includes the climate-forcing effects of CO_2 as well as other greenhouse gases (thus the assumed CO_2 level may be less than an actual doubling of current CO_2 levels). The effects of higher CO_2 levels on climate beyond an actual or effective doubling are generally not evaluated. For this reason, some economic evaluations have also considered a suite of temperature and precipitation changes from current levels, in order to analyze a wider range of potential climate consequences than suggested by the GCM-based forecasts. For example, Adams, et al. (1999b) evaluated the economic effects of temperature changes of 1.5 to 5.0 degrees C and precipitation changes of minus 15 to plus 15 percent, for major agricultural production regions of the U.S., along with two GCM-based forecasts.

5. Temperature increases lead to higher respiration rates, shorter periods of seed formation, and consequently lower biomass production. For example, higher temperatures result in a shorter grain filling period, smaller and lighter grains, and therefore lower crop yields and perhaps lower grain quality (i.e., lower protein levels).

6. Modeling studies have not included adjustments for improved water use efficiency that could result from increased CO_2 levels, which might reduce marginal irrigation and soil moisture needs.

References

Adams, D.M., R.J. Alig, B.A. McCarl, J.M. Callaway, and S.M. Winnett. 1999a. Minimum Cost Strategies for Sequestering Carbon in Forests. *Land Economics* 75(3): 360–374.

Adams, R.M. 1986. Agriculture, Forestry, and Related Benefits of Air Pollution Control. *American Journal of Agricultural Economics* 68: 885–894.

Adams, R.M., C.C. Chen, B.A. McCarl, and D.E. Schimmelpfenning. 2000. Climate Variability and Climate Change: Implications for Agriculture. In *The Long Term Economics of Climate Change, Volume 3, Advances in the Economics of Environmental Resources.* D. Hall and R. Howarth, eds. Elsevier Science Publisher, New York, NY.

Adams, R.M., R. Fleming, B.A. McCarl, and C. Rosenzweig. 1995. A Reassessment of the Economic Effects of Climate Change on U.S. Agriculture. *Climatic Change* 30: 147–167.

Adams, R.M., J.D. Glyer, B.A. McCarl, and D.J. Dudek. 1988. The Implications of Global Change for Western Agriculture. *Western Journal of Agricultural Economics* 13: 348–356.

Adams, R.M., B.H. Hurd, S. Lenhart, and N. Leary. 1998. Effects of Global Climate Change on Agriculture: An Interpretative Review. *Climate Research* 11: 19–30.

Adams, R.M., B.A. McCarl, K. Segerson, C. Rosenzweig, K.J. Bryant, B.L. Dixon, R. Conner, R.E. Evenson, and D. Ojima. 1999b. The Economic Effects of Climate Change on U.S. Agriculture. Chapter 2 in *The Economics of Climate Change*. R. Mendelsohn and J. Neumann, eds. Cambridge University Press, Cambridge, UK.

Adams, R.M., C. Rosenzweig, J. Ritchie, R. Peart, J. Glyer, B. McCarl, B. Curry, and J. Jones. 1990. Global Climate Change and U.S. Agriculture. *Nature* May: 219–224.

Brown, L. 1994. Facing Food Insecurity. In *State of the World*. L. Brown, et al., eds. W.W. Norton & Co., Inc., New York.

Callaway, J.M., R.F. Darwin, and R. Nesse. 1985. *Economic Valuation of Acidic Deposition: Preliminary Results from the 1985 NAPAP Assessment*. Battelle Pacific Northwest Laboratory, Richland, WA.

CAST. 1992. Preparing U.S. *Agriculture for Global Climate Change*. Task Force Report No. 119. Council for Agricultural Science and Technology, Ames, IA.

Chen, C.C., B.A. McCarl, and R.M. Adams. Agricultural Economic Effects of Shifts in ENSO Event Frequency and Strength. *Climatic Change*. In press.

Crosson, P., and J.R. Anderson. 1994. Demand and Supply Trends in Global Agriculture. *Food Policy* 19: 105–119.

Darwin, R., M. Tsigas, J. Lewandrowski, and A. Raneses. 1995. *World Agriculture and Climate Change: Economic Adaptations. Agricultural Economic Report No. 703*. U.S. Department of Agriculture, Natural Resources and Environment Division, Economic Research Service, Washington, DC.

Easterling, W.E. III, P.R. Crosson, N.J. Rosenberg, M.S. McKenny, L.A. Katz, and K.M. Lemon. 1993. Paper 2. Agricultural Impacts of and Responses to Climate Change in the Missouri-Iowa-Nebraska-Kansas (MINK) Region. *Climatic Change* 24: 23–61.

Fischer, G., K. Frohberg, M.L. Parry, and C. Rosenzweig. 1994. Climate Change and World Food Supply, Demand and Trade. *Global Environmental Change* 4(1): 7–23.

Hall, D. 1997. *Impacts of Global Warming on Agriculture*. Invited paper, International Association of Agricultural Economists biannual meeting. Sacramento, CA. August 3–7.

Hanson, J.D., B.B. Baker, and R.M. Bourdon. 1993. Comparison of the Effects of Different Climate Change Scenarios on Rangeland Livestock Production. *Agricultural Systems* 41: 487–502.

Heck, W.W., W.W. Cure, J.O. Rawlings, L. Zaragoza, A. Heagle, H. Heggestad, R. Kohut, L. Kress, and P. Temple. 1984. Assessing Impacts of Ozone on Agricultural Crops. *Journal of the Air Pollution Control Association* 34: 729–735.

Huffman, W.E., and R.E. Everson. 1992. Contributions of Public and Private Science and Technology to U.S. Agricultural Production. *American Journal of Agricultural Economics* 74: 751–756.

Hurd, B.H. 1994. Yield Response and Production Risk: An Analysis of Integrated Pest Management in Cotton. *Journal of Agricultural and Resource Economics* 19(2): 313–326.

Hurd, B.H., J.M. Callaway, J.B. Smith, and P. Kirshen. 1999. *Economic Effects of Climate Change on U.S. Water Resources. The Economic Impacts of Climate Change on the U.S. Economy*. R. Mendelsohn and J.E. Neumann, eds. Cambridge University Press, Cambridge and New York.

IPCC. 1996a. Chapters 13 and 23 in *Climate Change 1995: The IPCC Second Assessment Report, Volume 2: Scientific-Technical Analyses of Impacts, Adaptations, and Mitigation of Climate Change*. R.T. Watson, M.C. Zinyowera, and R.H. Moss, eds. Cambridge University Press, Cambridge and New York.

IPCC. 1996b. Chapter 2 in *Climate Change 1995: The Science of Climate Change*. Intergovernmental Panel on Climate Change. Cambridge University Press, Cambridge and New York.

Kaiser, H.M., S.J. Riha, D.S. Wilks, D.G. Rossier, and R. Sampath. 1993. A Farm-Level Analysis of Economic and Agronomic Impacts of Gradual Warming. *American Journal of Agricultural Economics* 75: 387–398.

Marland, G., B.A. McCarl, and U. Schneider. 1999. Soil and Carbon Policy and Economics. *Carbon Sequestration in Soils: Science Monitoring and Beyond*. N.J. Rosenberg, R.C. Isurralde, and E.L. Malone, eds. Battelle Press, Columbus, OH, pp. 153–169.

Mendelsohn, R., W.D. Nordhaus, and D. Shaw. 1994. The Impact of Global Warming on Agriculture: A Ricardian Analysis. *The American Economic Review* 84(4): 753–771.

Mendelsohn, R., W.D. Nordhaus, and D. Shaw. 1999. The Impact of Climate Variation on U.S. Agriculture. In *The Impacts of Climate Change on the U.S. Economy*. R. Mendelsohn and J. Neumann, eds. Cambridge University Press, Cambridge, UK.

NASS. 1998. See http://www.hqnet.usda.gov/nass.

Norse, D. 1994. Multiple Threats to Regional Food Production: Environment, Economy, Population. *Food Policy* 4: 133–148.

Office of Technology Assessment (OTA). 1993. *Preparing for an Uncertain Climate*. Office of Technology Assessment, U.S. Congress, Washington, DC.

Pope, R.D. 1982. Empirical Estimation and Use of Risk Preferences: An Appraisal of Estimation Methods that Use Actual Economic Decisions. *American Journal of Agricultural Economics* 63(1): 161–163.

Reilly, J., and K.O. Fuglie. 1998. Future Yield Growth in Field Crops: What Evidence Exists? *Soil and Tillage Research* 47: 279–290.

Reilly, J., and N. Hohmann. 1993. Climate Change and Agriculture: The Role of International Trade. *American Economic Association Papers and Proceedings* 83: 306–312.

Reilly, J., F. Tubiello, B. McCarl, and J. Melillo. 2000. Climate Change and Agriculture in the United States. Chapter 13 in *Public Comment Draft, National Assessment Synthesis Team Foundation Report*.

Rosenberg, N. 1992. Adaptation of Agriculture to Climate Change. *Climate Change* 21: 385–405.

Rosenzweig, C., and A. Iglesias, eds. 1994. *Implications of Climate Change for International Agriculture: Crop Modeling Study*. EPA 230-B-94-003. U.S. Environmental Protection Agency, Office of Policy, Planning, and Evaluation, Climate Change Division, Adaptation Branch, Washington, DC.

Rosenzweig, C., A. Iglesias, X.B. Yang, P. Epstein, and E. Chivian. *2000. Climate Change and U.S. Agriculture: The Impacts of Warming and Extreme Weather Events on Productivity, Plant Diseases, and Pests*. Center for Health and the Global Environment, Harvard Medical School, Cambridge, MA, p. 47.

Rosenzweig, C., and M.L. Parry. 1994. Potential Impact of Climate Change on World Food Supply. *Nature* 367: 133–138.

Rosenzweig, C., M.L. Parry, and G. Fischer. 1995. *World Food Supply. As Climate Changes: International Impacts and Implications*. K.M. Strzepek and J.B. Smith, eds. Cambridge University Press, Cambridge and New York, pp. 27–56.

Schimmelpfennig, D., J. Lewandrowski, J. Reilly, M. Tsigas, and I. Parry. 1996. *Agricultural Adaptation to Climate Change: Issues of Long Run Sustainability. Agricultural Economic Report, No. 740*. U.S. Department of Agriculture, Natural Resources and Environment Division, Economic Research Service, Washington, DC.

Smit, B., D. McNabb, and J. Smithers. 1996. Agricultural Adaptation to Climatic Variation. *Climatic Change* 33(1): 7–29.

Smith, J.B., S. Huq, S. Lenhart, L.J. Mata, I. Nemesova, and S. Toure, eds. 1996. *Vulnerability and Adaptation to Climate Change: A Synthesis of Results from the U.S. Country Studies Program*. Kluwer Academic Publishers, Dordrecht, The Netherlands.

U.S. Country Studies Program. 1994. *Guidance for Vulnerability and Adaptation Assessment*, Washington, DC.

U.S. EPA. 1996. *National Air Quality and Emissions Trends Report, 1995*. EPA 454/R-96-005. U.S. Environmental Protection Agency, Office of Air Quality Planning and Standards, Research Triangle Park, NC.

Wigley, T.M.L. 1999. *The Science of Climate Change: Global and U.S. Perspectives*. Pew Center on Global Climate Change, Arlington, VA.

Wolfe, D.W., and J.D. Erickson. 1993. *Carbon Dioxide Effects on Plants: Uncertainties and Implications for Modeling Crop Response to Climate Change. Agricultural Dimensions of Global Climate Change*. H.M. Kaiser and T.E. Drennen, eds. St. Lucie Press, Delray Beach, FL.

ten-fold increase in the rate of global sea-level rise over the 21st century. The mid-range estimate is a two- to five-fold increase relative to historical trends.[4] Regional-scale sea-level rise will differ from the global rise due to factors such as changes in ocean circulation and wind and pressure patterns (Gregory, 1993; IPCC, 1996a). These effects are still being quantified in modeling experiments.

Low, medium, and high relative sea-level rise scenarios for the period 1990 to 2050, combining the global sea-level rise estimates from IPCC (1996a) with expected land elevation changes, are shown in Figure 2 for the East and Gulf coasts of the United States. These estimates correspond to: (a) a continuation of observed 20th century trends (i.e., an 11-cm global rise), (b) a 20-cm global rise, and (c) a 40-cm global rise. The "hot-spots" of coastal Louisiana/east Texas and the mid-Atlantic region are apparent, with relative sea-level rises of up to 100 cm in the first region and 60 cm in the latter region. Elsewhere, the rise could be up to 50 cm, with a higher uncertainty in central Texas owing to uncertainties concerning the rate of subsidence. Note that even the low-rise global scenario produces significant rises in sea level.

Figure 2

Future Projections of Relative Sea-Level Rise (cm) on the U.S. Coast; 1990-2050

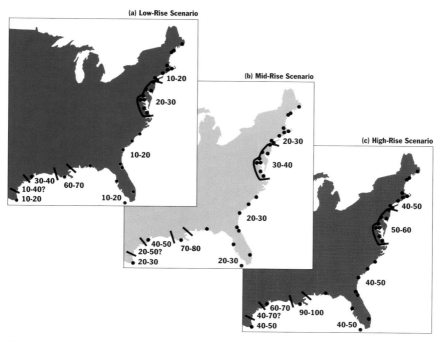

Source: Nicholls and Leatherman (1996).

Factors Affecting the Vulnerability of the U.S. Coastal Zone

Coastal regions of the United States differ in their vulnerability to the risks of sea-level rise. Important local and regional factors that affect vulnerability include variations in physical characteristics (e.g., the slope, elevation, and composition of coastal lowlands), rates of projected population growth and investment, and storm damage trends.

Geographic Characteristics

Regions of the United States that are most susceptible to the impacts of sea-level rise include the Gulf Coast, because of its relatively low-lying coastal topography and high rate of land subsidence, and the mid-Atlantic and south Atlantic areas, where low-lying coastal topography allows marine influence and hence the impacts of sea-level rise to penetrate large distances inland. Extensive coastal lowlands in Louisiana (see Box 1) and south Florida as well as eastern Texas, North Carolina, and the Chesapeake Bay, particularly parts of Maryland, would be affected by sea-level rise (Titus and Richman, 2000).

Overall, the coastal zones of the Northeast and West are least susceptible to sea-level rise impacts because of steeper average coastal profiles, geologic substrates composed of less erodible rock or glacial and riverine till, and lower rates of land subsidence. Coastal barrier islands and spits in the Northeast (e.g., outer Cape Cod, Massachusetts) and low-lying salt marshes (e.g., in San Francisco Bay) are exceptions in these regions; these areas are especially susceptible to erosion from storm surges associated with higher sea-levels (Heinz Center, 2000).

Demographic and Development Rates

Physical factors may determine whether an area is susceptible to flooding, erosion, or storm damage, but the overall magnitude of vulnerability in economic terms also depends on population patterns and economic investment. More than half of the U.S. population lives in counties located along the 20,000 km (12,000 miles) of coastline (Miller and Auyong, 1991). Growth projections suggest that by 2010 the coastal population will have grown by 60 percent from 1960 levels (Culliton et al., 1990; Miller and Auyong, 1991).

Florida is experiencing unprecedented population shifts as baby boomers enter retirement age and depart northern population centers for the southwest coast and the Miami-Ft. Lauderdale metropolitan region. Similarly, coastal resort communities such as Hilton Head and Myrtle Beach, South Carolina; the Outer Banks of North Carolina; and various communities in Georgia and along the Gulf of Mexico in Mississippi and Alabama are experiencing dramatic population growth. From 1950 to 1985, the coastal population of Texas increased 250 percent (Landsea, 1993); Southern

Louisiana: Case Study in Coastal Vulnerability and Potential Sea-Level Rise Impacts

The deltaic coast of Louisiana exhibits the greatest vulnerability to sea-level rise of the entire U.S. coast. In this area, relatively high rates of vertical marsh accretion cannot offset high rates of natural and human-induced land subsidence. Extensive losses of wetlands are expected in the future as salt marshes convert to open or sheltered water (perhaps 1,800 km^2 over the next 50 years by one estimate, Louisiana Legislative Study Group, 1999), although many of these losses would occur even in the absence of accelerated sea-level rise (DeLaune et al., 1992). If there are extensive wetlands losses as sea level rises, concomitant economic impacts could be significant. Louisiana contains approximately 40 percent of the coastal wetlands in the continental United States. These wetlands and their associated estuarine ecosystems support natural resources with an estimated value of more than $1 billion annually (Stone and McBride, 1998).

Studies of various sites in coastal Louisiana show that annual rates of accretion of organic matter and mineral sediment to coastal marshes are insufficient to keep pace with increasing sea-level rise and submergence. A study of the Barataria Bay ecosystem, for example, found that widespread conversion of marshes to open water cannot be forestalled unless sediment from the Mississippi River system is reintroduced to counteract the effects of coastal submergence (Baumann et al., 1984).

The state has adopted several measures to restore wetlands, but they cannot keep up with current pressures, let alone the increased land loss expected with accelerated sea-level rise. By one estimate, the combined effects of all restoration projects currently being implemented will slow the rate of land loss by only about 23 percent. Nonetheless, the state is actively considering more aggressive land restoration efforts, utilizing freshwater and sediment diversions from the massive Mississippi and Atchafalaya Rivers in a strategic manner (Louisiana Legislative Study Group, 1999).

California's population is expected to increase by 5.6 million people (almost 20 percent) over the next 20 years (NOAA, 1999).

The increase in coastal population has spurred a concomitant increase in infrastructure and property values that also contribute to vulnerability to sea-level rise. Each week, about 8,700 new single-family homes are constructed along the U.S. coast (NOAA, 1999). From 1988 to 1993, the total value of insured property in coastal counties from Maine to Texas increased 69 percent, from $1.9 trillion to $3.15 trillion (Pielke and Landsea, 1998).[5] Table 1 shows changes in the rates of population and property value growth in some coastal regions. In general, the trend showing an increase in insured property is good news, as financial risks to coastal properties are being spread across a wider base. At the same time, it is important to

Table 1

Key Regional Trends in the U.S. Coastal Zone Affecting Vulnerability to Sea-Level Rise

Coastal Region	Change in Coastal Population (millions of people/percent change), 1980–1993[a]	Change in Value of Insured Coastal Property (billions of 1993 dollars/percent change) 1988–1993[b]
Northeast (ME, NH, MA, RI)	0.4/+8%	$211/+74%
Mid-Atlantic (CT, NY, NJ, DE, MD, VA)	1.0/+7%	$591/+79%
South Atlantic (NC, SC, GA)	0.3/+23%	$62/+88%
Florida	2.8/+37%	$306/+54%
Gulf Coast (MS, AL, LA, TX)	0.1/+3%	$145/+62%

Notes: a) Includes population in coastal counties only. b) Includes insured commercial and residential property in coastal counties only.

Source: Socioeconomic data from Insurance Institute for Property Loss Reduction and Insurance Research Council (1995).

continue to provide insurers with accurate information on the full range of coastal hazards, including sea-level rise, if insurance is to provide an appropriate private incentive to mitigate risks to property in the coastal zone.

Storm Damage Trends

Sea-level rise can lead to increased vulnerability to storm damage and flooding (i.e., storm surges) in coastal areas.[6] Recent tropical storms and hurricanes provide powerful evidence of the vulnerability of people and property in U.S. coastal areas. Total damages from Hurricane Andrew in 1992, for instance, equaled about $30 billion, making it the most costly hurricane in U.S. history.

One reason for increased vulnerability to storms may be the interaction of the natural variability in storm activity and its link to trends in coastal development. In the two or three decades prior to 1990, the eastern United States experienced a period of relatively mild hurricane activity. Rapid coastal development, especially toward the latter portion of the period of lessened hurricane activity, left coastal regions much more vulnerable to storms, which have increased in frequency and intensity in the 1990s.

Overall, storm activity along the East Coast shows no clear trends over the 20th century, but great variability (Zhang et al., 2000). However, sea-level rise has exacerbated storm damage as flooding (i.e., storm surges) and waves penetrate further inland. This trend will continue through the 21st century, with the potential to increase storm damages over time.

Analysis of landfalling hurricanes since 1925 indicates that seven hurricane seasons similar to the seasons experienced between 1940 and 1969 would have resulted in damages of $10 billion or more if they had occurred with 1995 patterns of coastal development (Pielke and Landsea, 1998). Future hurricane damages, projected from past storms, could average $5 billion per year (Pielke and Landsea, 1998). If climate change results in more frequent or more powerful storm events, damage estimates could be even higher. A more accurate estimate of the risks imposed by storms, however, would not only reflect anticipated storm patterns and projections of future development patterns based on historical trends but also the continuing effects of sea-level rise and storms on future development patterns.

Physical Impacts of Sea-Level Rise on Coastal Resources

The physical impacts of relative sea-level rise will vary by location and will depend on a range of physical and socioeconomic factors, including human response (Turner et al., 1996).

Major physical impacts of sea-level rise include the following:

- Erosion of beaches, bay shores, and tidally influenced river deltas (Heinz Center, 2000);

- Permanent inundation or wetland colonization of low-lying uplands;

- Increased flooding and erosion of marshes, wetlands, and tidal flats, potentially resulting in net degradation and losses;

- Increased flooding and storm damage in low-lying coastal areas as episodic storm surges and destructive waves penetrate further inland; and

- Increased salinity in estuaries, marshes, coastal rivers, and coastal aquifers (Leatherman, 1989; Kana et al., 1984).

These primary impacts will trigger other impacts such as damage to settlements, industry, and other coastal infrastructure, including ports, ship channels, and bridges. Where hazardous waste landfills are affected, pollutants in the landfills may migrate because of flooding and water-table changes or may even be directly released via erosion. As sea levels rise, these impacts may become more severe, depending on individual site characteristics and response strategies.

There are already widespread problems on the East and Gulf coasts linked by varying degrees to sea-level rise in the 20th century and often exacerbated by poor management. For example, attempts to maintain navigation channels and other human activities in the Mississippi Delta have contributed to major wetland losses by removing the sediment supplies that help the wetlands maintain elevation and keep pace with relative sea-level rise, as well as severely altering their hydrology (Boesch et al., 1994). Most beaches are eroding and extensive tidal wetland losses are occurring, particularly in Louisiana and around the Chesapeake Bay (Heinz Center, 2000; Komar, 2000). Sea-level rise appears to contribute to these widespread problems (e.g, see Leatherman et al., 2000)

National assessments suggest that a one-meter rise in global sea levels could have significant impacts, including the possibility of inundating about 19,000 km^2 (6,400 mi^2) of upland (Smith and Tirpak, 1989; Titus et al., 1991). The 100-year coastal flood plain (i.e., the area subject to flooding during a storm that occurs on average once every 100 years) also could increase by 38 percent, or at least 18,000 km² (7,000 mi²) (FEMA, 1991). A 50-cm global sea-level rise would inundate about 12,000 km² (4,000 mi²) of upland, two-thirds of the potential inundation estimated under a one-meter rise because of the distribution of coastal elevation (Smith and Tirpak, 1989). In either case, major cities such as New Orleans, Tampa, Miami, Baltimore, Philadelphia, New York, Boston, and Washington, DC, will have to upgrade flood defenses and drainage systems or face adverse consequences. New Orleans is particularly threatened by the rapid loss of the surrounding wetlands in the Mississippi delta, which currently protect the city from storm surges.

The extent and geographic distribution of wetland losses depend on physical as well as human factors. Wetlands are not passive elements of the coastal landscape and, given sea-level rise, they will accrete vertically due to increased biological and sedimentary inputs and/or migrate inland (Reed, 1995). Drowning of wetlands only occurs if sea-level rise is too fast. Wetlands with limited sediment supplies and low tidal range (i.e., a small elevation difference between low and high tide) appear to be most threatened by sea-level rise (IPCC, 1996b). The availability of low-lying upland areas landward of the wetlands is also critical for wetland survival. If the uplands are protected by coastal defenses, landward migration is not possible. Titus et al. (1991) estimate that a one-meter rise in sea level would cause the loss of about 16,000 km² (6,000 mi²) of wetlands and, if no wetland migration were possible, this loss would rise to 22,000 km² (8,500 mi²). Loss estimates under a 50-cm rise are about two-thirds those under the one-meter scenario. Table 2 summarizes one assessment of the spatial distribution of projected wetland losses in the U.S. due to sea-level rise. Consistent with the notion that more rapid relative sea-level rise may outstrip rates of sediment accumulation, the data in Table 2 identify the U.S. Gulf Coast region as the area most likely to suffer extensive wetland losses due to sea-level rise.

Global analyses identify the Atlantic and Gulf coasts as the most vulnerable in North America to wetland losses from sea-level rise (Hoozemans et al., 1993; Nicholls et al., 1999). Where wetlands support commercial fisheries, that resource could be dramatically affected. For example, Browder et al. (1989) suggest that the present catch rate in the Louisiana brown shrimp fishery is unsustainable if Louisiana's wet-land losses continue, and the fishery could dramatically decline in the 21st century. Other functions of wetlands, such as concentrating nutrients, maintaining water quality, and protecting coasts from storm surges, would also be degraded or lost.

In addition to the gradual sea-level rise described above, the threat of rapid, abrupt rises in sea level associated with massive glacier and ice shelf melting has been raised in the scientific literature. If the West Antarctic ice shelf were destabilized by global warming and slid into the ocean, there would be a 5- to 7-meter rise in global sea levels (Mercer, 1978). There is geological evidence that this scenario occurred about 100,000 years ago. While the IPCC (1990; 1996a) has reviewed available evidence

Table 2

Projected Changes in U.S. Wetlands Acreage Attributed to Sea-Level Rise

Coastal Region	Coastal Wetlands, 1986 (thousands of acres)[a]	Projected Percent Change in Wetlands Acreage by 2100[b]
Northeast (ME, NH, MA, RI)	121	−3
Mid-Atlantic (CT, NY, NJ, DE, MD, VA)	733	−13
South Atlantic (NC, SC, GA)	1,377	+4
Florida	736	+29
Gulf Coast (MS, AL, LA, TX)[c]	3,885	−62
West Coast[d]	89	−40

Notes: a) Baseline wetlands estimates based on a 1986 estimate by NOAA; these are the latest national estimates available that are derived from a consistent method. b) Projected changes in wetlands acreage based on estimates by Armentano et al., assuming a 57-cm increase in sea level due to climate change. c) Louisiana accounts for the majority of wetlands in this region, as well as the majority of potential wetlands losses under a 57-cm increase in sea-level rise. d) Does not include Hawaii or Alaska.

Source: Armentano et al., 1988.

and has concluded that such an event is extremely unlikely in the 21st century, the impacts of catastrophic melting of ice sheets would be enormous. Large parts of the East and Gulf coasts, particularly in Louisiana and Florida, are beneath a 5- to 7-meter elevation.[7]

Human Response and Adaptation to Coastal Threats

Three generic options are available to decision-makers who contemplate responding to coastal threats: protection, accommodation, and planned retreat (IPCC, 1996b; Klein et al., 2000). Protection seeks to exclude the hazard, accommodation allows human activities and the hazard to coexist, while planned retreat removes human activity from the hazardous zone.

Protection

Protection options include hard structure responses (such as building dikes and sea walls) and "soft" engineering responses that utilize sediment (particularly beach nourishment, or replenishment of sand resources). These strategies could protect some of the resources vulnerable to sea-level rise (e.g., coastal property), but might sacrifice other resources in the process (e.g., natural wetlands or beaches). Building hard structures, for example, would limit the ability of beach and wetland resources to migrate inland as sea level rises. These losses are often termed "coastal squeeze" (IPCC, 1996b), and are already recognized as a major problem in some other parts of the world (see Nicholls, 2000).

Because of their relatively high cost, hard structures are most often applied in developed and urban areas where the value of protection is high. Hard structures do not have to be constructed until they are needed. Ideally, they would be built just in advance of the threat of inundation. Because the pace of greenhouse induced sea-level rise is unknown, some anticipation and monitoring is required to implement appropriate pre-emptive responses. Hard structures could eventually degrade seaside beaches, though, because they amplify the ability of waves to scour sand away. Hard protection can therefore cause recreational and ecological resources to decline. For example, in Galveston, Texas, a seawall constructed to protect the city has resulted in the loss of the natural beach.

The addition of sediment and/or improved sediment management is a "soft protection" alternative with the goal of increasing or at least maintaining the size of the beach or wetland (Davison et al., 1992; Boesch et al., 1994). Beach nourishment (i.e., the placement of sand usually dredged from offshore) is a popular and cost-effective option in highly developed areas with popular beaches and valuable beachfront real estate, especially during the early onset of erosion (American Society of Coastal Engineers, 1992). Florida, Delaware, and New Jersey, for example, have established funds for a variety of beach nourishment projects (Yohe and Neumann, 1997). However, the long-term effectiveness of beach nourishment remains uncertain. As sea levels rise more rapidly and cheaper sand resources are exhausted, the cost of nourishment will rise, and it may cease to be cost-effective in some coastal locations. Thus both hard and soft protection raise problems, and it is unlikely that the entire U.S. coast could ever be protected, even if that were desirable. Therefore, accommodation and planned retreat are likely to be a preferred response in many locations.

Accommodation and Planned Retreat

Accommodation and planned retreat are achieved primarily through land use and development planning. Accommodation might include elevation of buildings above flood levels, higher capacity drainage systems, or land-use planning. Planned retreat often involves restrictions on coastal development such as setback measures and post-disaster plans.

Setback measures, a zoning mechanism for planned retreat employed by some states, require that new structures be set back from the shore, usually by some multiple of the average annual rate of erosion (e.g., 20 to 60 times). According to a recent review of erosion control policies, 23 states make use of setbacks within erosion-prone areas, although implementation and enforcement of these policies varies considerably (Heinz Center, 2000; also see Klarin and Hershman, 1990; Titus, 1990; Yohe and Neumann, 1997; relevant Australian experience is discussed by Caton and Eliot, 1993). Maine's Dune Rules reference setbacks explicitly to sea-level rise by requiring demolition of large structures if the sea level rises by one meter or more (Titus, 1990). These types of measures can be controversial and legally contentious, but once implemented they are an effective means of communicating and enforcing a planned retreat strategy.

Post-disaster reconstruction plans are a second type of mechanism to encourage planned retreat (or accommodation). Post-disaster plans limit or prohibit reconstruction

Climate Change: Science, Strategies, & Solutions

of coastal property severely damaged by hurricanes, storms, or episodic flooding. For example, South Carolina's Beachfront Management Act of 1988 states that structures incurring damages of more than two-thirds of pre-storm value cannot be reconstructed (Yohe and Neumann, 1997). Table 3 describes four federal programs that encourage better management of coastal resources, including the National Flood Insurance Program, which requires elevation of structures damaged more than 50 percent of pre-storm value above the 100-year flood level plus wave heights. Effective post-disaster policies can actually reduce vulnerability to sea-level rise in areas where storm surge flooding provides an early indication of which structures will later be vulnerable to inundation.

One method of accommodating sea-level rise is to elevate structures. This option has been implemented by property owners in places such as the New Jersey and South Carolina shorefronts and provides an effective means of protecting property from episodic flooding associated with sea-level rise.

Table 3

Federal Land-Use Policies for U.S. Coastal Zones

Agency	Policy	Major Provision	Concern
Federal Emergency Management Agency	National Flood Insurance Program	Subsidize insurance for communities that adopt land-use regulations and building standards	Creates incentives to develop flood-prone areas (Klarin and Hershman, 1990)
	Upton-Jones Amendment	Encourage removal of unstable structures and conduct long-term planning	Few claims have been made (Platt et al., 1992)
Department of Commerce	Coastal Zone Management Act	State coastal programs must plan to minimize loss of life and damage caused by the destruction of natural features	States have wide latitude (Edgerton, 1991)
Department of the Interior	Coastal Barrier Resource Act	Creates national system of protected areas and disallows federal subsidies of insurance and the location of infrastructure	States have wide latitude

Economic Impacts on Coastal Property and Wetlands

Current estimates suggest that a 50-cm sea-level rise by 2100 could cause cumulative damage to coastal property in the United States of $20 billion to $150 billion. Estimates at the low end of the range reflect assumptions that the most economically efficient adaptation to sea-level rise will take place. Those at the high end assume that all currently developed vulnerable areas will be protected, regardless of cost.

The IPCC (1996b) suggests that impact assessments reflect at least five categories of impacts: inundation and erosion of property, inundation of wetlands, effects on recreation, effects on drinking water quality and quantity, and effects on port infrastructure. The potential costs associated with all of these impacts are interrelated, although they are often assessed separately. Property losses dominate cost estimates in the United States. The monetary value of wetland losses, which may prove to be large, currently are not included in most impact assessments. Likewise, national assessments of the effects of sea-level rise on recreation, drinking water, and ports have not been conducted. While these five categories are likely to be the most important for impact assessments in the U.S. context, studies in other countries may consider a broader range of effects (Box 2).

Box 2

International Impacts of Sea-Level Rise

The effects on tourism, subsistence fishing, agriculture, human settlements, and freshwater supplies may be of much greater relative importance in developing coastal countries than in the United States. Globally, the impacts of flooding from sea-level rise are expected to be most serious in Africa and South and South-East Asia (Nicholls et al., 1999). Delta areas of Bangladesh, Vietnam, and Myanmar, for example, support significant agricultural production in areas that would be threatened by sea-level rise. Small coral atolls and reef islands in the Pacific and Indian Oceans face a significantly greater risk of loss of freshwater supplies to saltwater intrusion than the United States. Inundation of some small island nations, whose economies may be largely or almost entirely dependent on coastal activities, may cause economic and social collapse. Although the contribution of these threatened countries to the global economy may be small, the loss of unique cultures and ecosystems would be significant (IPCC, 1996b).

In addition, the capacity to adapt to accelerated sea-level changes is likely to be more limited in developing countries than in the United States. The costs of protection or adaptation to sea-level rise may represent a much larger percentage of the total economy of a developing nation. One estimate for the Marshall Islands indicates that the costs of a one-meter sea-level rise could exceed 7 percent of gross national product (GNP) (Holthus et al., 1992). In the Maldives, costs could approach one-third of GNP (IPCC, 1996b). Considering impacts of this magnitude may require new frameworks of analysis. For example, a recent study of Fiji found that traditional impact assessment methods could undervalue economic and cultural assets of the coast and might overstate the resilience of the economy to sea-level rise (Yamada et al., 1995).

The Economic Cost to Developed Property

Estimates of the costs of sea-level rise have fallen in recent years as the ability to model adaptation and incorporate local conditions has increased. The magnitude of developed property impact estimates reflects two key factors: the sea-level rise scenario and the modeling of responses.

Some of the earliest results, such as those produced by Schneider and Chen (1980), reflected extreme sea-level rise scenarios — 450 to 750 cm by 2100 — that are now considered irrelevant (except with regard to the catastrophic Antarctic melting scenario discussed earlier). The first systematic national study of more moderate sea-level rise scenarios was the U.S. Environmental Protection Agency's 1989 Report to Congress (Smith and Tirpak, 1989). Results from this and other studies are summarized in Table 4.

Studies that take adaptation into account have produced more moderate cost estimates, indicating that adaptation can help minimize costs. A study by Fankhauser (1994) addressed whether protecting individual segments of coastal property reflected local conditions and considered the relative costs and benefits of protection.

Table 4

Potential Cost of Sea-Level Rise Along the Developed Coastline of the United States (billions of 1990 dollars)

Global Sea-Level Rise (source)	Measurement	Annualized Estimate	Cumulative Estimate	Annual Estimate in 2065
100 cm (Yohe, 1989)	Property at risk of inundation	n/a	321	1.37
100 cm (EPA, 1989)	Protection	n/a	73–111	n/a
100 cm (Nordhaus, 1991)	Protection	4.9	n/a	n/a
100 cm (Fankhauser, 1994)	Protection	1.0	62.6	n/a
100 cm (Yohe et al., 1996)	Protection and abandonment	0.16	36.1	0.33
50 cm (Yohe, 1989)	Property at risk of inundation	n/a	138	n/a
50 cm (Fankhauser, 1994)	Protection	0.57	35.6	n/a
50 cm (Yohe et al., 1996)	Protection and abandonment	0.06	20.4	0.07
50 cm (Yohe and Schlesinger, 1998)	Expected protection and abandonment	0.11	n/a	0.12
100 cm (Yohe and Schlesinger, 1998)	Expected protection and abandonment	0.38	n/a	0.40
41 cm (mean) (Yohe and Schlesinger, 1998)	Protection and abandonment	0.09	n/a	0.10
10 cm (10th percentile) (Yohe and Schlesinger, 1998)	Protection and abandonment	0.01	n/a	0.01
81 cm (90th percentile) (Yohe and Schlesinger, 1998)	Protection and abandonment	0.23	n/a	0.31

Note: All of the cumulative estimates but Fankhauser's are undiscounted; his estimates are discounted effectively by the annual rate of growth of per capita GNP (expected to average approximately 1.6% for the United States through 2100). The annual estimate in 2065 is available for the one study that estimates costs along the transient (i.e., estimates for each individual year); this estimate is undiscounted.

Fankhauser assumed gradual inundation patterns and reported annual protection costs of $570 million for a 50-cm rise and $1 billion for a 100-cm rise by 2100 (in 1990 dollars). This estimate is one-fifth of that from an earlier Nordhaus (1991) study, which did not include adaptation. A study by Yohe et al. (1996) provided a more thorough, site-specific assessment of the effects of adaptation. Yohe et al. found that consideration of the variability of local topography and baseline land use in decisions to protect or abandon property, based on assessments of the costs and benefits of available options, further lowered impact estimates.

Foresight is another factor that could influence costs. Yohe et al. (1996) and Yohe and Schlesinger (1998) found that costs are about 30 percent higher when property owners respond to sea-level rise as it happens, without the benefit of foresight, compared to response under the assumption of perfect foresight. Nordhaus (1999) described this difference as a first representation of the extra cost of climate surprise.

Wetlands Assessment

Existing assessments of lost wetland areas, while only first approximations, nonetheless point to the potentially large magnitude of damages that sea-level rise could cause to coastal wetland resources (Smith and Tirpak, 1989; Titus, 1992; Armentano et al., 1988; see Table 2 above). Coastal wetlands provide a wide range of amenities that are economically and ecologically valuable. These amenities include flood control, habitat for ecologically important and endangered species as well as commercially important fish and shellfish species, and groundwater recharge.

Current estimates of the impact of wetland losses suffer from two critical methodological and data gaps. First, as noted above, it is difficult to model the dynamic processes of wetland accretion and migration (Reed, 1995), which could mitigate wetland losses in some cases (Patrick and DeLaune, 1990). Significant challenges remain in determining which wetlands will be inundated, when the inundation will occur, what the impacts of natural and artificial barriers will be, and whether wetlands that migrate can replace existing ones. Second, there have been many efforts to assign a dollar value to wetland resources over the last two decades, but the overall economic value of wetlands (including aesthetic value) is not well-characterized by existing techniques.

Impacts of Recreation Losses

The use of coastal public lands and recreational resources has risen in step with population growth. Accelerated beach erosion caused by sea-level rise might lead to large losses in the value of these resources. By one estimate, recreational beach visits are valued at over $3 billion annually (Loomis and Crespi, 1999).

Although some coastal recreation losses could be mitigated through beach nourishment, the scale of nourishment required to hold the line for the next century could be vast, even if sea-level rise does not accelerate. Assessments of likely coastal changes and potential demands for nourishment over the next century for a range of sea-level rise scenarios would be helpful. In addition, recreational impacts should also be reflected in estimates of lost property value since coastal property values include a premium associated with access to the coast. Uneven erosion of recreational value would cause some property values to rise even as others fall, suggesting that distributional effects should be tracked carefully.

Implications for Future Assessments

In many cases, the impacts of sea-level rise could be mitigated by forward-looking state or local land-use policies. Consistent with this conclusion, several recent efforts have focused on defining coastal management policies that can enhance the ability of coasts to adapt to sea-level rise (Nicholls and Branson, 1998; Klein et al., 1998). The major challenges of future impact assessments include improving their comprehensiveness and accuracy and making their results more accessible and useful to state and local decision-makers who are most able to prepare coastal areas to respond to the threat of sea-level rise.

Some efforts are underway to improve impact assessments. These include U.S. Environmental Protection Agency work to better understand the impact of state and local policies on protect and retreat decisions, efforts to develop models of the local policy-making process and the process of learning in response to storm damage (Moser, 1999), and efforts to incorporate the combined effects of storm damage, erosion, and flooding in coastal areas. Such efforts may provide not only improved accuracy in impact assessment but also better ways to communicate their results to influence local decisions affecting the long-term capacity to adapt.

The IPCC, among others, has begun to put more emphasis on the study of adaptive measures themselves, as recognition grows that sea-level rise of some magnitude is inevitable regardless of the outcome of emissions reduction negotiations (for example, Klein et al., 2000). In particular, attention is now turning to the multiple determinants of adaptive capacity — that is, those factors that define the extent to which humans and natural systems can reasonably adapt to stress. There remains a need to further study the dynamics of adaptation in human systems, the processes of adaptive decision-making, the conditions that stimulate or constrain adaptation, and the role of non-climatic factors. In addition, because of the unequal distribution of the costs and benefits of actions to increase adaptive capacity to coastal stresses, there is a need for clarification of the roles and responsibilities in adaptation of individuals, communities, corporations, private and public institutions, governments, and international organizations.

Endnotes

1. There is evidence that an increase in air and sea surface temperatures and the resulting sea-level rise is already happening (IPCC, 1996a; Cane et al. 1997).

2. Note that this "post-glacial rebound" effect contributes to uplift in formerly glaciated areas, but also contributes to subsidence on the fringes of glaciated areas, such as the mid-Atlantic region.

3. During major earthquakes, rapid subsidence will occur, but the time interval between these events is typically hundreds of years.

4. An analysis based on a survey of the opinions of climate modelers provided a more moderate median estimate of a 34-cm rise from 1990 to 2100, with a 99 percent confidence interval falling between a decrease of 1 cm and a rise of 104 cm (Titus and Narayanan, 1995).

5. Estimates are in constant 1993 dollars, see Insurance Institute for Property Loss Reduction and Insurance Research Council (1995).

6. While the relationship between climate change and storms is quite uncertain, increases in coastal populations, infrastructure, and development (which have resulted in increasing damages from storms) suggest that even small growth in storm frequency or severity would create a disproportionate increase in damages (Pielke and Landsea, 1998).

7. Ice shelf collapse will become more likely in the 22nd or 23rd century, if global warming continues unabated. Recent research suggests that the melting of clathrates, sea floor water ice crystals that contain methane, may explain sea-level drops during warm periods of the geologic record (Bratton, 1999). Catastrophic melting of clathrates, however, is probably a very low probability event (Harvey and Huang, 1995).

References

American Society of Coastal Engineers. 1992. Effects of Sea-Level Rise on Bays and Estuaries. *Journal of Hydraulic Engineering* 118(1): 1–10.

Armentano, T.V., R.A. Park, and C.L. Cloonan. 1988. Impacts on Coastal Wetlands Throughout the United States. Chapter 4 in *Greenhouse Effect, Sea-Level Rise, and Coastal Wetlands.* J.G. Titus, ed. U.S. Environmental Protection Agency, Washington, DC.

Baumann, R., J. Day, and C. Miller. 1984. Mississippi Deltaic Wetland Survival: Sedimentation Versus Coastal Submergence. *Science* 224: 1093–1094.

Boesch, D.F., M.N. Josselyn, A.J. Mehta, J.T. Morris, W.K. Nuttle, C.A. Simenstad, and D.J.P. Swift. 1994. Scientific Assessment of Coastal Wetland Loss, Restoration and Management in Louisiana. *Journal of Coastal Research.* Special Issue No. 20.

Bratton, J.F. 1999. Clathrate Eustasy: Methane Hydrate Melting as a Mechanism for Geologically Rapid Sea-Level Fall. *Geology* 27(10): 915-919.

Browder, J.A., L.N. May, Jr., A. Rosenthal, J.G. Gosselink, and R.H. Naumann. 1989. Modeling Future Trends in Wetland Loss and Brown Shrimp Production in Louisiana Using Thematic Mapper Imagery. *Remote Sensing of Environment* 28: 45–49.

Cane, M.A., A.C. Clement, A. Kaplan, Y. Kushnir, D. Pozdnyakov, R. Seager, S.E. Zebial, and R. Murtugudde. 1997. Twentieth-Century Sea Surface Temperature Trends. *Science* 275: 957–960.

Caton, B., and I. Eliot. 1993. Coastal Hazard Policy Development and the Austrialian Federal System. In *Vulnerability Assessment to Sea-Level Rise and Coastal Zone Management.* R.F. McLean and N. Minura, eds. Proceedings of the IPCC/WCC'93 Eastern Hemisphere workshop, Tsukuba, Japan, 3-6 August 1993. Department of Environment, Sport and Territories, Canberra, Australia, pp. 417–427.

Culliton, T.J., M. Warren, T. Goodspeed, D. Remer, C. Blackwell, and J. McDonough III. 1990. *Fifty Years of Population Change Along the Nation's Coasts: 1960–2010.* National Oceanic and Atmospheric Administration, Rockville, MD.

Davison, A.T., R.J. Nicholls, and S.P. Leatherman. 1992. Beach Nourishment as a Coastal Management Tool: An Annotated Bibliography of Developments Associated with the Artificial Nourishment of Beaches. *Journal of Coastal Research* 8: 984–1022.

DeLaune, R.D., W.H. Patrick, Jr., and C.J. Smith. 1992. Marsh Aggradation and Sediment Distribution Along Rapidly Submerging Louisiana Gulf Coast. *Environmental Geology & Water Sciences* 20(1): 57–64.

Edgerton, L.T. 1991. *The Rising Tide: Global Warming and World Sea Levels.* Natural Resources Defense Council, Washington, DC.

Emery, K.O., and D.G. Aubrey. 1991. *Sea Levels, Land Levels and Tide Gauges.* Springer Verlag, New York, NY.

EPA. 1989. *The Potential Effects of Global Climate Change on the United States.* Report to Congress. U.S. Environmental Protection Agency, Washington, DC.

Fankhauser, S. 1994. *Protection vs. Retreat: Estimating the Costs of Sea-Level Rise.* CSERGE, London.

FEMA. 1991. *Projected Impact of Relative Sea-Level Rise on the National Flood Insurance Program.* Report to Congress. Federal Insurance Administration, Washington, DC.

Gregory, J.M. 1993. Sea-Level Changes under Increasing Atmospheric CO_2 in a Transient Coupled Ocean-Atmosphere GCM Experiment. *Journal of Climate* 6: 2247–2262.

Harvey, L.D.D., and Z. Huang. 1995. Evaluation of the Potential Impact of Methane Clathrate Destabilization on Future Global Warming. *Journal of Geophysical Research* 100(2): 2905–2926.

Heinz Center. 2000. *Evaluation of Erosion Hazards.* Prepared for the Federal Emergency Management Agency by the H. John Heinz III Center for Science, Economics, and the Environment, Washington, DC.

Holthus, P., M. Crawford, C. Makroro, and S. Sullivan. 1992. *Vulnerability Assessment for Accelerated Sea-Level Rise Case Study: Majuro Atoll, Republic of the Marshall Islands.* SPREP Reports and Studies Series No. 60. South Pacific Regional Environment Programme, Apia, Western Samoa.

Hoozemans, F.M.J., M. Marchand, and H.A. Pennekamp. 1993. *A Global Vulnerability Analysis: Vulnerability Assessment for Population, Coastal Wetlands and Rice Production on a Global Scale.* 2nd edition. Delft Hydraulics, The Netherlands.

Insurance Institute for Property Loss Reduction and Insurance Research Council. 1995. *Coastal Exposure and Community Protection: Hurricane Andrew's Legacy.* Boston, MA.

IPCC. 1990. *Sea-Level Rise. Climate Change: The IPCC Scientific Assessment.* J.T. Houghton, G.J. Jenkins, and J.J. Ephramus, eds. Cambridge University Press, Cambridge, UK, pp. 257–281.

IPCC. 1996a. *Changes in Sea Level. Climate Change 1995: The Science of Climate Change.* J.T. Houghton, L.G. Meira Filho, and B.A. Callander, eds. Cambridge University Press, Cambridge, UK, pp. 359–405.

IPCC. 1996b. *Coastal Zones and Small Islands. Impacts, Adaptations and Mitigation of Climate Change: Scientific-Technical Analyses.* R.T. Watson, M.C. Zinyowera, and R.H. Moss, eds. Cambridge University Press, Cambridge, UK, pp. 289–324.

Kana T., J. Michel, M. Hayes, and J. Jensen. 1984. The Physical Impact of Sea-Level Rise in the Area of Charleston, South Carolina. — Chapter 4 in *Greenhouse Effect and Sea-Level Rise — A Challenge for This Generation.* M.C. Barth and J.G. Titus, eds. Van Nostrand Reinhold, New York, NY, pp. 105–150.

Klarin, P., and M. Hershman. 1990. Response of Coastal Zone Management Programs to Sea-Level Rise in the United States. *Coastal Management* 18: 143–165.

Klein, R., M. Smit, H. Goosen, and C. Hulsbergen. 1998. Resilience and Vulnerability: Coastal Dynamics or Dutch Dikes. *The Geographical Journal* 164: 259–268.

Klein, R.J.T., J. Aston, E.N. Buckley, M. Capobianco, N. Mizutani, R.J. Nicholls, P.D. Nunn, and S. Ragoonaden. 2000. Coastal–Adaptation Technologies. In *IPCC Special Report on Methodological and Technological Issues in Technology Transfer.* B. Metz, O.R. Davison, J.W. Martens, S.N.M. van Rooijen, and L.L. Van Wie McGrory, eds. Cambridge University Press, Cambridge, UK, pp. 349–372.

Komar, P.D. 2000. Coastal Erosion: Underlying Factors and Human Impacts. *Shore and Beach* 68(1): 3–16.

Landsea. C. 1993. A Climatology of Intense or Major Atlantic Hurricanes. *Monthly Weather Review* 121: 1703–1713.

Leatherman, S. 1989. *National Assessment of Beach Nourishment Requirements Associated with Accelerated Sea-Level Rise. The Potential Effects of Global Climate Change on the United States, Appendix B: Sea-Level Rise.* J.B. Smith and D. Tirpak, eds. U.S. Environmental Protection Agency, Washington, DC, pp. 2–1 to 2–30.

Leatherman, S.P., K. Zhang, and B.C. Douglas. 2000. Sea-Level Rise Shown to Drive Coastal Erosion. *EOS Transactions American Geophysical Union* 81(6): 55–57.

Loomis, J., and J. Crespi. 1999. Estimated Effects of Climate Change on Selected Outdoor Recreation Activities in the United States. In *The Impact of Climate Change on the United States Economy.* R. Mendelsohn and J.E. Neumann, eds. Cambridge University Press, Cambridge, UK, pp. 289–314.

Louisiana Legislative Study Group. 1999. *Danger and Opportunity: Implications of Climate Change for Louisiana.* A Report for the Louisiana State Legislature to Fulfill House Concurrent Resolution 74, Regular Session, 1996 (May).

Mercer, J.H. 1978. West Antarctic Ice Sheet and CO_2 Greenhouse Effect: A Threat of Disaster. *Nature* 271: 321–325.

Miller, M.L., and J. Auyong. 1991. Coastal Zone Tourism: A Potent Force Affecting Environment and Society. *Marine Policy*, pp. 75–99.

Moser, S. 1999. *Selected Bibliography of Evaluation Studies of Federal Programs.* Contribution to a Workshop on Coastal Zone Impact, National Science Foundation, Charleston, SC.

National Research Council. 1987. *Responding to Changes in Sea Level: Engineering Implications.* National Academy Press, Washington, DC.

Nicholls, R., and J. Branson. 1998. Coastal Resilience and Planning for an Uncertain Future: An Introduction. *The Geographical Journal* 164: 255–257.

Nicholls, R.J., F.M.J. Hoozemans, and M. Marchand. 1999. Increasing Flood Risk and Wetland Losses due to Global Sea-Level Rise: Regional and Global Analyses — The Science of Climate Change. *Global Environmental Change* 9: 569-587.

Nicholls, R.J., and S.P. Leatherman. 1996. Adapting to Sea-Level Rise: Relative Sea-Level Trends to 2100 for the USA. *Coastal Management* 24(4): 301-324.

Nicholls, R.J. 2000. Chapter 14. Coastal Zones. In *Assessment of the Potential Effects of Climate Change in Europe.* M.L. Parry, C. Parry, and M. Livermore, eds. Jackson Environment Institute, University of East Anglia, in press.

National Oceanic and Atmospheric Administration (NOAA). 1999. *State of the Coast.* See http://state-of-coast.noaa.gov/bulletins.

Nordhaus, W.D. 1999. *The Economic Impacts of Abrupt Climatic Change.* Paper prepared for a Meeting on Abrupt Climate Change: The Role of Oceans, Atmosphere, and the Polar Regions, National Research Council, January 22.

Nordhaus, W.D. 1991. To Slow or Not to Slow. *Economics Journal* 5: 920–937.

Patrick, W.H., and R.D. DeLaune. 1990. Subsidence, Accretion, and Sea-Level Rise in South San Francisco Bay Marshes. *Limnology and Oceanography* 35(6): 1389–1395.

Pielke, R., and C. Landsea. 1998. Normalized Hurricane Damages in the United States: 1925–1995. *Weather and Forecasting* 13: 621–631.

Platt, R.H., H. Crane Miller, T. Beatley, J. Melville, and B.G. Mathenia. 1992. *Coastal Erosion: Has Retreat Sounded?* Institute of Behavioral Science, University of Colorado, Boulder.

Reed, D.J. 1995. The Response of Coastal Marshes to Sea-Level Rise: Survival or Submergence? *Earth Surface Processes and Landforms* 20: 39–48.

Schneider, S., and R. Chen. 1980. Carbon Dioxide Warming and Coastline Flooding: Physical Factors and Climatic Impact. *Annual Review Energy* 5: 107–140.

Smith, J.B., and D.A. Tirpak, eds. 1989. *The Potential Effects of Global Climate Change on the United States.* Report to Congress. U.S. Environmental Protection Agency, Washington, DC.

Stone, G., and R. McBride. 1998. Louisiana Barrier Islands and their Importance in Wetland Protection: Forecasting Shoreline Change and Subsequent Response of Wave Climate. *Journal of Coastal Research* 14(3): 900–915.

Titus, J.G. 1990. Greenhouse Effect, Sea-Level Rise, and Barrier Islands: Case Study of Long Beach Island, New Jersey. *Coastal Management* 18: 65-90.

Titus, J.G. 1992. The Costs of Climate Change in the United States. In *Global Climate Change: Implications, Challenges, and Mitigation Measures.* S.K. Majumdar, L.S. Kalkstein, B.M. Yarnal, E.W. Miller, and L. M. Rosenfeld, eds. Pennsylvania Academy of Science, Easton, PA.

Titus, J.G., and V.K. Narayanan. 1995. *The Probability of Sea-Level Rise.* U.S. Environmental Protection Agency, Washington, DC.

Titus, J.G., R.A. Park, S.P. Leatherman, J.R. Weggel, M.S. Green, P.W. Mausel, S. Brown, C. Gaunt, M. Trehan, and G. Yohe. 1991. Greenhouse Effect and Sea-Level Rise: Potential Loss of Land and the Cost of Holding Back the Sea. *Coastal Management* 19: 171–204.

Titus, J.G., and C. Richman. 2000. *Maps of Lands Vulnerable to Sea-Level Rise: Modeled Elevations along the U.S. Atlantic and Gulf Coasts.* Climate Research, in press.

Turner, R.K., S. Subak, and W.N. Adger. 1996. Pressures, Trends and Impacts in the Coastal Zones: Interactions between Socio-Economic and Natural Systems. *Environmental Management* 20: 159–173.

Yamada, K., P.D. Nunn, N. Mimura, S. Machida, and M. Yamamoto. 1995. Methodology for Assessment of Vulnerability of South Pacific Island Countries to Sea-Level Rise and Climate Change. Journal of Global Environment Engineering 1: 101–125.

Yohe, G. 1989. The Cost of Not Holding Back the Sea — Economic Vulnerability. *Ocean and Shore Management* 15: 233-255.

Yohe, G., and J. Neumann. 1997. Planning for Sea-Level Rise and Shore Protection under Climate Uncertainty. *Climatic Change* 37: 111–140.

Yohe, G., J. Neumann, P. Marshall, and H. Ameden. 1996. The Economic Cost of Greenhouse Induced Sea-Level Rise for Developed Property in the United States. *Climatic Change* 32: 387–410.

Yohe, G., and M. Schlesinger. 1998. Sea-Level Change: The Expected Economic Cost of Protection or Abandonment in the United States. *Climatic Change* 38: 447–472.

Zhang, K., B.C. Douglas, and S.P. Leatherman. 2000. Twentieth-Century Storm Activity along the U.S. East Coast. *Journal of Climate* 13: 1748–1761.

Potential Impacts on U.S. Water Resources

Kenneth D. Frederick, Peter H. Gleick

Water is critical to a society's welfare: It is vital for agriculture and industry, energy production, transportation, recreation, the functioning of natural ecosystems on which humans depend, and the disposal of wastes. Natural variability in water supply also affects society. Too much water results in floods and too little results in drought, with potentially large socioeconomic costs.

The United States, on average, is well endowed with water. Annual precipitation averages nearly 30 inches, or 4,200 billion gallons per day (bgd), throughout the continental 48 states. While two-thirds of this precipitation quickly evaporates or transpires back to the atmosphere, the remaining one-third replenishes the nation's streams and groundwater supplies. This runoff provides a renewable supply that is nearly 15 times larger than current consumptive use — water withdrawn from, but not returned to, a water source in a usable form. In addition, water stored in lakes, reservoirs, and groundwater aquifers within 2,500 feet of the surface is equivalent to more than 50 years of renewable supply (U.S. Water Resources Council, 1978).

But these averages hide important regional and temporal problems with distribution. Figures 1 and 2 show regional variations in average annual precipitation and runoff. Despite its apparent abundance and renewability, fresh water can be scarce virtually anywhere in the United States at some time, especially in the arid and semiarid West. Moreover, the availability of water to meet the demands of a growing and increasingly affluent population — while sustaining a healthy environment — has emerged as one of the nation's primary resource issues. These concerns are based in part on uncertainties about the availability of supplies stemming from the vicissitudes of the hydrologic cycle, growing populations, and, more recently, the prospect that human-induced climate changes will alter the cycle in uncertain ways.

The hydrologic cycle naturally consists of large seasonal, annual, and regional variations in supplies. Because of these variations, we have built a vast number of dams, reservoirs, canals, pumps, and levees to collect, control, and contain surplus flows and to distribute water on demand.

The nature of our current water-use patterns and the infrastructure to regulate and allocate supplies are the result of past hydrologic conditions. Even today, the design and evaluation of water investments and management strategies assume that future precipitation and runoff patterns will be similar to those of the past. The increasing

This paper is a condensed and updated version of a report published in September 1999 by the Pew Center: Water & Global Climate Change: Potential Impacts on U.S. Water Resources. The authors and the Pew Center are grateful for the input of John Boland, Barbara Miller, and Kenneth Strzepek.

Figure 1

Average **Annual Precipitation** in the United States and Puerto Rico

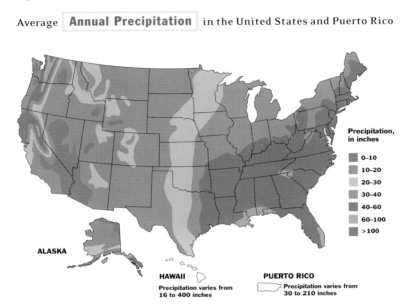

Note: The figures are based on 30-year average precipitation data.
Source: U.S. Water Resources Council, 1978.

likelihood that human-induced, or anthropogenic, climate change could affect water quality and availability, as well as demand, raises doubts about these assumptions and the most appropriate water policies for the future.

This chapter reviews what is known about the potential impacts of anthropogenic climate change on the supply and demand for water in the United States and the resulting economic and ecological implications of those impacts. A tremendous amount has been written about this subject.[1] This chapter reviews the most critical information and identifies the most important gaps in our knowledge.

Climate Change and its Impacts on Hydrology and Water Resources

Changes in Temperature and Precipitation

Changes in climate affect all aspects of the hydrologic cycle. Changes in tempera-ture affect cloud characteristics, soil moisture, snowfall and snowmelt regimes, and the rate at which water evaporates from the earth's surface or transpires from plants. Changes in precipitation affect the timing and magnitude of floods and droughts, shift runoff regimes, and alter groundwater recharge characteristics. Synergistic effects alter cloud formation and extent, vegetation patterns and growth rates, and soil conditions.

Climate Change: Science, Strategies, & Solutions

Figure 2

Average  in the Continental United States and Alaska

Note: The figures are based on 30-year average precipitation data.
Source: U.S. Water Resources Council, 1978.

Atmospheric concentrations of carbon dioxide may also affect the water cycle by increasing plant growth and thereby influencing evapotranspiration rates.[2]

On a larger scale, climatic changes can affect atmospheric circulation patterns and storm frequencies and intensities. All of these factors, in turn, have major implications for water availability and quality.

Projections of future climate based on computer-driven models are very uncertain and results from different models vary. Nevertheless, the best current estimates indicate that average global temperatures will rise 1°C–6°C over the next century and that average global precipitation will increase by as much as 15 percent, with increases in some regions and decreases in others. Although there is wide agreement among scientists that temperatures will increase and precipitation patterns will change, there is a great deal of uncertainty surrounding the timing and extent of future changes, particularly at the regional level.

A recently completed assessment of the impacts of climate change on the United States (see http://www.nacc.usgcrp.gov/) used two general circulation models, the Canadian Global Climate Model (CGCM) and the British Hadley2 (HADCM2) model, to examine the impacts of changes in greenhouse gas emissions on temperature and precipitation. Both models show warming over most of the North American continent in all seasons of at least 4°C and significant changes in precipitation patterns. While both models show increases in precipitation over the West Coast, the Canadian model predicts more extensive drying in the Great Plains and Southeastern United States (Doherty and Mearns, 1999).

The average surface temperature of the earth has increased by nearly a degree over the past century. The 1990s was the warmest decade of the entire millennium, and 1998 the warmest year (Mann and Bradley, 1999). The higher latitudes have warmed more than the equatorial regions (IPCC, 1996a; OSTP, 1997).

Recent research shows that climate-induced changes in the hydrologic cycle of the earth may already be occurring. A number of these changes are sufficiently different from the past record to be the result of something other than just natural variability. By the late 1980s, observers had noticed a general increase in precipitation outside of the tropics, with a tendency for rainfall declines in the subtropics (IPCC, 1990; 1996a). From 1900 to 1988, precipitation over land increased by 2.4 mm per decade and global mean rainfall rose by more than 2 percent (Dai et al., 1997a, 1997b). Precipitation has increased over land in the high latitudes of the Northern Hemisphere, particularly during winter (Karl et al., 1995). Similar results are being seen in the United States. An analysis by Karl and Knight (1998) shows more precipitation in the continental United States, due primarily to an increase in heavy and extreme daily precipitation events — a worrisome trend in regions where flooding is a problem.

Total annual snowfall in the far northern latitudes seems to be increasing. At the same time, snow and ice cover seem to be decreasing and melting earlier (Robinson et al., 1993; Groisman et al., 1994, Peterson et al., 2000). These changes are linked to higher temperatures. Other observed effects include earlier lake ice melting, earlier snowmelt-related floods in western Canada and the western United States, and earlier warming of Northern Hemisphere land areas in the spring (Nicholls et al., 1996).

Impacts on Water Supplies

Because of all the connections between climate and hydrology, climate changes could have major impacts on a region's water availability and quality. In the arid and semiarid western United States, relatively modest changes in precipitation can have disproportionately large impacts on runoff. While specific regional impacts will depend on future climatic changes as well as economic, institutional, and structural conditions, the following general impacts are expected to occur:

- Regions with snowfall, such as the Rocky Mountains and the Sierra Nevada, will experience seasonal shifts in runoff, with increases in winter and early spring runoff, decreases in late spring and summer runoff, and possible increases in flood intensity.

- Higher latitudes are more likely to receive increases in precipitation and runoff; lower latitudes are more likely to see decreases in runoff.

- In regions where water-quality problems are related to average temperatures or flows, problems could be exacerbated by warming.

Climate Change: Science, Strategies, & Solutions

- Coastal freshwater aquifers in places such as Cape Cod, Long Island, and Florida will be at greater risk of saltwater intrusion due to rising sea level.

- Midcontinental regions, particularly the semiarid and arid western United States, may experience drying of soils during the growing season or more variability in water availability.

Recent studies in the United States indicate that higher temperatures and changes in precipitation could cause major shifts in the amount and timing of river runoff, which is vital for water supply, navigation, hydroelectricity generation, and sustaining natural aquatic ecosystems. The climate forecasts of the Canadian and Hadley2 GCMs have strikingly different implications for mean annual runoff for the 18 major water resource regions in the continental United States from 1990 to 2030 and from 1990 to 2090. For example, the Canadian model suggests runoff would decline in all regions except California by 2030. In 12 of the 18 regions, runoff declines by more than 20 percent. In contrast, the Hadley2 model projects increases in runoff in most regions. (Wolock and McCabe, 1999).

These results illustrate that runoff is extremely sensitive to climatic conditions. Significantly higher temperatures (even with more precipitation) can lead to large reductions in regional runoff, while smaller temperature increases and large increases in precipitation can lead to much greater runoff.

Table 1 provides estimates of the impacts of a range of temperature and precipitation changes on annual runoff for several mountainous river basins in the western United States. The results of simulation studies summarized in Table 1 support the conclusion that runoff is sensitive to changes in temperature and precipitation. In studies with an increase in temperature and no change in precipitation, runoff decreases. With no change in precipitation, runoff declines by 2–12 percent with a 2°C increase in temperature, and by 4–21 percent with a 4°C increase in temperature.

Other studies simulating the impacts of climate change on particular river basins predict big changes in future hydrologic conditions relative to historical conditions, including shifts in snowfall and snowmelt dynamics in mountainous watersheds. Dozens of studies show increases in winter and early spring runoff, decreases in late spring and summer runoff, and higher peak flows in regions such as the Rocky Mountains and Sierra Nevada (see, for example, Gleick, 1987b; Lettenmaier and Gan, 1990; Nash and Gleick, 1993; Leung and Wigmosta, 1999). A shift in runoff timing is already evident in the Sacramento River basin, where the fraction of annual runoff that occurs in the April to July snowmelt season has been decreasing steadily over the past century (see Figure 3) (Gleick and Chalecki, 1999). While this may not be due to human-induced climate change, it is precisely the kind of effect seen in regional studies.

Studies based on available records of river runoff have begun to reveal trends that are consistent with modeling projections and that cannot be explained by natural variability. Three studies published in 1994 all found evidence that certain rivers are exhibiting runoff trends consistent with the effects of global warming. Burn (1994) found a statistically significant trend toward earlier spring runoff in several rivers in western Canada. Lins and Michaels (1994) reported statistically significant increases in autumn and winter streamflow in North America between 1944 and 1988, and Lettenmaier et al. (1994) detected clear increases in winter and spring streamflow across much of the United States between 1948 and 1988. More work looking at recent runoff records is needed.[3]

Impacts on Water Quality

Climate change will lead to changes in streamflow conditions, storm surges, and water temperatures, each of which in turn affects water quality. Stream quality is often most compromised during critical low-flow periods when there is less water to dilute pollutant discharges, maintain dissolved oxygen levels, and support aquatic life.

Current understanding of the hydrological impacts is insufficient to determine whether climate change would improve or worsen low-flow conditions in specific watersheds. Likewise, the direction as well as the magnitude of the climate impacts on lake quality from changes in precipitation and evaporation rates

Table 1

Changes in Mean Annual Runoff

in Mountainous River Basins (in percent)

Precipitation Change	River Basin [Source]	Temperature Change +2°C	Temperature Change +4°C
−25	Carson[7]	−25	−25
	American[7]	−51	−54
−20	Upper Colorado[3]	—	−41
	Animas River[3]	−26	−32
	White River[3]	−23	−26
	East River[3]	−19	−25
	East River[8]	—	−30
	Sacramento[2]	−31	−34
−12.5	Carson[7]	−24	−28
	American[7]	−34	−38
−10	Great Basin Rivers[1]	−17 to −28	—
	Sacramento River[2]	−18	−21
	Inflow to Lake Powell[3]	−23	−31
	White River[3]	−14	−18
	East River[3]	−19	−25
	Upper Colorado[4]	−35	—
	Lower Colorado[4]	−56	—
	Colorado River[5]	−40	—
	Animas River[3]	−17	−23
0	Sacramento River[2]	−3	−7
	Inflow to Lake Powell[3]	−12	−21
	White River[3]	−4	−8
	East River[3]	−9	−16
	East River[8]	—	−4
	Animas River[3]	−7	−14
	Animas River[6]	−2	—
+10	Great Basin Rivers[1]	+20 to +35	—
	Sacramento River[2]	+12	+7
	Inflow to Lake Powell[3]	+1	−10
	White River[3]	+7	+1
	East River[3]	+1	−3
	Colorado River[5]	−18	—
	Animas River[3]	+3	−5
+12.5	Carson[7]	+13	+7
	American[7]	+20	+19
+20	Upper Colorado[3]	—	+2
	Animas River[3]	+14	+5
	East River[3]	+12	+7
	East River[8]	—	+23
	White River[3]	+19	+12
	Sacramento[2]	+27	+23
+25	Carson[7]	+39	+32
	American[7]	+67	+67

Notes: Some of these models also evaluated the impacts of climate changes from general circulation models. Refer to the original references for details.

Sources: [1]Flaschka et al., 1987 (all Great Basin Rivers results); [2]Gleick 1986, 1987a, 1987b (all Sacramento River results); [3]Nash and Gleick, 1993 (all Lake Powell, White, East, and Animas River results); [4]Stockton and Boggess, 1979; [5]Revelle and Waggoner, 1983; [6]Schaake, 1990; [7]Duell, 1992, 1994 (Carson and American Rivers results); [8]McCabe and Hay, 1995.

is uncertain. However, cli-mate-induced changes in storm surges and water temperatures are likely to have a negative impact on water quality. Dissolved oxygen levels tend to decline as temperatures rise because warmer water holds less oxygen. More intense precipitation in some regions is likely to increase the amount of agricultural and urban wastes washed by storm flows into rivers and lakes, but could improve water quality where pollutants are diluted by increased flow.

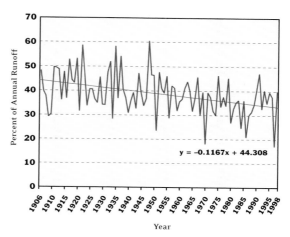

Figure 3

Sacramento River Runoff April to July
(as percent of Annual Runoff)

$y = -0.1167x + 44.308$

Note: The figures are based on the Four-River Index—unimpaired runoff records from the American, Feather, Sacramento, and Yuba River gauging stations.

Source: Gleick and Chalecki, 1999.

Impacts on Ecosystems

Healthy ecosystems depend on receiving appropriate amounts of water of a certain quality at certain times. Since ecosystem processes are affected by changes in tempera-ture and flow regimes, they will be affected by changes in climatic conditions. Studies in the United States have identified a wide range of possible impacts, including extinction of endemic fish species already close to their thermal limits, declining area of wetlands with reductions in waterfowl populations, and major habitat loss (Eaton and Scheller, 1996; Covich et al., 1997; Hauer et al., 1997; Meyer, 1997; Schindler, 1997). Actual impacts will depend on the nature of the climatic changes, the regional characteristics of the ecosystems, and the nature and scope of intentional interventions by humans.

Lake levels, water residence time, and mixing regimes could change, with profound effects on ecosystems. Dissolved oxygen concentrations and ice cover are predicted to decrease (Hostetler and Small, 1999; Stefan and Fang, 1999). Warm-water fish pop-ulations will increase, while cold-water species and wetlands will be disrupted and possibly lost (Meyer et al., 1999).

Impacts in northern latitudes may be particularly severe. Studies for Alaska indi-cate dramatic decreases in permafrost, draining of existing lakes, creation of new ones, and alteration of nutrient exchanges and food web structures. The Rocky Mountain and Sierra Nevada regions would see greater fragmentation of cold-water habitats, shorter duration of ice cover for lakes, and a greater likelihood of late summer channel drying.

Changes in the southeastern United States could include increased rates of production and nutrient cycling, more extensive summer deoxygenation, more drying of coastal wetland soils resulting in greater fire threat, and expansion of subtropical species northward (Meyer et al., 1999; Hostetler and Small, 1999; Burkett and Kusler, 2000).

Some climate change impacts may help ecosystems: riverine, lake fringe, and other wetland areas may benefit from increased precipitation; vegetation biomass may increase due to rising CO_2 levels; and lower lake levels may permit colonization by wetland vegetation (Burkett and Kusler, 2000). But researchers who have noted potential benefits from climate change also express concern about the lack of practical options for protecting wetlands and other aquatic ecosystems from uncertain but potentially large changes. The limited ability of natural ecosystems to adapt to or cope with those changes over the short time frame in which the impacts are likely to occur may lead to irreversible impacts such as species extinction.

Another concern is the lack of formal water rights held by ecosystems. As a result, competition for water has often come at the expense of aquatic systems. A variety of measures including legislation to preserve wild and scenic rivers and protect endangered species and private and public purchases of water rights for instream use have been introduced to guarantee some minimum water level for sensitive ecosystems.

Implications of Climate Change for Water Use

Precipitation, temperature, and carbon dioxide levels affect the demand for water as well as the supply. Water withdrawals in the United States — overall and on a per capita basis — have been declining over the past two decades due to rising costs, environmental concerns, efficiency improvements, technological change, and scarcity. But higher temperatures are expected to increase future demand for irrigation, household, and industrial uses. Current withdrawal and consumptive uses (see Figures 4 and 5) and the potential impacts of the climate on water use are discussed below.

Agricultural, Household, and Industrial Uses

Irrigation, which accounts for almost 40 percent of all U.S. water withdrawals and over 80 percent of consumptive use, is particularly sensitive to climatic conditions. Irrigation becomes more critical for crop production as conditions become hotter and drier. Consequently, in areas with available and affordable water supplies, these conditions would increase both the area under irrigation and the amount of water applied per irrigated acre. However, any increases in water-use efficiency attributable to higher atmospheric CO_2 levels would tend to counter the need to apply more water as temperatures rise (see endnote 2). The net effect on the demand for irrigation water is uncertain.

developed, subsequent increases in storage require ever larger investments. The social and environmental costs of storing and diverting water increase as the number of free-flowing streams declines and society attaches more value to water left in a stream.

Other factors likely to contribute to higher future costs of water are the threats to existing supplies posed by contamination and groundwater depletion. Although billions of dollars have been spent in recent decades to improve their quality, 36 percent of the nation's surveyed rivers and streams and 39 percent of the surveyed lakes, reservoirs, and ponds are impaired for one or more of their designated uses (U.S. EPA, 1998).

Groundwater accounted for 22 percent of total U.S. freshwater withdrawals in 1995 (Solley et al., 1998). In some areas, current levels of groundwater use are unsustainable. For example, declining aquifer levels, higher pumping lifts, and falling well yields have increased water costs in western Texas, causing farmers to take several million acres out of irrigation in recent decades. In California, groundwater overdrafts averaged nearly 1.5 maf yearly, equivalent to 4 percent of the state's total agricultural and urban use in 1995 (California Department of Water Resources, 1998). And pumping from some coastal aquifers in California, Cape Cod, Long Island, New Jersey, and Florida has exceeded natural recharge, resulting in saltwater intrusion into the aquifers.

The socioeconomic impacts of climate change look very different depending on whether the change brings less water as projected by the Canadian climate model or more water as projected by the Hadley2 model. Frederick and Schwarz (1999) examined a set of socioeconomic impacts associated with changes in the long-term availability of water based on the hydrologic implications of the Canadian and Hadley2 models for the 18 water resource regions in the conterminous United States. Under the least-cost management scenario, the Canadian model projects that estimated annual costs of balancing supplies and demands in 2030 in these 18 regions would increase by nearly $105 billion relative to the costs without climate change. In contrast, the increase in water supplies projected by the Hadley2 model would reduce the annual costs by nearly $5 billion relative to the no-climate-change case. These estimates do not include any costs or benefits from either increases or decreases in flooding that might occur.

Adaptation Strategies

Climate is just one of many factors challenging future water planners and managers. Indeed, changes in population, economic conditions, technology, policies, and how society values various water uses may be more important determinants of future supply and demand conditions than those attributable to climate change (IPCC, 1996b; Frederick and Major, 1997). Climate changes will be imposed on top of these other long-term changes. Thus, even if the impacts of climate change are less than the other combined impacts — which is by no means certain — the marginal effect could be substantial and costly.

The socioeconomic impacts of climatic and non-climatic factors affecting water supply and demand will depend in large part on how society adapts. However, there is no strong consensus yet about the effectiveness of different coping and adaptation approaches to deal with future climate changes.

One view holds that little needs to be done now because climate changes are highly uncertain, will manifest themselves slowly, and will be swamped by the many demographic, economic, and societal changes that will occur simultaneously. This view also notes that a wide variety of tools are already available to water managers for dealing with risk and uncertainty, and that these tools will prove sufficient for coping with plausible impacts from future climate changes (Schilling and Stakhiv, 1998).

There is merit to this position, but some problems as well. The first problem is that along with the many remaining uncertainties about the details of climate change come potentially large risks. We are uncertain whether some climate changes will be rapid or of such a large magnitude that they will overwhelm existing systems. As noted above, regional modeling studies suggest that even modest changes in climate can lead to changes in water availability outside the range of historical variability. Because water-supply systems were designed on the assumption that future climate will look like past climate, existing systems and management methods may not be adequate to deal with these changes. And even if adaptation to different climatic conditions is possible, it may be very costly. There are also likely to be surprises that analysts are unable to predict.

Finally, water managers tend to be reactive, not proactive. As Schilling and Stakhiv (1998) note, water managers are technical and empirical pragmatists. They are trained to react to real events, and their tools of choice are physical rather than economic or institutional. The uncertainties about future climate are significant barriers to action. Only a few water agencies or river basin commissions have evaluated their vulnerabilities to possible future climatic changes (Boland, 1998; Steiner, 1998). The American Water Works Association (AWWA), the nation's largest organization of water utilities, recommends that water managers re-examine design assumptions, operating rules, and contingency planning for a wider range of climatic conditions than traditionally used (AWWA, 1997) (see Box 1).

Water-use efficiency improvements are a major tool for meeting future water demands in water-scarce regions where extensive infrastructure already exists. Such improvements can be made faster and more cheaply, with fewer environmental and ecological impacts, than continued investment in new supply. Some studies have recently begun to explore how effective such improvements might be for addressing climate-related impacts. In an assessment of urban water use, Boland (1997, 1998) shows that water conservation measures such as education, industrial and commercial reuse, modern plumbing standards, and pricing policies can be extremely effective at mitigating the impacts of climate change on regional water supplies.

Proper prices and well-designed markets are also increasingly important for balancing supply and demand. Higher prices provide incentives to use less and produce

Climate Change: Science, Strategies, & Solutions

Box 1

Summary of Recommendations for Water Managers from the American Water Works Association's Public Advisory Committee

- While water management systems are often flexible, adaptation to new hydrologic conditions may come at substantial economic costs. Water agencies should begin now to reexamine engineering design assumptions, operating rules, system optimization, and contingency planning for existing and planned water-management systems under a wider range of climatic conditions than traditionally used.

- Water agencies and providers should explore the vulnerability of both structural and non-structural water systems to plausible future climate changes, not just past climatic variability.

- Governments at all levels should re-evaluate legal, technical, and economic approaches for managing water resources in the light of possible climate changes.

- Water agencies should cooperate with leading scientific organizations to facilitate the exchange of information on the state-of-the-art thinking about climatic change and impacts on water resources.

- The timely flow of information from the scientific global change community to the public and the water-management community would be valuable. Such lines of communication need to be developed and expanded.

Source: AWWA, 1997.

more, and markets enable resources to move from lower- to higher-value uses as conditions change. In spite of their potential advantages, effective pricing structures have been slow to develop as tools for adapting to changing supply and demand conditions.

Water marketing — the voluntary transfer of water rights to new uses and users — has great potential to increase the efficiency of both water use and allocation (National Research Council, 1992; Western Water Policy Review Advisory Commission, 1998). However, both the nature of the resource and the institutions established to control it have inhibited water marketing. Efficient markets require that buyers and sellers bear the full costs and benefits of transfers. But when water is transferred, third parties, such as seasonal farm laborers and local agricultural equipment suppliers, are likely to be affected. The challenge for developing more effective water markets is to develop institutions, such as water banks, that can expeditiously and efficiently take third-party impacts into account (Loh and Gomez, 1996; Gomez and Steding, 1998).

Water banks can facilitate temporary transfers by providing a clearinghouse of water rights for rental. The temporary nature of such a transfer blunts a principal third-party concern that a transfer will permanently undermine the economic and social viability of the water-exporting area. Emergency water banks helped mitigate the impacts of California's prolonged drought by facilitating water transfers among willing buyers and sellers. Idaho and Texas have established permanent water banks, and other states are considering establishing them as well.

Long-term imbalances that result from changing demographic and economic factors, social preferences, or climate might at some point warrant a permanent transfer of water rights. The prospect that neighboring watersheds and states will be affected very differently by climate change could increase the potential benefits of interbasin and interstate transfers. Such transfers have occurred, but the process of resolving third-party issues and eliminating other constraints to moving water across hydrologic and political boundaries remains slow, costly, and contentious.

It is increasingly apparent that climatic changes are coming with a complex mix of impacts on all aspects of the water resources of the United States. Some of these impacts can be foreseen with some certainty; many others are unpredictable. Additional research will help resolve some of the uncertainties and expand opportunities for adaptation to change. Meanwhile, maintaining options and building in flexibility are important for designing efficient water programs in the context of climate change.

Endnotes

1. An extensive bibliography of this literature can be found at http://www.pacinst.org/CCbib.html. In addition, a new national assessment report on climate and U.S. water resources can be found at http://www.pacinst.org/naw.htm.

2. The level of carbon dioxide (CO_2) in the atmosphere may affect water availability through its influence on vegetation and evapotranspiration rates. CO_2 is the most important greenhouse gas resulting from human activity. Higher CO_2 levels have been shown to increase plant growth. A larger area of transpiring tissue and the corresponding increase in transpiration would tend to reduce the runoff associated with a given level of precipitation. On the other hand, higher CO_2 levels increase the resistance of plant stomata to water vapor transport, resulting in decreased transpiration per plant unit. The net effect on water supplies is uncertain, but would depend on factors such as vegetation, soil type, and climate.

3. Although long records of runoff are essential to determining whether runoff is changing over time, very few rivers have reliable records longer than several decades. Records longer than a century are extremely rare. Moreover, human interventions in the form of water withdrawals, the construction of dams and reservoirs, and land-use changes in watersheds have already caused significant changes in runoff regimes, greatly complicating the use of past runoff records to detect climate changes or even trends in natural variability.

References

American Water Works Association (AWWA). 1997. Climate Change and Water Resources. Committee Report of the AWWA Public Advisory Forum. *Journal of the American Water Works Association* 89(11): 107–110.

Boland, J.J. 1997. Assessing Urban Water Use and the Role of Water Conservation Measures under Climate Uncertainty. *Climatic Change* 37: 157–176.

Boland, J.J. 1998. Water Supply and Climate Uncertainty. Water Resources Update 112. In *Global Change and Water Resources Management*. K. Schilling and E. Stakhiv, eds. Universities Council on Water Resources, Carbondale, IL.

Burkett, V., and J. Kusler. 2000. Climate Change: Potential Impacts and Interactions in Wetlands of the United States. *Journal of the American Water Resources Association* 36(2): 313–320.

Burn, D.H. 1994. Hydrologic Effects of Climatic Change in West-Central Canada. *Journal of Hydrology* 160: 53–70.

California Department of Water Resources. 1998. *Update to the California Water Plan*. Bulletin 160–98. Sacramento, CA.

Covich, A.P., S.C. Fritz, P.J. Lamb, R.D. Marzolf, W.J. Matthews, K.A. Poiani, E.E. Prepas, M.B. Richman, and T.C. Winter. 1997. Potential Effects of Climate Change on Aquatic Ecosystems of the Great Plains of North America. *Hydrological Processes* 11: 993–1021.

Dai, A., I.Y. Fung, and A.D. Del Genio. 1997a. Surface Observed Global Land Precipitation Variations during 1900–1988. *Journal of Climate* 10: 2943–2962.

Dai, A., A.D. Del Genio, and I.Y. Fung. 1997b. Clouds, Precipitation and Temperature Range. *Nature* 386: 665–666.

Doherty, R., and L.O. Mearns. 1999. *A Comparison of Simulations of Current Climate from Two Coupled Atmosphere-Ocean Global Climate Models Against Observations and Evaluation of their Future Climates.* Report to the National Institute for Global Environmental Change. National Center for Atmospheric Research, Boulder, CO (February).

Duell, L.F.W., Jr. 1994. The Sensitivity of Northern Sierra Nevada Streamflow to Climatic Change. *Water Resources* Bulletin 30(5): 841–859.

Duell, L.F.W., Jr. 1992. Use of Regression Models to Estimate Effects of Climate Change on Seasonal Streamflow in the American and Carson River Basins, California-Nevada. In *Managing Water Resources During Global Change.* Proceedings of the 28th American Water Resources Association Annual Conference, Bethesda, MD.

Eaton, J.G., and R.M. Scheller. 1996. Effects of Climate Warming on Fish Thermal Habitats in Streams of the United States. *Limnology and Oceanography* 41(5): 1109–1115.

Flaschka, I.M., C.W. Stockton, and W.R. Boggess. 1987. Climatic Variation and Surface Water Resources in the Great Basin Region. *Water Resources Bulletin* 23: 47–57.

Frederick, K.D., and D.C. Major. 1997. Climate Change and Water Resources. *Climatic Change* 37: 7–23.

Frederick, K.D., and G.E. Schwarz. 1999. Socioeconomic Impacts of Climate Change on U.S. Water Supplies. *Journal of the American Water Resources Association* 35(6): 1563–1583.

Gleick, P.H. 1986. Methods for Evaluating the Regional Hydrologic Impacts of Global Climatic Changes. *Journal of Hydrology* 88: 97-116.

Gleick, P.H. 1987a. The Development and Testing of a Water-Balance Model for Climate Impact Assessment: Modeling the Sacramento Basin. *Water Resources Research* 23(6): 1049-1061.

Gleick, P.H. 1987b. Regional Hydrologic Consequences of Increases in Atmospheric Carbon Dioxide and Other Trace Gases. *Climatic Change* 10(2): 137–161.

Gleick, P.H. 1998. *The World's Water 1998–1999.* Island Press, Washington, DC.

Gleick, P.H., and B. Chalecki. 1999. The Impacts of Climatic Changes for Water Resources of the Colorado and Sacramento-San Joaquin River Basins. *Journal of the American Water Resources Association* 35(6): 1563–1583.

Gomez, S., and A. Steding. 1998. *California Water Transfers: An Evaluation of the Economic Framework and A Spatial Analysis of the Potential Impacts.* Pacific Institute for Studies in Development, Environment, and Security, Oakland, CA.

Groisman, P. Ya., T.R. Karl, R.W. Knight, and G.L. Stenchikov. 1994. Changes in Snow Cover, Temperature, and the Radiative Heat Balance over the Northern Hemisphere. *Journal of Climate* 7: 1633–1656.

Guldin, R.W. 1989. *An Analysis of the Water Situation in the United States: 1989–2040: A Technical Document Supporting the 1989 USDA Forest Service RPA Assessment.* General Technical Report RM-177. U.S. Department of Agriculture, Forest Service, Rocky Mountain Forest and Range Experiment Station, Fort Collins, CO.

Hauer, F.R., J.S. Baron, D.H. Campbell, K.D. Fausch, S.W. Hostetler, G.H. Leavesley, P.R. Leavitt, D.M. McKnight, and J.A. Stanford. 1997. Assessment of Climate Change and Freshwater Ecosystems of the Rocky Mountains, USA and Canada. *Hydrological Processes* 11: 903–924.

Hostetler, S., and E. Small. 1999. Response of North American Lakes to Simulated Climate Change. *Journal of the American Water Resources Association* 35(6): 1625–1638.

Interagency Floodplain Management Review Committee. 1994. *Sharing the Challenge: Floodplain Management into the 21st Century.* Report to the Administration Floodplain Management Task Force. Washington, DC.

Intergovernmental Panel on Climate Change (IPCC). 1990. *Climate Change: The IPCC Scientific Assessment.* J.T. Houghton, G.J. Jenkins, and J.J. Ephrauns, eds. Cambridge University Press, Cambridge, UK.

IPCC. 1996a. Climate Change 1995: *The Science of Climate Change. Contribution of Working Group I to the Second Assessment Report of the Intergovernmental Panel on Climate Change.* Cambridge University Press, New York, NY.

IPCC. 1996b. *Climate Change 1995: Impacts, Adaptations and Mitigation of Climate Change: Scientific-Technical Analyses. Contribution of Working Group II to the Second Assessment Report of the Intergovernmental Panel on Climate Change.* Cambridge University Press, New York, NY.

Karl, T.R., and R.W. Knight. 1998. Secular Trends of Precipitation Amount, Frequency, and Intensity in the United States. *Bulletin of the American Meteorological Society* 79(2): 231–241.

Karl, T.R., R.W. Knight, and N. Plummer. 1995. Trends in High-Frequency Climate Variability in the Twentieth Century. *Nature* 377: 217–220.

Lettenmaier, D.P., and T.Y. Gan. 1990. Hydrologic Sensitivities of the Sacramento-San Joaquin River Basin, California, to Global Warming. *Water Resources Research* 26(1): 69–86.

Lettenmaier, D. P., E.F. Wood, and J.R. Wallis. 1994. Hydro-Climatological Trends in the Continental United States 1948–1988. *Journal of Climate* 7: 586–607.

Leung, L.R., and M.S. Wigmosta. 1999. Potential Climate Change Impacts on Mountain Watersheds in the Pacific Northwest. *Journal of the American Water Resources Association* 35(6): 1463–1472.

Lins, H.F., and P.J. Michaels. 1994. Increasing U.S. Streamflow Linked to Greenhouse Forcing. *EOS, Transactions, American Geophysical Union* 75(281): 284–285.

Loh, P., and S. Gomez. 1996. *Water Transfers in California: A Framework for Sustainability and Justice.* Pacific Institute for Studies in Development, Environment, and Security. Oakland, CA.

Mann, M., and R. Bradley. 1999. *Geophysical Research Letters* (March 15).

McCabe, G.J., and L.E. Hay. 1995. Hydrologic Effects of Hypothetical Climate Changes on Water Resources in the East River Basin, Colorado. *Hydrological Sciences* 40: 303–318.

Meyer, J.L. 1997. Stream Health: Incorporating the Human Dimension to Advance Stream Ecology. *Journal of the North American Benthological Society* 16: 439–447.

Meyer, J.L., M.J. Sale, P.J. Mulholland, and N.L. Poff. 1999. Impacts of Climate Change on Aquatic Ecosystem Functioning and Health. *Journal of the American Water Resources Association* 35(6): 1373–1386.

Nash. L. 1993. *Environment and Drought in California 1987–1992: Impacts and Implications for Aquatic and Riparian Resources.* Pacific Institute for Studies in Development, Environment, and Security, Oakland, CA (July).

Nash, L.L., and P.H. Gleick. 1993. *The Colorado River Basin and Climatic Change: The Sensitivity of Streamflow and Water Supply to Variations in Temperature and Precipitation.* U.S. Environmental Protection Agency, EPA230-R-93-009, Washington, DC.

National Research Council. 1992. *Water Transfers in the West: Efficiency, Equity, and the Environment.* National Academy Press, Washington, DC.

National Weather Service. 1999. *Hydrologic Information Center.* http://www.nws.noaa.gov/oh/hic/ flood_stats/.

Nicholls, N., G.V. Gruza, J. Jouzel, T.R. Karl, L.A. Ogallo, and D.E. Parker. 1996. Observed Climate Variability and Change. In *Climate Change 1995: The Science of Climate Change. Contribution of Working Group I to the Second Assessment Report of the Intergovernmental Panel on Climate Change.* Cambridge University Press, New York, NY.

Office of Science and Technology Policy (OSTP). 1997. *Climate Change: State of Knowledge.* Executive Office of the President, Washington, DC.

Peterson, D.H., R.E. Smith, M.D. Dettinger, D.R. Cayan, and L. Riddle. 2000. An Organized Signal in Snowmelt Runoff over the Western United States. *Journal of the American Water Resources Association* 36(2): 421–432.

Revelle, R.R., and P.E. Waggoner. 1983. Effects of a Carbon Dioxide-Induced Climatic Change on Water Supplies in the Western United States. In *Changing Climate.* National Academy of Sciences. National Academy Press, Washington, D.C.

Robinson, D.A., K.F. Dewey, and R.R. Heim, Jr. 1993. Global Snow Cover Monitoring: An Update. *Bulletin of the American Meteorological Society* 74: 1689–1696.

Schaake, J.C. 1990. From Climate to Flow. In *Climate Change and U.S. Water Resources.* P.E. Waggoner, ed. J. Wiley and Sons, New York, NY.

Schilling, K.E. 1987. *Water Resources: The State of the Infrastructure.* Report to the National Council on Public Works Improvement. Washington, DC.

Schilling, K.E., and E.Z. Stakhiv. 1998. Global Change and Water Resources Management. *Water Resources Update* 112: 1–5 (Summer).

Schindler, D.W. 1997. Widespread Effects of Climatic Warming on Freshwater Ecosystems in North America. *Hydrological Processes* 11: 1043–1067.

Solley, W.R., R. Pierce, and H.A. Perlman. 1998. *Estimated Use of Water in the United States in 1995.* U.S. Geological Survey Circular 1200. Denver, CO.

Stefan, H.G., and X. Fang. 1999. Simulation of Global Climate-Change Impacts on Temperature and Dissolved Oxygen in Small Lakes of the Contiguous U.S. In *Potential Consequences of Climate Variability and Change to Water Resources of the United States.* D.B. Adams, ed.

Steiner, R.C. 1998. Climate Change and Municipal Water Use. Water Resources Update 112, In *Global Change and Water Resources Management*. K. Schilling and E. Stakhiv, eds. Universities Council on Water Resources, Carbondale, IL.

Stockton, C.W., and W.R. Boggess. 1979. *Geohydrological Implications of Climate Change on Water Resource Development*. U.S. Army Coastal Engineering Research Center, Fort Belvoir, VA.

U.S. Army Corps of Engineers. 1991. *The National Study of Water Management during Drought: A Research Assessment*. IWR Report 91-NDS-3.

U.S. Army Corps of Engineers. 1994. *Executive Summary of Lessons Learned during the California Drought (1987–1992)*. IWR Report 94-NDS-12.

U.S. Army Corps of Engineers. 1996. *National Inventory of Dams*. See http://crunch.tec.army.mil/.

U.S. Environmental Protection Agency. 1998. *National Water Quality Inventory: 1996 Report to Congress*. EPA841-R-97-008. Washington, DC.

U.S. Water Resources Council. 1978. *The Nation's Water Resources 1975–2000: Second National Water Assessment*. Government Printing Office, Washington, DC.

Western Water Policy Review Advisory Commission. 1998. *Water in the West: Challenge for the Next Century*. National Technical Information Service, Springfield, VA.

Wolock, D.M., and G.J. McCabe. 1999. Estimates of Runoff Using Water-Balance and Atmospheric General Circulation Models. *Journal of the American Water Resources Association* 35(6): 1341–1350.

+

+

+

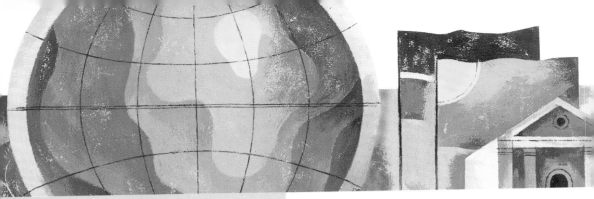

GLOBAL STRATEGIES

Contents

Action Around the World

Eileen Claussen

Many promising and proven strategies exist for reducing greenhouse gas emissions. Indeed, national governments and businesses already are adopting these strategies to positive effect. They range from domestic trading schemes and revenue-neutral "eco-taxes" to innovative energy and industry-sector policies.

However, too little attention has been paid to "what works" in mitigating climate change, leaving the public and many politicians with the idea that we don't have the slightest idea how to solve the problem. This lack of awareness that we can indeed make progress only contributes to further delays in the implementation of a determined effort to address this problem at the global and national levels.

Undoubtedly the path ahead will be difficult, but there are smart, cost-effective steps that countries and businesses can take now to put their greenhouse gas emissions on a downward path, or at least reduce their growth. And there are sensible approaches that can be taken in the longer term as well.

Growing scientific consensus about the problem of climate change, combined with increasing public support for solutions, have prompted national governments throughout the world to devote more attention to this issue and to some of the mitigation steps that they can and should pursue.

A Look at What Nations are Doing

In an effort to get a better handle on what various countries are doing to mitigate climate change, and what lessons might be learned from these activities, we have asked several experts to take a closer look at this issue on a country-by-country basis. In this chapter, we present their findings.

Looking broadly at the global picture conveyed in our experts' assessments, we find that action to address climate change is very uneven throughout the world. There are certainly pockets of progress. For example, the United Kingdom (UK) already has exceeded the emissions reduction target it agreed to in the 1997 Kyoto Protocol and is moving ahead with a commitment for even greater reductions by 2010. But the UK is an exception. Most of the European Union (EU) countries, together with Japan and the United States, are not close to meeting their Kyoto targets, let alone on a course toward significant long-term reductions in emissions. Indeed, the new administration in Washington has expressed an unwillingness to even consider proceeding with U.S. commitments under the Kyoto Protocol. In the developing world, current efforts to address this issue are overshadowed by the fact that emissions growth in Argentina,

Brazil, China, India, and Korea will increase by two-thirds by 2015 under the most optimistic scenarios.

EU member states have been at the forefront of international efforts to reduce greenhouse gas emissions. As John Gummer and Robert Moreland note, the political commitment to address this issue remains strong in the EU. To date, however, the major factors contributing to emissions reductions achieved by EU countries are a switch from coal to natural gas in the production of electricity in the United Kingdom (an action taken without regard to climate change), together with dramatic decreases in emissions in the former East Germany. Nevertheless, a number of more "intentional" strategies to mitigate climate change are producing clear results. These include energy-efficiency incentives and high gasoline prices. In the future, Gummer and Moreland note, EU states are likely to rely more on renewable energy, combined heat-and-power schemes, eco-taxes, voluntary agreements with industry, and other strategies to reduce emissions.

Other industrialized countries are not nearly as far along as the EU member states in investigating and pursuing the policy avenues that are necessary to achieve long-term emissions reductions. Nevertheless, there are still signs of hope. For example, Kenichiro Yamaguchi and Naoki Matsuo report that the Japanese government is "highly motivated" to have the Kyoto Protocol enter into force, despite the fact that Japan faces the highest marginal cost among industrialized countries in reducing its emissions. Even in the United States, which has yet to forge an aggressive national strategy to mitigate climate change, there is increasing interest in expanding research, demonstrating and deploying new, cleaner-energy technologies, and implementing more stringent energy efficiency standards. There is also a growing consensus among many of the nation's largest companies that this issue merits their attention.

Obstacles to Progress

The growing acceptance and understanding that changes are needed does not make those changes easy to make, particularly since they involve decisions with economic consequences that will be made by politicians. It is very clear from our experts' reports that countries are having a difficult time achieving consensus on how to reduce their emissions and move forward with new strategies and policies for doing so.

In developing nations, the problem in many respects is a lack of technical options. As William Chandler reports, energy sources such as wind, biomass-fired power, nuclear power, and hydroelectric power simply are not cost-competitive with coal. And, despite the fact that many developing countries have embraced natural gas as a fuel source, reducing carbon dioxide emissions significantly will require much more dramatic changes in the mix of power generation — changes that would greatly increase the costs of power.

In developed countries, the obstacles to significant reductions in emissions are less technical and more political and cultural in nature. In Japan, for example, incidents such as a September 1999 accident at a nuclear facility have proved a setback to the country's nuclear power plan, which has long been considered a linchpin of Japanese efforts to decouple economic growth from rising carbon dioxide emissions.

The United States, for its part, faces daunting cultural and political obstacles to the adoption of far-reaching strategies for reducing emissions, as Henry Lee affirms. "Coordination within the U.S. government over the past decade has been harder, not easier, to achieve, and these difficulties have developed around issues that are far less complex than climate change," Lee writes.

Several factors have played a part either in stalling or putting the brakes on progress. They are:

- **The role and status of national governments.** Governments around the world have entered the 21st century in a position of weakness. In the United States, government has been divided along partisan lines for the better part of two decades, and it is more divided than ever as a result of the 2000 elections. As Henry Lee points out, the pluralistic system of government in the United States does not allow for quick movement on multi-dimensional problems in the absence of a clear public mandate. While some might consider this a blessing, it is still an obstacle to concerted action on environmental and other issues. The same can be said of the small parliamentary majorities that are a fact of political life in Europe and elsewhere.

- **Failure to value the future.** We are a world preoccupied with the here and now. The poor are concerned with fulfilling their basic needs for food, water, and shelter. And, for those who are able to meet these needs, other day-to-day pressures begin to dominate, with the result that there is little time or inclination to consider the interests of future generations. Adding to the problem is the fact that politicians operate on short election cycles, making it hard to shape and adopt forward-looking policies, particularly if those policies incur short-term costs.

- **The state of the science.** Recent years have seen the emergence of a sizeable body of scientific consensus supporting the need for serious action to address global climate change. Nevertheless, the precise impacts of climate change are the subject of much scientific debate. This debate, combined with a tendency to focus on uncertainties rather than certainties, has made it difficult for policymakers to take strong or effective action on these issues. The unfortunate result is that science — which is the necessary underpinning for action — too often is employed in the cause of those who wish to take no action at all.

- **National sovereignty.** Sovereignty concerns are at the heart of much of the debate about the appropriate strategies for mitigating climate change at both the national and international levels. European countries, according to Gummer and Moreland, have been unable to reach agreement on an EU-wide energy products tax. There has been much debate within the United States about whether to surrender policy independence in order to gain global agreement, and this debate is likely to continue to be contentious and vigorous, as indicated by Henry Lee. And the Brazilians say it is up to them and not the rest of the world to decide the extent to which greenhouse gas-eating rainforests are preserved. In each of these instances, national sovereignty concerns present a clear barrier to progress.

As this discussion shows, countries around the world will have to grapple with — and, in many senses, overcome — a number of underlying political and cultural issues if they are to achieve progress in addressing the issue of climate change. With few exceptions, national leaders and the citizens they govern simply have not been ready to embrace the strong measures needed to achieve significant reductions in their emissions.

Political Will

Countries that have taken strong action on this issue did so because their elected leaders had the political will to push hard for strategies that show promise in reducing emissions. The EU countries, as Gummer and Moreland note, became convinced early in the global warming debate that the growing scientific concern about this issue should be taken seriously. And, in adopting the "EU bubble" to apportion emissions reductions among member states, European leaders arrived at a practical solution for distributing the burden among the continent's wealthier nations and their not-so-wealthy neighbors.

Among the European countries, the United Kingdom is the only one that is on track to achieve its Kyoto commitments and move beyond them. In fact, when recent projections showed that emissions in the UK could begin to increase again between 2008 and 2012 unless additional actions were taken, the government issued a range of new proposals for intensifying current measures and launching additional ones. The proposals included everything from new energy taxes and the encouragement of renewables to a combined effort with industry to implement a domestic emissions trading regime.

The same type of focused, results-oriented policy-making is still a pipe dream in most other industrialized nations. The United States, for example, despite being responsible for nearly one-quarter of global emissions of greenhouse gases, may be less ready than ever to seriously address the issue. Twelve years after former President George Bush stated that we "must have a clear commitment to emissions reductions," the United States has barely started discussing exactly what such a commitment would entail.

The recalcitrance of U.S. leaders is even harder to excuse given the phenomenal strength of the nation's economy over much of the last decade. In Japan at least, political leaders can explain their unwillingness to deal seriously with this issue by pointing to the country's recent economic weakness, as well as the difficulties in nuclear plant siting noted above. For either of these countries to move forward, political leaders need to unite their citizens behind concrete and practical actions.

In the developing world, the problem of political leadership at the national level is eclipsed by the need for developed country leaders to deliver on their promises of outside assistance. As William Chandler reports, international cooperation will be critically important to the ability of developing countries to see even minimal reductions in the growth of their emissions in the coming years. This is particularly true in the introduction of advanced, low-emission electric power technologies. However, in yet another instance of flagging political will, industrialized countries have not yet fulfilled commitments they made in the United Nations Framework Convention on Climate Change to support developing countries through capacity building and technology transfer.

Looking Ahead

Meeting the challenge of global climate change calls for no less than a second industrial revolution. We need to promote new technologies, regulatory strategies, and investments that will put the entire world on a path to clean economic development. Completing work on an international regime for reducing atmospheric concentrations of greenhouse gases is obviously important. But we will win or lose this battle to the extent that we craft sound policies and strategies at the national level.

As the remainder of this chapter shows, the countries of the world are a long way from doing what's needed to win. The challenge that national leaders face in the months and years ahead is to move from a vague acceptance that climate change is a problem to committing their countries and citizens to the hard work that will be required to achieve long-term, sustainable reductions in greenhouse gas emissions.

European Union:
A Review of Five National Programs

John Gummer, Robert Moreland

Member states of the European Union (EU)[1] have been at the forefront of international efforts to mitigate global climate change. In 1990, the EU ministers of environment and energy agreed that carbon dioxide (CO_2) emissions of member states would be no higher in 2000 than in 1990. Seven years later in Kyoto, Japan, EU ministers agreed to reduce greenhouse gas (GHG) emissions 8 percent below 1990 levels during the period 2008 to 2012. This overall reduction was apportioned among the 15 member states. The wealthier nations accepted higher-percentage reductions, while the less economically developed nations agreed to moderate increases in emissions growth.

Action taken to meet the Kyoto targets is primarily a national responsibility. Although the report describes the EU context in which national actions are taking place, it concentrates on a detailed analysis of five member states: Germany, the United Kingdom, the Netherlands, Austria, and Spain, which in total contributed 60 percent of the EU's emissions in 1990. The report describes their progress in achieving national targets and assesses their chances of success.

Germany and the United Kingdom have the highest levels of emissions in the EU — together they accounted for 46 percent of the total in 1990 — and have assumed by far the largest commitment for absolute emissions reductions. The Netherlands and Austria have strong reputations for environmental leadership and are often regarded as role models by other EU member states. Austria has been taking a strong line on environmental issues with its Eastern European neighbors, which are currently applying for EU membership. The progress of these two countries in meeting their obligations will be carefully watched by other countries. Spain is less economically developed than the other four countries and is only responsible for limiting its increase in emissions.

The European Commission estimates that GHG emissions levels in the EU as a whole will be approximately the same in 2000 as in 1990. Although the EU has achieved reductions in some sectors, they have been offset by increases in others. The Commission estimates that emissions will increase by 6 percent between 2000 and 2008–12 if further measures to reduce them are not implemented.[2]

This paper is a condensed and updated version of a report published in June 2000 by the Pew Center: The European Union & Global Climate Change: A Review of Five National Programmes. The authors and the Center gratefully acknowledge Tom Burke, Jos Delbeke, Hermann Ott, Karl Steininger, Pier Vellinga, Hauke von Seht, and Anne Weir for reviewing early drafts of the report.

Major factors contributing to reductions that have been achieved in the EU to date are a switch from coal-powered to natural gas-fired electricity production and rehabilitation policies in the former East Germany. Other measures, most notably energy efficiency incentives and high gasoline prices (relative to the United States), played a part. However, in the future, member states are likely to rely more on renewable energy, combined heat and power (co-generation) schemes, eco-taxes, voluntary agreements with industry, and curbs on emissions from traffic. Despite political difficulties over coal, nuclear energy, eco-taxes, and road transport, the political commitment to reduce emissions remains strong.

The European Union Context for National Action

The Kyoto Protocol targets were not adopted unilaterally by each member state. Instead, each state negotiated with other member states in the effort to reduce GHG emissions from the EU as a whole by 8 percent over the period 1990 to 2008-12. Consequently, the position of each country has to be set in the overall EU context. (See Box 1.)

Background

Early in the global warming debate, the EU was convinced that scientific analysis of global warming should be taken seriously and decisive action was required. This conviction was matched by growing environmental concern in Europe, which led to the growth of "Green" parties and even — as in the case of Germany since 1998 — their formal inclusion in government.

There had already been considerable debate about the effects of some of the sources of GHG emissions, particularly coal use. The debate then was about the effects of acid rain rather than global warming. The issue assumed international importance in the 1970s and 1980s, with the Scandinavian countries blaming the UK and Germany for damage to their lakes and forests. The argument highlighted the need for Europe to reduce emissions from coal burning. Germany and the UK were not blind to the problem. Prior to reunification, there was concern in West Germany about air pollution (and CO_2 emissions) from East Germany. In the UK, there was legislation as early as the 1950s to reduce domestic coal burning that had caused the winter fog immortalized by Charles Dickens and Sir Arthur Conan Doyle.

While there was concern, there was also caution in a number of member states about reducing coal burning significantly. Since coal was once the main source of electricity generation in Europe, change meant huge infrastructure costs and finding inexpensive alternative energy sources. Coal and lignite production was also an important source of employment, particularly in the UK and Germany, but also in France, Belgium, and Spain. Any reduction in production was fiercely resisted by trade unions.

The European Union

Any examination of the EU requires an understanding of its role vis-à-vis its 15 member states. The EU is a supranational level of government with common law established primarily by agreement among its member state governments. Decisions about EU action, as distinct from member state action, lie primarily with the member state governments, which act together through the Council of Ministers, made up of ministers from each of the member state governments. Decision-making in the EU has a variety of mechanisms.[3] The two main mechanisms are described below.

- **Decisions taken by majority voting** (usually weighted to reflect the population of the individual member states) and with the agreement of the European Parliament (elected by popular vote).

  Majority voting is used primarily for those legislative proposals that ensure trade between member states operates on an open and fair basis. Most decisions on environmental issues (other than eco-taxes) are taken by majority voting. However, the agreement on target GHG levels under the Kyoto Protocol required negotiations between member states and, therefore, was taken without recourse to majority voting.

  The European Commission, which consists of nominees from the member states (two from each large member state and one from each smaller member state) as well as an administrative staff, has the sole right to make proposals for legislation but has a limited role in decision-making. However, if it disagrees with a decision by the Council of Ministers, the Commission can withdraw its proposals unless overruled by a unanimous vote of all EU governments. The European Parliament has had an increasing role in EU decision-making and now has power to veto many decisions of the Council of Ministers.

- **Decisions taken only by the unanimity of all member state governments.** By definition, this mechanism covers all matters not covered by majority voting. In particular, any EU tax — even if it affects the movement of trade in the EU — must be decided by unanimity.

  In nearly all cases for voting (either by majority or by unanimity), a legislative proposal is drafted by the European Commission and advice is obtained from the European Parliament regardless of its powers to alter legislation. Advice usually must be obtained from the Economic and Social Committee (which includes representatives from business, trade unions, consumers, farmers, and other interested parties) and the Committee of the Regions (which includes representatives from regional and local governments). The European Commission has the responsibility to ensure that member governments follow EU decisions. If necessary, the Commission can take governments to the European Court of Justice for failure to comply.

Twice in the 1970s, the European economy had been severely affected by restrictions in the supply of oil from the Middle East. The oil crises led to continuing concern about losing domestic sources of energy. Furthermore, publicity over the effects of accidents at nuclear power plants such as Three Mile Island and Chernobyl had damaging effects on public support for nuclear energy production. The public's fears were enhanced by environmental campaigns highlighting the problems of nuclear waste disposal.

The economics of coal production changed these positions. European coal was largely extracted from deep mines — a costly process. Both Belgium and France abandoned nearly all coal mining in the 1970s and 1980s, and France made a substantial and generally successful investment in nuclear power. In the UK, Margaret Thatcher successfully withstood a coal miners' strike in 1984 and, by deregulating the energy market, made possible a significant transfer from coal to natural gas. It was deregulation that cut the link between the power generators and the coal mines and finally ended the reign of "Old King Coal."

Coal from the remaining mines and imported coal — primarily from South Africa, Australia, and the United States — continued to be used. However, the discovery of natural gas under the North Sea and the construction of pipelines to natural gas reservoirs in Eastern Europe and North Africa supported the transition to gas. These developments opened up the possibility of substantial expansion in the use of natural gas, especially for electricity.

Concern about the environment in general — and climate change in particular — began to dominate the political debate during the 1980s in several countries in Northern Europe. Sweden, Austria, and Finland sought early action in reducing GHG emissions; they were joined by the UK, Germany, Denmark, and the Netherlands. Poorer countries — Spain, Portugal, Greece, and Ireland — were less convinced because they feared GHG emissions restrictions would inhibit economic development. However, each country came to support emissions reductions on the condition that the burden fell upon the main emitters and not upon the more fragile economies. The 1990 joint meeting of EU energy and environment ministers set the stage for an early and strong lead on international efforts to reduce greenhouse gases.

The level of the EU's concern was evident at the Rio Earth Summit in 1992, where the United Nations Framework Convention on Climate Change (UNFCCC) was signed. In March 1997, the EU adopted a negotiating position for Kyoto, which included a 15 percent reduction in emissions of CO_2, methane, and nitrous oxide for all industrialized countries by 2010 from 1990 levels.[4] In June 1997, the EU also agreed to propose at Kyoto an intermediate reduction of at least 7.5 percent by 2005.[5] When the Kyoto Protocol was signed in December 1997, the EU settled on a target of reducing total GHG emissions by 8 percent between 1990 and 2008–12.

EU Bubble

A unique aspect of the EU's Kyoto arrangement was the use of Article 4 of the Kyoto Protocol, which establishes that groups of countries may redistribute their emissions commitments in ways that preserve the collective goal (usually referred to as the "bubble"). Each member state has a specific target within the bubble. The targets

Table 1

Current | **EU Bubble Allocations** | (CO_2 equivalent, million metric tons)

Member State	Percent Change 1990–2008/12	Absolute Change	Total 1990	Total 2008–12
Austria	−13.0	−10	74	64
Belgium	−7.5	−10	139	129
Denmark	−21.0	−15	72	57
Finland	0.0	0	73	73
France	0.0	0	637	637
Germany	−21.0	−252	1,201	949
Greece	25.0	26	104	130
Ireland	13.0	7	57	64
Italy	−6.5	−36	542	506
Luxembourg	−28.0	−4	14	1
Netherlands	−6.0	−12	208	196
Portugal	27.0	18	69	87
Spain	15.0	46	301	347
Sweden	4.0	3	69	72
United Kingdom	−12.5	−97	775	678
Total	−8.0	−336	4,334	3,998

Source: *Preparing for Implementation of the Kyoto Protocol (Brussels 19.05.1999) (COM (99) 230).*

range from a 28 percent reduction in Luxembourg to a 27 percent increase in Portugal. Viewed in absolute terms, Germany agreed to a reduction of 252 million metric tons (MMT), while Spain was granted an increase of 46 MMT. Table 1 lists the current allocations for GHG emissions by country.

Allocation of targets within the bubble is guided by the principle that the more industrialized countries should take a greater share of the burden. The EU also supports development in the less economically developed regions and countries of the EU. In practice, the EU position meant that, for example, Germany would take the major share of the burden whereas Spain, which had a standard of living a third less than that of Germany, would be able to sustain a rise in emissions. The EU's argument was comparable to stating, for example, that in fulfilling its Kyoto commitments, the United States would not place the same burden on the states of West Virginia and Mississippi as it would on California and New York.

The EU also took into account other factors, such as whether a member state had already taken action on GHGs and/or had a low per capita GHG level. This particularly benefited France because of its switch to nuclear energy.

Implementing the Kyoto Protocol [6]

The EU is second only to the United States in gross domestic product and in the volume of GHG emissions. In 1990 — the base year for the targets agreed under the Kyoto Protocol — its GHG emissions were approximately 4,334 MMT.[7] The major contributors were Germany (28 percent of total emissions), the UK (18 percent), France (15 percent), Italy (13 percent), and Spain (7 percent). Member states with above average per capita emissions were Luxembourg, Ireland, Germany, Finland, the Netherlands, Belgium, Denmark, and the UK. In 1990, the main sources of GHG emissions were electricity and heat production.

The European Commission has indicated that the EU's GHG emissions may be roughly the same in 2000 as they were in 1990.[8] In its 1999 Communication, the Commission noted "with concern" that a decline in total GHG emissions was achieved in the first half of the decade, but emissions rose during the latter half, when economic growth in the EU was slightly higher.

It appears that the EU's target of stabilizing CO_2 in 2000 at 1990 levels will be achieved, largely due to a decline in emissions from Germany and the UK. Economic restructuring in the former East Germany and fuel switching from coal to natural gas, particularly in the UK, were the most significant factors in the emissions reductions. However, several factors are working in the opposite direction. The most significant of these is GHG emissions from the transport sector, which rose by about 22 percent over the decade.

Means of achieving the targets are left to the member states. EU-wide measures are essentially complementary. All member states have taken or are planning measures in all sectors producing GHGs. However, member states believe that such measures will only moderate the growth of GHG emissions from transport. Further action taken in the residential sector appears to leave GHG emissions from the residential and service sectors about the same between 2008 and 2012 as in 2000. Therefore, there is a heavy burden on industry and power generation to achieve the Kyoto target.

The EU has agreed on a number of EU-wide measures to meet the target. The following measures are applied across all member states:

- Agreement in 1998 with the automobile industry to improve fuel efficiency. Under this voluntary agreement, the average fuel efficiency of new cars will improve by 25 percent between 1995 and 2008.[9]

- Compulsory permit procedures for large installations that emit GHGs.

- European Commission guidelines to reduce state aid for coal between 1994 and 2002.

- Energy-saving initiatives and research programs.

- Legislation on waste (the "Landfill" Directive) to limit the biodegradable content of waste.

- The Strategy and Action Plan on Renewables, which aims to increase the share of renewables in EU primary energy production to 12 percent by 2010, saving 400 MMT of CO_2 per annum by 2010.[10]

Other measures are still under discussion, including a proposal for an energy products tax. The previous concept of a CO_2 tax with high rates of taxation and a high degree of harmonization across the member states has been replaced with a more pragmatic approach that foresees an extension of excise duties and a gradual increase in taxation. The proposal is deadlocked in the Council of Ministers largely because of the opposition of some countries to taxation at the EU level. However, most member states, including those with reservations on EU taxation, are introducing legislation for some form of national energy taxation. In general, the objective is to raise taxes on GHG-producing fuels and, at the same time, lower other business-related taxes in order to avoid increasing the overall tax burden on industry.

Box 2

Methodology

This analysis draws on data and reports produced by the EU member states, the European Commission, and the European Environment Agency (EEA). The analysis also draws on the independent external assessments of the "In-Depth Review" teams (IRT) established under the UNFCCC to review the progress of national plans.[11] Information has also been obtained from government ministers and officials and other interested parties, such as energy producers and business and environmental organizations. The authors note that these national forecasts of emissions trends have sometimes diverted from reality. Nevertheless, they have generally accepted the data provided by governments and well-known organizations as being reliable.

Member states are obligated under EU law to provide regular data related to GHG emissions to the European Commission and the EEA. In practice, the adequacy of information since 1997 varies. A lack of information, particularly for Spain, limits the ability to assess progress. Limitation of data also partly explains why this review does not go more deeply into the costs — particularly in terms of public expenditure — of reducing GHG emissions. Costing is also difficult because reducing GHG is not the only objective of many countries' programs.

Before the Kyoto Protocol, both EU and national GHG targets tended to emphasize CO_2. However, the EU accepts that the obligation to reduce emissions covers six GHGs.[12] Primarily because of difficulties in obtaining accurate measurement of hydro-fluorocarbons (HFCs), perfluorocarbons (PFCs), and sulfur hexafluoride (SF_6), the European Commission typically produces figures for total levels of GHG emissions in CO_2 (not carbon) equivalents based on CO_2, methane (CH_4), and nitrous oxide (N_2O) levels. For consistency, this report follows that procedure. Unless otherwise specified, figures for GHG emissions exclude the contribution of HFCs, PFCs and SF_6 — a difference of probably 1 to 2 percent. However, the analysis does consider actions being taken to reduce these three minor gases.

In discussing the programs of the five member states, the report highlights significant obstacles to action and the extent to which governments are prepared to face them. Many obstacles are political, rather than technical or financial. The authors analyze the political realities of reducing emissions and the political commitment of governments to achieve their targets. The factual position in 1990 is taken as the starting point for comparison as it is the base date for targets under the Kyoto Protocol. The report ends with some conclusions on the likelihood of the targets being reached.

Other European Commission proposals include: a tax on aircraft fuel "as soon as the international legal situation allows the Community to levy such a tax on all carriers including those from third countries,"[13] amendment of existing EU legislation on establishing common rules for certain types of combined transport of goods between member states (i.e., road/rail links and links to waterways), and revitalization of rail transport.

The European Commission published a discussion paper in 2000 on emissions trading by industry within the EU. The paper cites research indicating that emissions trading across all economic sectors could reduce the annual cost of compliance with the Kyoto Protocol by one-third.[14] Several member states are either examining plans for national emissions trading or devising such plans. Denmark has already passed a plan for national emissions trading for the electric power industry. The European Commission reviewed the plan in the context of EU competition rules and approved it in March 2000.[15] No decision has been made on the EU approach to international emissions trading. However, the EU does not regard international emissions trading as a substitute for domestic action, as it made clear at a meeting of the parties to the Kyoto Protocol in November 2000 (COP6).

The use of "sinks" is supported by the European Commission and the member states, but they are generally skeptical as to the extent sinks can make a major contribution to emissions reductions. In a presentation at COP6, the EU expressed concern about the scale and permanency of sinks, and envisioned their use as a "loophole" leading to the avoidance of necessary domestic action.

The EU's ability to meet its obligations under the Kyoto Protocol is dependent on actions taken by its member state governments. So far the commitment of these governments remains strong. However, as will be further demonstrated in the sections on the five individual countries, there are obstacles to be overcome and political nettles to be grasped. Some member states have underestimated the difficulties in achieving the target. Some — most notably the Netherlands — will fall short of their targets.

Germany

The Federal Republic of Germany[16] has the largest gross domestic product (GDP) and the largest volume of GHG emissions (1,201 MMT in 1990) in the EU.[17] It ranks third in total carbon emissions within the G-8 (group of eight major industrialized nations) after the United States and Japan. In 1990, Germany's GHG emissions were approximately 28 percent of the total of the entire EU and 55 percent more than the next highest EU emitter, the UK. Germany also has the third highest per capita level of emissions in the EU.

Germany will be making around two-thirds of the EU's emissions reductions under the Kyoto Protocol: a 21 percent reduction below 1990 levels by 2008–12. Germany also set a 25 percent target for CO_2 reductions between 1990 and 2005. Its willingness to play a pivotal role has enabled the EU to set its ambitious target for reducing GHG emissions. Environmental conditions were poor in the New Länder (the former

East Germany) when it was reunited with West Germany in 1989. Action to reduce emissions has been a major part of the overall German plan to improve the environment in the New Länder.

Progress to Date. The German government estimates that between 1990 and 1998, GHG emissions fell 16 percent overall, CO_2 fell 13 percent, CH_4 37.5 percent, and N_2O 28 percent.[18] The government projected a further reduction in overall GHG emissions of around 1.5 percent by 2000, making the total reduction about 17 percent. This reduction reflects a dramatic decrease — as much as 50 percent — in emissions from the former East Germany as a result of the government's emphasis on renovation, modernization, and the reduction of air pollution since reunification. There was a 2 to 3 percent increase in emissions in West Germany between 1990 and 2000. This increase is partly explained by the immigration of about one million people from the former East Germany and an additional one million from other Eastern European countries since reunification. An increase in manufacturing in West Germany to compensate for the decline in production in the New Länder is another factor.

Emissions of CH_4 and N_2O have been declining due to improvements in technological processes, the implementation of regulatory measures, and the decline in coal mining. The German government estimates that in 1998, non-CO_2 greenhouse gases made up about 13 percent of GHG emissions.[19] The German IRT reported in 1997 that "the measures implemented under the CO_2 reduction program will cut emissions of other GHGs ... back by 40 to 50 percent by the year 2005 compared to 1990."[20] There is some evidence of an increase in the three minor gases (HFC, PFC, SF_6). The German government is undertaking further research into these emissions and future trends.

Means of Achieving Targets. Emissions forecasts indicate that the government's targets cannot be achieved without additional measures. The government estimates that without any such measures since 1990, CO_2 emissions would increase by 19 percent by 2005 if economic growth continues at the current rate.[21] The government's program to reduce GHG emissions has been updated regularly. The emphasis is on reducing current emissions from industry, buildings, and energy production, and reducing the increase in emissions from transport that would have occurred without policy measures.

The government has announced a "package of 150 measures" that it has implemented or is planning to implement to meet its target.[22] The package involves direct government intervention through education and training programs, new economic instruments and tax reform, new legislation, and cooperation with Länder (state governments). An integral part of the program involves voluntary agreements with industry backed by possible tax penalties for failure to achieve targets. The package covers action in different areas, particularly energy savings, energy generation, and transport. Germany puts a high priority on savings in the production and use of energy in industry and in energy use in buildings. District heating systems and co-generation are promoted particularly in the New Länder.

The main elements of the program are:

- **Voluntary Agreements with Industry.** Both the current and previous German governments have emphasized persuasion rather than legal measures in obtaining lower GHG emissions from industry. In March 1996, 15 German industry associations signed a declaration on preventing global warming. These associations represented over 70 percent of industrial energy consumption and 99 percent of the public power generation capacity. The declaration requires industry participants to make an overall 20 percent reduction in CO_2 emissions between 1990 and 2005. The agreement also contains specific sectoral commitments.

- **Eco-Taxes.** Increases in petrol and diesel taxes are substantially above the rate of inflation, with further increases envisaged annually to 2002 (and, if necessary, beyond). In November 1999, the Bundestag or lower house of Parliament adopted "The Law on Continuing the Ecological Tax Reform," which provides for a further four-step increase in taxation on various fuels from 2000 to 2003. At the time of writing, the government is completing all the legislative and constitutional processes required to introduce further taxation. The proposals include a package of higher taxation on energy use in industry. However, the cost to industry will be more than offset by payments to reduce employer and employee contributions to the state pension schemes.

- **Promotion of Renewable Energy.** In 1999, renewables contributed about 5 percent of electricity generation. The government plans to double this share by 2010.[23] In February 2000, the "German Renewable Energy Law" was adopted by the Bundestag to change the method of pricing, which has not been sufficient to promote renewables to the desired extent, and also to exempt renewables from energy taxation. Government expenditure on renewables is increasing annually and in 1999 was around $21 million per year.

- **Promotion of Combined Heat and Power, and District Heating.** The Combined Heat and Power, and District Heating program works to promote and modernize district heating systems and to encourage use of co-generation. The government aims to double the share of energy production from co-generation and district heating by 2010.[24]

- **Road Traffic Measures.** Between 1992 and 1998, the German government made four major increases in petrol and diesel fuel taxes. The petrol tax was increased another 10 percent in 1999 by Chancellor Gerhard Schröder's government, which has committed to annual increases until 2002. Several measures, largely implementing EU vehicle type approval legislation, will be taken on fuel and emissions efficiency of vehicle engines.

- **Switch to Cleaner Burning Fuels.** The government cut coal production by almost one-half and ended the exclusive long-term contract between the coal industry and electric utilities. Coal mining subsidies will gradually be cut in half from 1997 to 2005. Use of natural gas is increasing, partly because of access to new supplies.

The government is reviewing its policy on exports to the developing world in relation to the Clean Development Mechanism (CDM).[25] Particular attention is being given to products sold in countries where Germany's high GHG emissions standards for products sold in Germany do not apply. The government is undertaking a study with the iron and steel industry on the contribution that industry can make to CDM. Although the government will take part in EU and international discussions on emissions trading, plans for domestic emissions trading mechanisms, which so far have little support from German industry, still have to be developed.

Obstacles to Achieving Targets. Both the current Social Democrat/Green Party coalition government and the previous Christian Democrat government have been firmly committed to the need to reduce GHG emissions. In its comments to the IRT review of July 1997, the German government stated that "the measures taken so far within the framework of the CO_2 reduction program do not suffice to achieve the CO_2 reduction target."[26] The government is committed to additional measures to meet the target, but it faces the following obstacles:

- *Controversy over Nuclear Energy.* After reunification, all nuclear power stations in the New Länder were shut down for safety reasons.[27] Following the election of the Social Democrat/Green Party coalition in 1998, commitments were made to phase out nuclear energy, which Chancellor Schröder described as "socially unacceptable." However, nuclear energy contributes 30 percent of electricity supply in Germany and its removal would place Germany's commitments on climate change in severe jeopardy. Chancellor Schröder later described progress in phasing out nuclear energy as "step-by-step."[28] Following negotiations with the Green Party and the nuclear energy industry, he announced in May 2000 that nuclear energy production would be phased out over the next 20 years.

- *Unpopularity of Tax Increases.* German authorities see a need to increase petrol and diesel charges and road haulage charges to achieve their objectives under the Protocol. However, German taxes in these areas are already higher than in neighboring countries and three times higher than in the United States. As a result, there is hardly a petrol station for miles on the German side of the Luxembourg border because Luxembourg's fuel tax is so much lower. Germany is striving for greater tax harmonization at the EU level in this area. Further energy taxes are being introduced, although some compromises have been required. The government has to steer between the demands of environmental groups and members of government who would like higher taxes, and the business community, which argues that further energy taxation breaches the spirit of industry's voluntary agreement with government.

- *The Coal Industry.* The Social Democrat Party has managed to obtain a reluctant agreement for drastic reductions in coal production from its own party supporters and trade unions, but this agreement may not hold for further closures.

Climate Change: Science, Strategies, & Solutions

Nevertheless, liberalization of the energy market (largely as a result of EU legislation, which Germany has strongly supported) has inevitably led to greater pressure to use cheaper energy resources — particularly natural gas.

Monitoring Procedures. The German government has implemented a two-part monitoring system. An inter-departmental ministerial committee was formed to review the progress of the overall program on climate change and take decisions (subject to legislative approval) on future plans. In addition, the Rhineland-Westfalian Institute for Economic Research scrutinizes the agreements with industry annually on the government's behalf. Two monitoring reports have been published on developments in 1995–96 and 1996–97. One example of how these reports will be used by government is that if reports of the monitoring mechanism indicate that the voluntary agreements with industry may not be reached, the government will introduce further taxes on industry's energy consumption.

Political Commitment. Germany sees long-term commercial advantages in climate change, particularly as it has a 20 percent share of the global export market for environmental technology. Nevertheless, its support for the objectives of the Kyoto Protocol is largely based on the concern of its government about the effects of climate change. Greenhouse gas emissions are expected to be around 17 percent lower in 2000, compared to 1990 levels (and CO_2 around 13 percent lower). This leaves Germany only a few years to reach its domestic target of 25 percent in CO_2 emissions reductions by 2005.

The reductions to date have been in the former East Germany, but a decline in emissions is now expected in the former West Germany as well. At present, the government plans to achieve its targets through action within Germany and has not developed ideas on the use of international emissions trading. Consequently, the domestic target of a 25 percent reduction in CO_2 emissions and the Kyoto Protocol target of a 21 percent reduction in GHGs are steep — but not necessarily impossible — tasks. Parts of the government's plan are only now taking effect, and others have yet to be implemented. Germany's commitment, which is more than rhetoric, goes across the political party divide, as does the determination to change the policies and measures if they are not delivering. There is no reason to believe that this commitment would be changed by any likely successor government.

Table 2

Greenhouse Gas Emissions	Germany (1990)
Greenhouse Gas	**Percent**
CO_2 — carbon dioxide	83.9
CH_4 — methane	9.7
N_2O — nitrous oxide	5.7
HFCs — hydrofluorocarbons	0.2
PFCs — perfluorocarbons	0.2
SF_6 — sulphur hexafluoride	0.3

Source: The 1999 Report of the Federal Republic of Germany for a Monitoring Mechanism of Community CO_2 and Other Greenhouse Gas Emissions Pursuant to the Council Decision, 1999/296/EC.

The United Kingdom

Although not as pivotal as Germany in reducing EU emissions, the UK's role in the bubble will be significant. The UK is committed to reducing GHG emissions by 12.5 percent relative to 1990 levels by 2008–12 and will make approximately 29 percent of the EU's emissions reductions — second only to Germany.

When the government of Tony Blair came to power in May 1997, it adopted a target of a 20 percent reduction in CO_2 by 2010 compared to 1990 emissions levels. Based on the government's assumption that non-CO_2 gases would fall faster than CO_2, this reduction is in effect almost twice the Kyoto commitment. The government reaffirmed its commitment to the 20 percent CO_2 target in February 2000.[29]

Progress to Date. The UK has already exceeded its Kyoto target by reducing GHG emissions an estimated 14.6 percent between 1990 and 2000.[30] Changes in emissions by sector between 1990 and 2000 are presented in Table 4. Fuel switching from coal to natural gas for power generation has yielded about half of the reductions. Coal's share fell from 65 percent to about 38 percent, and natural gas' share increased from 1 percent to 28 percent within that period. The government publication "Energy Report 1998" predicted that the share of natural gas-fired generation might grow to as much as 60 percent by 2003 and the share of coal might fall to less than 10 percent.[31]

Table 3

CO₂ Emissions Sources	United Kingdom (1990)
Sector	**Percent**
Power Stations	32
Industry	21
Transport	21
Residential	13
Services	5
Land Use Change	5
Refineries	3

Source: "Climate Change: Draft UK Programme," Department of the Environment, Transport and the Regions, February 2000, Annex C.

Nevertheless, government projections show an increase in GHG emissions between 2008 and 2012 unless further measures are taken.[32] In February 2000, the government issued proposals for continuing or intensifying current policies and for implementing new measures.[33] Further emphasis will be placed on increasing the share of renewables in electricity generation and improving the productivity of nuclear energy. The program also includes energy-saving measures such as voluntary agreements with industry, district heating and co-generation, and residential energy savings. These measures have already contributed almost one-quarter of the reduction in CO_2 emissions between 1990 and 2000.

Climate Change: Science, Strategies, & Solutions

Table 4

| Changes in Greenhouse Gas Emissions | by End User, |

United Kingdom (1990-2000)

	Percent Change	Total Share of Emissions	
	1990–2000	1990	2000
Business	−17	43	37
Transport	6	18	23
Domestic	−12	23	23
Agriculture, Forestry, and Land Use	−10	12	13
Public Sector	−14	5	5

Source: Climate Change: Draft UK Programme, Department of the Environment, Transport and the Regions, February 2000, Table 2, page 6.

The main elements of the program are:

- *Liberalizing the Energy Market.* Government policy from the mid-1980s has transferred energy generation to the private sector and deregulated the energy market. The tie between mines and power stations was broken. Coal-fired power generation is gradually being replaced by cheaper natural gas-fired generation.

- *Introducing Energy Taxation.* Revenue from an energy tax on industry will be recycled to reduce other industrial costs. This "climate change levy" will also enable heavy energy users to gain 80 percent reductions if they sign an agreement to deliver significant emissions cuts. The funds generated will be recycled into industry through reductions in contributions businesses make for employee state pensions and other benefits. The government has strongly resisted the proposal for an EU-wide energy products tax, largely on the grounds that such taxation should be left to member states.

- *Examining Domestic Emissions Trading.* The government is working with industry to find ways of introducing a domestic carbon emissions trading system. Around 40 companies and trade associations have formed the UK Emissions Trading Group to develop such a system, which is expected to be implemented in 2001.

- *Encouraging Renewables, Combined Heat and Power (Co-generation), and District Heating.* The government plans to double the capacity of co-generation between 2000 and 2010. Production from renewables was about 2 percent in 2000 and is forecasted to rise to 5 percent by 2003. The government estimates that this increase will cut GHG emissions by about 1.5 percent from 2000 levels.[34] The government's policy is to ensure that renewables contribute 10 percent of electricity generation by 2010, although this scenario is "subject to the cost to consumers being acceptable."[35]

- **Increasing Fuel Duty.** Government policy has been to increase fuel duty by 5 percent more than inflation each year (called the "fuel duty escalator"). This has been supported by the introduction in 1999 of a lower tax on private automobiles of 1,100 cylinder capacity (cc) or less (about 8 percent of vehicles in 1999). However the "escalator" will not apply for the tax year 2000–01 and public opinion appears strongly opposed to further increases.

- **Introducing Value-Added Taxes on Fuel for Residential Use.** Emissions from households account for 25 percent of UK CO_2 emissions.[36] The value-added tax (sales tax) on domestic fuel, which was introduced at 8 percent in 1993, was raised to 17.5 percent in April 1995. Following the change in government in 1997, the tax was reduced to 5 percent.

Obstacles to Achieving Targets. The European Environment Agency (EEA) states in its 1999 report that the UK "is exceeding largely its national objective to stabilize GHG emissions by the year 2000 at 1990 emissions levels."[37] However, there are potential obstacles for 2000–12 that could frustrate efforts to achieve the Kyoto Protocol targets. The UK's success to date has been due largely to the considerable reduction in coal burning, but continued fuel switching will have diminishing returns, leaving less scope for future reductions. Further, the government faces a difficult task in moderating the growth in emissions from road transport, particularly in the light of increasing opposition to its policies to discourage use of the automobile.

Political Commitment. The United Kingdom has been a long-time proponent of action on climate change and has been willing to take a significant share of the GHG reductions within the EU bubble. The government underlined its commitment to action on climate change by giving Deputy Prime Minister John Prescott responsibility for this area. However, the government will be under pressure from opposition parties to maintain its momentum and produce detailed monitoring figures. The only definitive commitment is a formal review in 2004.[38]

Like Germany, the UK faces some political difficulties concerning nuclear energy, the decreased use of coal, the growth of domestic energy use, and the containment of traffic growth. The government's optimism regarding savings of emissions from domestic use is not universally shared because of its cut in tax on domestic energy use.[39] However, these difficulties do not, as yet, threaten the fulfillment of its commitment. Although the target of 10 percent of electricity generation from renewables is ambitious, the detailed program now produced suggests it may well be achieved.

The UK has made a clear political commitment to the Kyoto Protocol, and meeting its target of a 12.5 percent reduction seems realistic. Success, however, will depend on the government carrying out its February 2000 program.

The Netherlands

Given its size and its small share (5 percent) of overall EU emissions, the Netherlands may not seem at first glance to be central to the EU's objective to reduce GHGs. However, the Netherlands is one of the six founding countries of the Common Market (formed in 1957 and now the EU) and, by virtue of having both a successful economy and being one of the strongest advocates of a more integrated Europe, the Dutch government is expected to take a lead on many EU issues. In particular, it is considered an environmental leader both within the EU and internationally. The Netherlands held the EU presidency in 1992 and 1997 and gave strong leadership during climate change negotiations. However, with relatively high economic growth and an economic base geared to high energy use, the Dutch government's ambitions to reduce GHG emissions may be more difficult to realize than it envisaged.

The main commitment of the Dutch government within the EU bubble is a 6 percent (about 12 MMT) reduction in CO_2 emissions by 2008–12 compared to 1990 levels. Before making this commitment, the Dutch government established a national target of a 3 percent reduction in CO_2 emissions between 1990 and 2000. The Dutch target for CH_4 reductions is 10 percent for the same period. The 2000 target for N_2O is a level no higher than that of 1990.[40] There are no specific targets for HFCs, SF_6, or PFCs, but the Dutch government plans to include these gases in its overall reduction program.

Progress to Date. The EEA stated in its 1999 report that the Netherlands' CO_2 emissions in 2000 are expected to be 17 percent higher than in 1990.[41] Not only will the Netherlands have failed to meet its own target of a 3 percent reduction, but it will also fail to meet the overall EU objective of stabilizing CO_2 emissions at 1990 levels by 2000.[42] The Dutch government expects that CH_4 emissions will be reduced by about 30 percent in 2000, but N_2O emissions are expected to be 14 percent higher.[43]

Emissions have grown significantly in the transport sector, from 31 MMT in 1990 to 36 MMT in 1997.[44] There also has been no significant reduction in GHG emissions from the horticultural sector despite a 48 percent improvement in energy efficiency in this sector between 1980 and 1998 due to the use of small-scale co-generation, improvements to greenhouses, use of residual heat, and improved productivity (i.e., higher crop production per square meter).

One reason for the failure of the Netherlands to achieve its emissions targets is the policy of cheap energy for industry relative to other EU countries, which goes back to the development of the natural gas reservoirs in the North Sea and Groningen. Relatively low oil prices between 1990 and 2000 and liberalization of the energy market in the EU have exacerbated this situation. In addition, the Netherlands has had one of the more successful economies in the European Union, with an economic growth rate above the average for the EU.

Despite its disappointing performance, particularly on CO_2 emissions reductions, the Dutch government has reported two achievements. The first is an agreement with industry to save energy and increase production efficiency, which has saved about 6 MMT of GHG emissions from 1990 to 1997. The second is the doubling of co-generation capacity to 7,800 megawatts (MW) (covering in 2000 up to 40 percent of installed electricity generation capacity) between 1990 and 1998.

Means of Achieving Targets. The Dutch government has devised several scenarios to reduce emissions based on differing levels of economic growth. Using the current forecasts of economic growth, the government predicts that GHG emissions would increase by about 20 percent between 1990 and 2008–12 without existing policy measures.[45] Given the considerable increase in Dutch emissions from 1990 to 2000, this forecast could be an underestimate.

To address the situation, the Dutch government has devised a policy package in three parts:

- *Domestic Measures.* Domestic measures, some planned and some already in effect, are designed to halve the shortfall by 2008–12. This is to be achieved primarily through energy conservation in the residential and business sectors and reductions in the carbon intensity of fuel, with greater reliance on renewable energy sources and transport measures. The government's transport program includes tax incentives for fuel-efficient passenger cars, road pricing (i.e., charging vehicles for the use of roads liable to congestion, particularly at peak periods), further taxes on private cars, stricter control of speed limits, and tax incentives to encourage fuel-efficient driving behavior. The government also has voluntary agreements with industry to improve energy efficiency at coal-fired power stations and to reduce emissions from offshore gas production from gas venting.

- *Measures Outside the Netherlands, to be Taken at a Later Stage.* The Netherlands plans to achieve 50 percent of its commitment through emissions trading. Although the government has made a preliminary allocation of funds for this program, no plan has yet been devised, pending clarification of the overall EU policy toward emissions trading.

- *Reserve Measures.* Reserve measures will be prepared if emissions reductions fall short of achieving the domestic target. However, reserve measures will not automatically go into operation in the event of a shortfall. The government's plans for 2002 and 2005 include a full evaluation of Dutch progress in reducing GHG emissions. Following each review, the government will decide if additional measures, which might include tax hikes, should be taken.

Austrian public appears to believe that the country has no great problem from emissions from sources other than transport and could react against tougher measures.

Spain

Spain has a gross domestic product (GDP) equal to 8 percent of the EU total and has the third lowest per capita income in the EU. In 1990, Spain's GHG emissions were 301 MMT, approximately 7 percent of the EU total, making Spain the fifth highest emitter.[57] However, Spain's emissions level was the second lowest after Portugal in per capita terms (7.6 MT) and approximately 42 percent below the average of the EU member states.

The low level of per capita emissions reflects the relatively low level of economic development and a per capita GDP approximately 20 percent below the EU average. Reducing the disparities in economic development among member states has always been a central policy in the EU and is important to the development of a single currency in Europe. The other member states accept that a reduction in Spain's emissions levels could threaten its economic development and worsen its already high level of unemployment.

Thus, Spain's main commitment within the EU bubble is to restrict the growth of GHG emissions to no more than 15 percent between 1990 and 2008–12. Before Kyoto, Spain's 1991–2000 Energy Plan (which assumes an average annual GDP growth rate of 3.6 percent in the 1990s) limited the growth of its energy-related CO_2 emissions to 25 percent for the period of the plan. Without such measures, CO_2 emissions could have increased by 45 percent.[58] Although the share of emissions from non-CO_2 GHGs is higher than for most industrialized countries, Spain has no specific emissions reduction targets for these gases. The government predicts that the level of non-CO_2 emissions will increase at a lower rate than that of CO_2.

Progress to Date. Between 1990 and 1995, CO_2 emissions rose by around 10 percent, CH_4 emissions rose by 9 percent, and N_2O emissions fell by 4 percent. The government's forecast for 2000 is that the overall increase in emissions will be around 11 to 13 percent over 1990 levels, while CO_2 emissions will be around 14 percent higher than in 1990.[59] Various Spanish environmental organizations have criticized these estimates, claiming that from their research the percentage is higher.[60] Increased use of natural gas and renewables instead of coal, improved energy efficiency in industry, and the promotion of co-generation have helped restrict Spain's growth in GHG emissions.

The availability of statistics on Spain's GHG emissions is more limited than for other member states covered in this report. Spain has been slow in supplying statistics relating to emissions that it is obliged to produce under EU law. There is also little up-to-date information available on emissions monitoring. The lack of data limits the extent of this analysis.

Means of Achieving Targets. The government's program is largely restricted to reducing emissions from the energy sector. The central thrust of Spanish GHG policy is increasing the use of natural gas, which in the 1970s was only 2 percent of primary energy supply. Other aspects of the Spanish policy are as follows:

- All new government housing must meet high energy-saving standards.

- The government is increasing investment in public transport, especially rail infrastructure, which has been generally obsolete and non-competitive with road transport.

- The government has limited parking places in new buildings, increased parking space close to public transport stations, created preferential bus lanes in some cities, introduced liquefied natural gas as fuel for buses, and promoted biofuel through EU subsidies.

- Where coal mines remain open, the government maintains subsidies for coal production. However, the sale price of coal must reflect costs before subsidies. In practice, the government is paying for the financial losses of coal mines. In addition, duties have been abolished on imported coal.

- Several measures are enhancing the use of sinks. Forest coverage will increase by about 8 percent over the period 1990 to 2000 to absorb carbon dioxide.[61]

- Government funding for the Energy-Saving and Efficiency Plan is supported by EU money. However, the need to restrict GHG emissions competes with funding for other environmental priorities, particularly the need to increase the water supply in southern Spain.

A plan to promote renewables was announced by the Spanish National Institute for Energy in May 1999 and a national climate change strategy was expected in 2000. These plans would include tax breaks, subsidies, and incentives for renewables. However, the Spanish press has reported that the government's Finance Ministry considers the cost of these proposals to be excessive.[62]

Obstacles to Progress. The main obstacle to restricting emissions is Spain's desire to improve its economic position vis-à-vis other member states. There is a general reluctance to pressure Spain to take any action that could inhibit economic progress. The other significant limitation relates to the potential for domestic energy production. Since 1984, there has been a moratorium on increasing nuclear power production, which has limited its share as a primary energy source to 14 percent. This share is likely to decline, as old plants being phased out will not be replaced. Pressure from trade unions and others has restricted the government's ability to eliminate coal subsidies.

Political Commitment. Although it accepts the need for limiting GHG emissions, Spain believes that, given its low per capita emissions and need for economic development, the main responsibility for action lies with other member states. Also, environmental concern has been generally lower in Spain than elsewhere in Europe.

Motorway tolls were recently lowered in Spain as an anti-inflationary measure despite criticism of its environmental effects. However, attitudes may be changing. Southern Spain is extremely susceptible to land being overtaken by desert and water scarcity. The public recognizes that this situation could worsen without action to combat climate change.

In terms of the gap between expenditure and payments, Spain receives the largest net contribution from the EU budget. Those member states that are net contributors to the EU budget — particularly Germany, the Netherlands, and the UK — have shown increasing reluctance to support increases in the EU budget and could pressure Spain to meet its Kyoto obligations as a quid pro quo for continued financial support.

Spain's goals of accelerating economic growth and reducing unemployment may be incompatible with achieving its Kyoto objectives. Its emissions during the 1990s appear to have increased by 11 to 13 percent. Consequently, the task of achieving its Kyoto objectives will not be an easy one.

Nevertheless, it appears that the rate of growth of emissions in the period 1995 to 2000 slowed considerably. The increasing use of natural gas and renewables, replacing oil and coal, should help the task, but much will depend on the political will in Spain and on pressure from other member states of the EU. Expressions of public concern about the need to take action on climate change are emerging. A combination of this concern and EU pressure might ensure that the Spanish government takes further action.

Conclusions

Despite difficulties in achieving targets, the political commitment to reduce GHG emissions in each of the five member states is strong. There is general public support for the need to reduce emissions and governments are planning further measures toward that end. However, there is a gap between commitment and performance and there are real questions regarding whether these countries will deliver on their Kyoto obligations because of political opposition to implementing some measures.

The UK seems likely to achieve its target, with the possibility of exceeding it substantially. Germany has a harder task and has obstacles to overcome. Austria and the Netherlands have underestimated the task ahead. The Netherlands is unlikely to meet its target. The position in Spain is difficult to assess because of the lack of relevant data. Nevertheless, it will require a greater effort to meet its target.

The report highlights the following points:

- Two of the five countries covered — Germany and the UK (the highest emitters) — are on track to achieve their individual targets to stabilize CO_2 emissions in 2000 at 1990 levels. In contrast, estimates point to a 17 percent increase in emissions from the Netherlands, which had a national target of a 3 percent reduction over this period.

- Increasing co-generation schemes and the share of renewables in the total energy supply are major features of all five member state programs. To achieve these targets, governments are providing renewables suppliers with financial support and favorable tax and regulatory treatment.

- Concerns about safety and costs have generally ruled out increasing the share of electricity produced from nuclear energy. There is especially strong hostility to nuclear energy from the governments of Germany, Austria, and Spain. However, the German government has stated that reducing its GHG emissions is a "pre-condition" to closing nuclear power plants and this is a major factor in the government's acceptance of a 20-year period before a complete phase-out of nuclear energy production.

- The use of coal for power generation has diminished in all five countries and further reductions are expected. Reduced coal use has been particularly pronounced in the UK and Germany — the two main producers of coal in the European Union.

- Generally member states strongly emphasize the role of voluntary agreements with industry in achieving targets. Government incentives and taxes have usually supported these agreements.

- Agreement on an EU-wide energy products tax has not been reached. However, although some governments — especially the Netherlands — and the European Commission believe this failure is an obstacle toward progress in reducing emissions, the lack of agreement has not stood in the way of the introduction of energy taxes by member states using national legislation. In general, the objective of these taxes is to promote better environmental practices, rather than to increase the overall tax burden on industry. However this is not always the public perception.

- All five countries recognize that GHG emissions from transport will continue to rise and they aim to modify — rather than eliminate — this growth. Most member states have increased petrol and vehicle taxes and plan further increases. However, there are signs that public opposition to such measures will result in some modification of measures that discourage vehicle use, placing more emphasis on improving public transport.

- The public expenditure required to reduce GHGs is not sufficiently clear from any of the five countries' plans — partly because GHG reductions result from programs not solely related to climate change. Lack of public funding appears less of an obstacle than other issues in meeting emissions reduction targets.

Endnotes & References

1. The European Union (EU) has fifteen member states: Austria, Belgium, Denmark, Finland, France, Germany, Greece, Ireland, Italy, Luxembourg, the Netherlands, Portugal, Spain, Sweden, and the United Kingdom. Austria, Sweden, and Finland did not join the EU until 1995, but adopted the EU measures referred to in this document.

2. *Preparing for Implementation of the Kyoto Protocol* (Brussels 19.05.1999) (COM(99)230), Policy-Makers Summary (hereafter "1999 Communication").

3. In December 2000, member state governments approved certain revisions to the decision-making processes of the EU, largely to take into account the enlargement of the EU's membership. The revisions, embodied in the Treaty of Nice, will not materially alter the processes described in Box 1. They await ratification by the national parliaments.

4. See Council Press Release (03-03-1997) Number 6309/97. One of the authors, John Gummer, represented the United Kingdom as Secretary of State for the Environment at this meeting.

5. See Council Press Release (19-06-1997) Number 9132/97.

6. This section is based largely on three "Communications" from the European Commission to the Council, the European Parliament, the Economic and Social Committee, and the Committee of the Regions: *Climate Change — The EU Approach for Kyoto* (Brussels 01.10.1997) (COM(97)481); 1999 Communication; and *EU Policies and Measures to Reduce Greenhouse Gas Emissions: Towards a European Climate Change Programme (ECCP)* (Brussels 08.3.2000) (COM(00)88).

7. 1999 Communication, Annex 1.

8. 1999 Communication, Policy-Makers Summary.

9. Environmental Agreement with ACEA (European Car Manufacturers Association), *Proposal for Council Decision Establishing a Scheme to Monitor the Average Specific Emissions of Carbon Dioxide from New Passenger Cars* (COM(95)689). See also Monitoring System, European Commission, Brussels (COM(98)348) for reports on negotiations of emissions agreements with Japanese, Korean, and other firms outside ACEA.

10. Altener 2 Programme (Brussels), see *Energy for the Future: Renewable Sources of Energy,* White Paper for a Community Strategy and Action Plan (Brussels 26.11.97) (COM(97)599).

11. Each Annex I party to the Kyoto Protocol must produce a communication every three years on its progress in controlling GHG emissions for the UNFCCC. This communication is reviewed by a team of experts who then produce a "Report on the In-Depth Review of the National Communication." (An "In-Depth Review Team" or its report is hereafter "IRT.")

12. The six gases covered in the Kyoto Protocol are: carbon dioxide (CO_2), methane (CH_4), nitrous oxide (N_2O), hydroflurocarbons (HFCs), perflurocarbons (PFCs), and sulphur hexafluoride (SF_6).

13. *Report from the Commission to the Council and the European Parliament on Excise Duty Reductions and Exemptions* (Brussels 14.11.96) (COM(96)549 final).

14. *Green Paper on Greenhouse Gas Emissions Trading within the EU,* European Commission (Brussels 08.03.2000) (COM(00)87). Emissions trading is "the buying and selling of emissions allowances. Article 17 of the Kyoto Protocol establishes trading of assigned amounts between Annex B Parties. It is expected that domestic and international schemes will be set up for industrial emissions trading" (Grubb, M., with C. Vrolijk and D. Brack. 1999. *The Kyoto Protocol: A Guide and Assessment.* The Royal Institute of International Affairs, London).

15. The objective of the Danish government is to reduce CO_2 emissions in the energy sector from 28.9 MMT in 1997 to 20 MMT in 2003. The allocation principle is "grandfathering," i.e., allowances based on share of emissions from the relevant companies in 1994 to 1998. Given the implications of this allocation principle on competition — particularly within the energy industry — the system was examined, and subsequently approved, by the European Commission. For further information, contact the Ministry of Environment and Energy, Denmark. Tel: 45-33-92-67-00; Fax: 45-33-11-47-43; E-mail: ens@ens.dk; http://www.energy-istyrelsen.dk.

16. Unless otherwise stated, all statistics on Germany refer to the area of the country after reunification. References to the "New Länder" refer to the former East Germany and references to "Länder" cover the state governments throughout Germany.

17. *Annual European Community Greenhouse Gas Inventory 1990–1996,* prepared by the European Environment Agency for the European Commission, April 1999 (hereafter "EEA 1999 Report"). The three major greenhouse gases are CO_2, CH_4, and N_2O and are expressed in CO_2 equivalents.

18. *The 1999 Report of the Federal Republic of Germany for a Monitoring Mechanism of Community CO_2 and Other Greenhouse Gas Emissions Pursuant to the Council Decision 1999/296/EC* (hereafter "1999 FRG Monitoring Mechanism"), Table 2. In its annual report (1999), the EEA estimates that CO_2 reduction between 1990 and 2000 will be 11.5 percent. However, current, as yet unpublished, estimates of German reductions provided by the Federal Ministry of the Environment in January 2000 indicate that this figure might be an underestimate of 1 to 2 percent.

19. 1999 FRG Monitoring Mechanism, Tables 1 and 2.

20. IRT for Germany, 1997, paragraph 5.

21. Ibid., paragraph 45.

22. *Fifth International Climate Protection Conference in Bonn: A Guide Published by the Federal Environment Ministry,* Berlin, November 1999.

23. Speech delivered by Jürgen Trittin on November 16, 1999 to the Royal Institute of International Affairs, London (http://www.riia.org).

24. Ibid.

25. Article 12 of the Kyoto Protocol provides a mechanism to allow assistance to less developed countries to achieve sustainable development and to contribute to the ultimate objective of reducing greenhouse gases.

26. IRT for Germany, 1997, paragraph 9.

27. In Germany, as elsewhere, nuclear energy is a controversial subject. The authors take no side on this issue beyond recognizing that nuclear energy does not create GHG emissions. There is the view that phasing out nuclear energy would not jeopardize a reduction in emissions in a recent publication (Hennicke, P., and D. Wolters. *2000: Klimaschutz und Atomaussteig-eine weltweite Perspektive,* Energiewirtschaftliche Tagesfragen, Nr. 3, 140–145).

28. "Because we trust in Germany's vitality...." Policy Statement by Gerhard Schröder, Chancellor of the Federal Republic of Germany, to the Bundestag, 10 November 1998.

29. *Climate Change: Draft UK Programme,* Department of the Environment, Transport and the Regions, February 2000 (hereafter "Draft UK Programme"). This is a consultation document with a final policy document envisaged for later in 2000. It is unlikely that there will be significant changes to the main lines of action on GHG emissions.

30. Ibid., page 46, paragraph 4. The government has deliberately taken 1995 as the base year for the three minor gases as allowed under the Kyoto Protocol.

31. *Energy Report 1998,* HMSO 1998, Volume 1, page 15.

32. Draft UK Programme, page 176.

33. Draft UK Programme.

34. Ibid., page 60.

35. Ibid., Introduction, under Policies to Deliver Cuts in Emissions.

36. Ibid., page 25.

37. *Overview of the National Programmes to Reduce Greenhouse Emissions* (Topic Report 08/1999) from the Topic Center on the Emissions of the European Environment Agency (1999) (hereafter "EEA National Programmes, 1999").

38. Draft UK Programme, page 162.

39. Ibid., page 48. The government forecasts a reduction in GHG emissions by end users from domestic sources of about 5 percent over the period 1990 to 2010. However, Cambridge Econometrics (a UK economics consultancy company and think tank) is forecasting a 16 percent increase in emissions from households over the same period (*UK Energy and Environment,* Cambridge Econometrics, January 2000). This difference relates to the government's apportionment of the decline in emissions from electricity production between end users and Cambridge Econometrics' emphasis on increasing fuel use by domestic consumers. Another area of policy that is open to debate is the extent to which renewable energy can contribute 10 percent of energy production by 2010 without considerable input of public investment.

40. IRT for the Netherlands, July 1996, paragraph 5.

41. EEA National Programmes, 1999.

42. *The Netherlands' Climate Policy Implementation Plan: Part 1 Measures in the Netherlands,* Ministry of Housing, Spatial Planning and the Environment, June 1999, page 23 (hereafter "Implementation Plan: Part 1").

43. EEA National Programmes, 1999.

44. Measures in the Traffic Sector from the government's fact sheet series, *The Netherlands' Climate Policy Implementation Plan* (hereafter "Series"). This is a series of fact sheets produced in June 1999. At the same time the Ministry of Housing, Spatial Planning and the Environment published Implementation Plan: Part 1, which covers the same information as the fact sheets.

45. Summary and Readers' Guide, page 4, from Series.

46. Measures in the Sector Energy and Waste Disposal, from Series.

47. Implementation Plan: Part 1, page 21.

48. Ibid.

49. International Conference on the Changing Atmosphere: Implications for Global Security, Toronto, Canada, 1988.

50. Data in this section are largely provided by IRT for Austria, December 1996, paragraph 4. Updated information provided by the Austrian Ministry of Environment.

51. Schleicher, S., K. Kraten, and K. Radunsky. 1999. *The Austrian CO$_2$ Balance 1997*. Austrian Council on Climate Change (March) (http://www.accc.gov.at/bilanz/bil97).

52. Schleicher, S., K. Kraten, and K. Radunsky. 2000. *The Austrian CO$_2$ Balance 1998*. Austrian Council on Climate Change (January).

53. The EEA notes in its 1999 annual report (EEA National Programmes, 1999) that if past action and future plans are considered, "it seems possible for Austria to reduce its CO$_2$ emissions...well below the Toronto target."

54. IRT for Austria, 1996.

55. *Second National Climate Report from the Austrian Government* (for UNFCCC), 1997 (http://www.unfccc.de/resource/country/austria.html), paragraph 5.3.1.4.

56. Industriewissenschaftliches Institut, Bioenergie-Cluster, Vienna, 1998.

57. Figures are from EEA 1999 Report.

58. IRT for Spain, paragraph 2.

59. EEA National Programmes, 1999, Appendix on Spain.

60. *ENDS Daily*, 15/11/99, "Comisiones Obreras (CCOO), one of Spain's biggest trade union confederations, and environmental NGO 'Ecologists in Action,' independently calculates that Spain's carbon dioxide emissions rose by 22.8 percent or 23.2 percent, respectively, between 1980 and 1998." According to Carlos Martínez, CCOO, "technical experts" in the industry ministry "accept these figures."

61. IRT for Spain, paragraph 48. A sink is a remover of atmospheric CO$_2$.

62. *ENDS Daily*, 15/11/99, Enrique José Vicente, Deputy Director General for Electric Power, Spanish Industry Ministry, said that "approximately $2.5 billion tax incentives for renewable energy producers included in the plan are considered 'excessive' by the finance ministry. Meanwhile, the finance ministry last week announced a 7 percent reduction in tax on motorway tolls as an anti-inflationary measure, which, critics say will encourage car use and therefore fossil fuel consumption."

U.S. Climate Policy: Factors & Constraints

Henry Lee

As a candidate for president in 1988, George Bush called for federal action to meet the emerging challenge of climate change, stating, "Those who think we are powerless to do anything about the 'greenhouse effect' are forgetting about the White House effect. As president I intend to do something about it."[1] Yet, by the final year of his presidency, his resolve had evaporated as he found the climate change issue to be fraught with economic and political peril. Buffeted by conflicting opinions within his own administration, he wavered for several weeks before agreeing to attend the 1992 United Nations Conference on the Environment and Development in Rio de Janeiro. His position on climate became more cautious — emphasizing "economically sensible action," as opposed to decisive actions.[2]

President Clinton was not appreciably more successful in tackling the problem. As a candidate, he repeatedly claimed that he would do what former President Bush would not — commit the United States to reduce its emissions of greenhouse gases and assume a position of global leadership in response to this problem. In fact, his vice president, Albert Gore, had publicly called for a massive program to reduce the threat of global warming, modeled in scale and scope after the Marshall Plan, to rebuild Europe after World War II.[3] But the policy finally proposed by President Clinton consisted of 47 programs, almost all of which were voluntary.[4] Seven years into the Clinton presidency, U.S. greenhouse gas emissions had increased 11.2 percent from 1990 levels — the target each developed country had volunteered to meet.[5]

As former President Bush's son, George W. Bush, assumes the presidency, the United States finds its government divided and gridlocked on the climate issue and its public confused. Both the Congress and the Executive Branch acknowledge the seriousness of the problem, but have been unable or unwilling to construct a political constituency for action. Worried that a proactive response may be too costly, result in regional economic dislocations, and place the United States at a competitive disadvantage, U.S. officials are hesitant to act. Interest groups such as environmentalists, business groups, and labor have found little common ground. Thus, as the country enters a new millennium, there is no clear consensus on how it should respond to this problem, and no indication that a consensus is imminent.

The author and the Pew Center are grateful for the research and input provided by Ilana Brito and the helpful comments received from Sheila Cavanagh, Karen Filipovich, and Robert Stavins, and for the editorial assistance provided by Kate Kennedy.

Some blame the lack of action on an absence of leadership, inferring that if the United States only had stronger presidents, it would have pursued more stringent policies. Others argue that presidents must be more cautious and circumspect than candidates, pointing out that rhetorical excesses on the campaign trail often turn into harsh political realities once the candidate finds himself in the Oval Office. Still others ascribe the inertia to the Republicans' capture of the Congress in 1994 and the widening of the ideological gap between the two branches.

In truth, assigning blame is too simple an answer. The political realities are much more complex. Elected officials, be they Republican or Democrat, respond to the general public and the interests that elect them. Presidents and members of Congress who get too far in front of their political support are likely to lose that base of power and, with it, their credibility with the American people. Interest groups are an important part of this base. The two that have been most prominent in the climate debate — the business community and the environmental interest groups — have rarely spoken with one voice.

Despite the aggressive rhetoric of the Global Climate Coalition, one of the most visible of the industry groups, the business community has been divided, and this is more true in 2000 than it was in 1992. Many U.S. companies have expressed their concern about possible job losses from a carbon reduction program. They have argued against "unrealistic targets and timetables," warning of their negative consequences to the "U.S. economy and all American families, workers, seniors, and children."[6] Simultaneously, some companies have concluded that there is a strong probability that future governments will respond actively to the climate problem. It would be folly for companies whose businesses require them to invest in assets with a 40 to 50-year life to ignore this likelihood. Their support for early reduction credit legislation, joint implementation initiatives, and accelerated energy research and development programs are indications of their desire to reduce the cost of whatever regulatory burdens may be imposed upon them and to frame the debate in a way that is both proactive and in their long-term self-interest. Some of the largest companies, such as BP and Monsanto, have actually set greenhouse gas (GHG) reduction targets for themselves that are far more ambitious than those agreed to by the U.S. government. Yet, other companies see higher costs and very little benefits in making meaningful reductions, and they continue to vigorously oppose government intervention.

The responses of the major environmental groups also have been far from uniform. Most have fought vigorously for stringent reduction targets at the international level, but this has been a relatively costless position for them to take. The differences at this level have been more strategic than substantive, with certain groups pushing for an immediate compromise in order to obtain an agreement prior to the end of the Clinton Administration, and others holding out for stronger implementation language and more rigid requirements. On the domestic front, groups such as Environmental Defense and the Natural Resources Defense Council (NRDC) have developed cap and trade proposals, which, if enacted, would impose limits on GHG emissions and allow for carbon trading. Other groups have chosen to take a less proactive role while continuing to advocate issues such as increased automobile efficiency standards.

One might argue that the domestic debate has yet to be joined and the differences to date have been stylistic and tactical. Yet, if one looks back on the debate surrounding President Clinton's proposed Btu (British thermal unit) tax in 1993, one does not find a vigorous and united front of environmental interests pushing for its approval. Advocating increased fuel prices can put some environmental groups in an uncomfortable positions vis-à-vis their members. Most are not organized around single issues, and in crafting their positions, they are careful not to alienate the supporters needed to meet their other goals, such as biodiversity, land preservation, and pollution abatement.

The lack of consensus on this topic is not surprising. Climate change is a difficult issue. It is complex and replete with scientific uncertainties as to timing and impact. It involves multiple emissions sources from almost every sector of the U.S. economy. Thus, the list of actors is large. It is global in scope and requires international coordination on an unparalleled scale. Moreover, the cost of most response mechanisms must be borne by the present generation, but the benefits lie with future generations. Finally, it requires changing our present fossil fuel-based energy system to one more reliant on less carbon-intensive sources. Historically, transitions from heavy reliance on one form of energy to another require between 30 and 50 years and are marked by fierce political battles between entities that might lose their markets and power bases and those that might gain them.[7]

It would be unreasonable to expect simple solutions and processes to emerge easily or quickly. Strategies that may seem reasonable at one point in time often turn out to be imperfect in hindsight. If it takes 50 years to change global fuel use patterns and the Rio Framework Convention was the first step, then the global community is in year eight of a global transition process, or about 16 percent of the way down an uncharted path.

Are there broad cultural and political factors that have shaped U.S. policy on climate change and are those factors unique to this issue, or are they embodied in the fabric of the American political psyche and the U.S. system of governance? The answer seems to be both. A review of these may give the reader a better understanding of why U.S. efforts to date have accomplished relatively little and may provide an indication of what can be expected from future administrations.

Fundamental Themes

Certain themes re-emerge throughout the 20th century and play an important role in shaping U.S. policy. They include: 1) a government constrained by pluralistic governance processes and institutions; 2) a strong trust in technology; 3) an aversion to government intervention; 4) a tendency to be influenced by events and images as opposed to arguments and analysis; and 5) tension between a desire to assert global leadership and a desire to protect U.S. sovereignty.

Pluralistic Governance

When the founding fathers gathered in Philadelphia in 1776, they were intent on protecting America against the potential for an individual or a single body of government to abuse its power. Hence, they established layers of government and then made sure that policy could only be developed and implemented through cooperation between these layers. The U.S. government structure was formed with checks and balances so that no one body of government would become too powerful.

Effective policy in the United States requires cooperation across branches of government (legislative, executive, and judicial), across levels of government (federal, state, and local), and across agencies within each level. No meaningful GHG reduction strategy will be possible without close cooperation between the Congress and the Executive Branch. Within the Executive Branch, such a strategy would require the involvement of at least seven major agencies — the departments of State, Energy, Commerce, Agriculture, Treasury, and Justice, and the Environmental Protection Agency (EPA). Within Congress, it would require the involvement of multiple committees and subcommittees. Further, many of the programs and policies crafted at the federal level will have to be implemented in concert with the 50 state governments.[8] It is not easy to get such a pluralistic structure of governance to embrace new concepts or ideas. Instead, the U.S. government was deliberately organized to avoid the tyranny of big government, even if it came at the expense of efficient government.[9]

In 1989, Harvard University professors Ray Vernon and Roger Porter wrote a monograph on U.S. foreign economic policy-making. In it they made two major points.[10] First, they argued that foreign economic problems almost always arise in forms that affect more than one agency or more than one congressional committee, and second that the United States has historically had great difficulty coordinating these agencies and committees. These difficulties arise because most international economic policies substantially affect domestic economic issues — issues around which strong and powerful interest groups are arrayed. As a result, initiatives that are developed unilaterally and independently by the Executive Branch or by a small group within the Executive Branch are rarely sustainable.[11]

Climate change has strong implications for the economy and it clearly falls within the broad definition of "international economic" issues. Certainly, negotiated responses will have an impact on the U.S. domestic economy. Therefore, the warnings expressed by Vernon and Porter are especially relevant. Further, coordination and cooperation between the executive and congressional branches of government has been poor or non-existent.

Throughout the past 12 years the United States has formed numerous interagency committees to coordinate the development of climate policies and provide recommendations to the president and his Cabinet. The effectiveness of this decision mechanism has been inconsistent. In the administrations of Bush senior and Clinton, there were significant differences of opinion on what policies should be pursued. The open disagreements between the U.S. EPA and senior staff in the Bush White House made it difficult to develop a coherent negotiation stance. In the Clinton Administration, there were differences between the State Department and EPA on one side and the Treasury Department and the Council of Economic Advisors on the other. While the new Bush Administration has only been in office three months, there are early signs that a consensus within his Cabinet may be as elusive as it was for previous administrations.

Congress is equally constrained by its division of responsibility and jurisdiction. It is divided into two branches — the Senate and the House — and each branch has numerous committees in which legislation is developed for consideration by the full membership. The breadth of the climate problem ensures that as many as a half-dozen committees in each branch will claim jurisdiction for overseeing or developing legislative responses to the problem. Each of these committees has its own personality, priorities, and constituencies.[12]

As it has with other transboundary air pollution issues, Congress perceives climate change as a regional economic issue.[13] That is, legislators are keenly sensitive to how an action or policy will affect the economy of the state or district they represent. Officials from states that rely more on fossil-fueled industries, such as coal-fired electricity generation, are more apt to doubt the need to reduce GHG emissions, since such reductions will have competitive ramifications for their states. This focus on regional economic competitiveness is not new; in fact, it has characterized almost every major air pollution and energy policy debate since 1970. The fear that a climate protocol might affect the United States' competitive advantage in international markets, and that this might translate into lost jobs in their home state, was the major motivation for the passage of the Byrd-Hagel resolution.[14] Congress has been determined to ensure that any response to the climate problem does not constrain the nation's economic future. In this regard, Congress's position is no different than positions espoused by elected officials in China, India, and other developing countries — ironically, many of the same countries whose climate policies members of Congress would like to change.[15]

One of the unique aspects of the American political scene is the influence of interest groups that have unprecedented access to the media, the courts, and, through the country's penchant for expensive and lengthy political campaigns, the candidates for elective office. Department heads understand that without the support of a coalition of interest groups, it is almost impossible to obtain passage of a new program or gain approval for a new policy. Hence, major initiatives require not only coordination between the Executive Branch and Congress and between agencies, but also coordination and support from interest groups — a requirement that has grown in importance over the past 20 years.

Climate Change: Science, Strategies, & Solutions

be of much greater import. International oversight institutions will develop guidelines and rules that will likely impact domestic programs and policies. Political tension will be inevitable as the United States tries to walk the narrow line between international cooperation and domestic political pressure not to relinquish its sovereign right to shape its domestic policy free of external constraints.

Historically, when the United States has decided to involve itself in an international organization with the potential to influence U.S. economic or security interests, it has demanded a certain degree of control over that organization and its decisions. Will the United States be able to demand a similar amount of control or influence in new institutions established to manage and monitor an international climate agreement? Given that the United States is only one of many votes in a future international body, what guarantee will it have that its interests will be treated fairly? Will the United States agree to put even a small portion of its economic flexibility into the hands of an international governing board over which it has no control, dominance, or priority status? Given the U.S. track record, these questions will prove to be contentious and, at a minimum, will trigger a vigorous political debate.

On the other hand, if the price of attaining an agreement is surrendering some flexibility on the sovereignty issue, there is some precedent that suggests the United States might compromise. For example, the United States agreed to arbitration provisions in the North American Free Trade Agreement — provisions similar to those we had resisted in prior trade negotiations.[23] However, concern over any infringement on U.S. sovereignty will remain and be a major point of contention both within the Executive Branch and between it and the Congress.

The Specter of Free Riders

These five characteristics have played a major role in shaping U.S. responses to the climate problem. But climate is a unique issue in terms of its size, its spatial and temporal dimensions, and the scientific uncertainty surrounding its impacts. These characteristics make it significantly more complex than conventional issues, such as reducing acid rain or cleaning up rivers and streams.

Climate is also a classic global commons problem. Technically, it does not matter where the emissions reductions occur. If country A invests billions in reducing its emissions and country B makes no investments, country B gets the same benefits as country A. Thus, there is a huge incentive to get other countries to bear a majority of the costs. Issues with these characteristics are very difficult to solve. There is a continuing suspicion that each country is trying to manipulate the international negotiations to impose greater costs on others and less on themselves. This suspicion is magnified in the domestic political arena. Climate agreements negotiated internationally will be scrutinized carefully to ensure that the burdens agreed to are truly "fair."

As one might expect in a pluralistic society, what seems fair to the international negotiator may not seem fair to an individual senator or interest group. This tension between the desire to reach an international accord and the desire to protect against any loss in competitive advantage will present a major obstacle to reaching an international agreement that is saleable both domestically and internationally.

Do Presidents Matter?

While it may be tempting to lay responsibility for the United States' cautious response to the climate problem on the doorstep of an individual, organization, or branch of government, forces are at work that have more to do with U.S. political culture and government structure.

Yet, presidents do make a difference, if only at the margin. It is, thus, useful to examine the strategies and tactics adopted by both former presidents Bush and Clinton and ask whether the legacy of their actions will affect the country's ability to negotiate an agreement either internationally or domestically.

As a candidate, Bush senior portrayed himself as an environmentalist — in part to contrast himself with the Reagan Administration and in part because he believed that there was public support for a strong environmental program. Once in office, he found that while the public understood the benefits of cleaner air and a reduction in acid rain, such was not the case with climate change. His aides were unable to show that reductions in carbon would provide tangible economic benefits, nor were they able to put together a constituency within his administration in favor of stringent reduction goals. To put this dilemma in perspective, one might ask what would have happened if Bush had gone to Rio and committed the United States to mandatory targets and timetables? Upon his return, would his program have received a receptive welcome in Congress? Would candidate Bill Clinton have applauded the commitment, despite the inevitable opposition from the industrial states in the Midwest and organized labor? The answers to these questions are not readily apparent. What is clear is that in 1992 the American public, as well as its government, was not prepared for emissions reduction timetables and targets.

President Clinton faced a similar dilemma. Congress made it clear that it would put any climate program — either domestic or international — through a stiff economic filter. If the program would result in serious harm to the economy, Congress would not approve it.[24] Clinton, thus, had three choices and, in an amazing demonstration of political dexterity, managed to do a little of all three: 1) develop a "comprehensive program" that had almost no costs, but did very little — the Clinton Action Plan of 1993;[25] 2) agree to an international protocol, which was strongly supported by environmental interest groups, but could not pass Congress's economic filter;[26] and 3) avoid a confrontation with Congress — one that his administration would inevitably lose — by not submitting either a climate program or a treaty for ratification.

Climate Change: Science, Strategies, & Solutions

Two strategic decisions were made — one in each administration — that are likely to inhibit future presidents. The first was the willingness of former President Bush's Administration to allow environmental interest groups to focus the Rio Conference on climate change as opposed to its original theme of the environment and economic development. A treaty on sustainable development with a strong climate section might have provided a much broader menu of opportunities for cooperation among countries and allowed U.S. negotiators to craft protocols that attracted developing countries to actively participate, possibly avoiding the executive-legislative gridlock that has stymied the debate. Changing the definition of the problem after eight years of international negotiations will be difficult, but it may be the only option that will persuade many of the developing countries to buy into the solution.

The second was President Clinton's decision to allow his negotiators at the Berlin meeting of the Conference of Parties in 1995 to commit to mandatory timetables and reduction targets. Once that commitment was made, his administration gave up substantial negotiating flexibility. Its success or failure at subsequent international negotiations inevitably would be judged by the length of the timetable and the stringency of the selected targets. The administration raised an expectation among other countries and constituency groups that it then had to meet. It also focused the subsequent domestic debate on the reasonableness of these timetables and targets — a focus that may have had a greater effect in stimulating detractors to slow the process down than in empowering countries to take stronger actions.

As Vernon and Porter argued, a major foreign economic policy negotiated without the cooperation of key players in both the legislative and executive branches is rarely successful. The result was a three-year period in which the relationship between Congress and the Executive Branch around this issue can at best be characterized as gridlock, and at worst as overt hostility. Restoring this relationship will be a challenge for the new Bush Administration.

The Future

As a new presidential administration begins its term, is there any reason to believe that the United States will change course and embrace a more proactive strategy to reduce its GHG emissions? While each of us may have an opinion about the direction of future political initiatives, such predictions are fraught with uncertainty. Will the factors influencing the United States' desire and ability to act grow stronger or weaker? The answer is complicated by their sheer number and breadth; some factors may become stronger, while others may dissipate.

Should we expect to see dramatically improved coordination and cooperation among branches, agencies, and levels of government around an issue as complex and far-reaching as climate? Will the public's hostility to fossil fuel price hikes diminish or remain as virulent as it has over the past three decades? Will the United States be willing to surrender even a small amount of its sovereignty to an international organization, such as the United Nations, in order to attain a meaningful and workable GHG reduction accord? Is it likely that an event or phenomenon will emerge similar to the hole in the ozone layer, providing the media with a compelling image to rouse and focus public opinion?

Recent trends do not bode well for any major shifts in the answers to these questions. Coordination within the U.S. government over the past decade has been harder, not easier, to achieve and these difficulties have developed around issues far less complex than climate change.

Raising taxes, in the best of circumstances, is a challenge, but with no discernible fiscal crisis to open a debate on this topic, it will be difficult for either Congress or the Executive Branch to take explicit actions that will result in higher energy prices.

While the most recent Intergovernmental Panel on Climate Change report only strengthens the conclusion that the planet is warming and will grow increasingly warmer over the next century,[27] climatologists will continue to find it difficult to point to a single natural disaster and say that it occurred because of climate change, as opposed to natural weather variations.

But, one could pose a different set of questions. Will the United States be willing to aggressively invest in new energy technologies? Will business become more open and less hostile to programs and policies that will reduce emissions? Will developing countries find it more in their self-interest to participate in international efforts? The answers to these questions may be more reassuring.

Spurred in part by the reports of the President's Advisory Committee on Science and Technology coordinated by Harvard Professor John Holdren and in part by congressional leaders from both parties, interest in expanding research and in demonstrating and deploying new cleaner energy technologies is growing.[28] The vision of an economy powered by fuel cells, automobiles that get 80 miles per gallon, or a hydrogen-based power system is enticing to the American public. It integrates easily into the country's longstanding trust in its capacity to solve problems through advancing technology. Further, Congress perceives investments in R&D, either directly or through tax credits, as creating jobs and spurring economic growth. Hence, these programs are easier to sell to the voters than those that restrict behavior or increase prices.

The downside is that the history of U.S. energy policy is replete with failed R&D initiatives. The government often selects the wrong technologies and support for these programs is rarely sustained over a sufficiently long period. It is unlikely that R&D programs by themselves will prove to be sufficient to stimulate the reductions in GHGs necessary to meet international targets. But they could be an important beginning.

Second, an increasing number of the nation's largest companies are questioning whether a strategy of outright opposition to GHG reductions is sustainable or sensible over the long term. These companies are investing in assets with an economic life of

30–40 years and are coming to the realization that it might be in their self-interest to: 1) begin to reduce their GHG emissions now; and 2) push government to think seriously about GHG emissions, so that future investments can be made with greater certainty that the regulatory rules of today will not be totally changed tomorrow. For example, a cap on CO_2 emissions could substantially reduce the need to invest in SO_2 and NO_x abatement equipment. Conversely, billions of dollars of SO_2 and NO_x abatement equipment could be stranded by a subsequent CO_2 reduction cap. Therefore, attaining greater regulatory certainty on GHG reductions is not an inconsequential issue for many companies.

Certainly not all companies embrace this view and even its proponents have reached no conclusion about how hard and fast to push or what type of emissions regime they would want. However if the gridlock experienced in the past between the Executive Branch and the Congress is to be broken over the next four years, the business community will be a likely catalyst.

Third, there is a growing awareness that climate change is not the only threat to sustainable development and that different countries give more or less weight to particular threats depending on their geographic and economic circumstances. Many developing countries will find it politically more palatable to broaden the international agenda to include issues such as biodiversity, desertification, and water pollution. If Congress will not accept a climate treaty without participation by developing countries and these countries are unable to convince their people that clear benefits will accrue from such participation, then broadening the agenda may turn out to be an attractive option for both sides.

None of these three factors overcome the public education hurdles, nor do they resolve the coordination problems or the sovereignty concerns. If climate change requires countries to change the way they provide and use energy, their rate of economic growth, their pattern of trade or the way they interact with the global community, then expecting rapid change is unrealistic. A Chinese philosopher wrote that a journey of a thousand miles begins with a single step. He did not suggest that it began with a gigantic leap. In the last two years, forces have emerged both inside and outside the government that, under certain circumstances, could start the United States on a steady path toward a workable GHG reduction strategy — one that is acceptable to all the branches of its government. Only time will tell whether the intellectual seeds now planted will bear fruit.

Endnotes & References

1. Bush, George H.W. September 24, 1988. From Afar, Both Candidates Are Environmentalists....*The New York Times.* Section 1, Column 4, p. 24.

2. Bush, George H.W. June 12, 1992. Address to the United Nations Conference On Environment and Development in Rio de Janeiro, Brazil.

3. Vice-President Albert Gore's proposal for sustainable development that is funded by the developed countries, primarily the United States, is described in his book: Gore, Albert. 1992. *Earth in the Balance: Ecology and the Human Spirit.* Houghton Mifflin, Boston, MA, pp. 295–307.

4. *The Climate Change Action Plan,* White House. (October 1993).

5. Total U.S. GHG emissions as measured in million metric tons of carbon equivalents. U.S. Environmental Protection Agency, Office of Policy. April 2000. *Inventory of United States Greenhouse Gas Emissions and Sinks: 1990–1998.* EPA Report 236-R-00-001. National Center for Environmental Publications and Information, Washington D.C., pp. ES-2–ES-3.

6. Letter signed by 117 CEOs sent to President Clinton by the Global Climate Coalition in July 1997.

7. Schurr, Sam H., and Bruce C. Netschert et al. 1977. Chapter 2 in *Energy in the American Economy: 1850–1975.* The John Hopkins Press, Baltimore, MD.

8. Lee, H. *Designing Domestic Carbon Trading Systems: Key Considerations.* Discussion Paper, Environment and Natural Resources Program, John F. Kennedy School of Government, Harvard University, Cambridge, MA, E-1998-21.

9. Kann, M.E. 1986. Environmental Democracy in the United States. Chapter 11 in *Controversies of Environmental Policy,* Sheldon Kamieniechi et al., eds. Albany, NY, pp. 252–274.

10. Porter, R.B., and R. Vernon. 1989. *Foreign Economic Policy-making in the United States: An Approach for the 1990s.* Center for Business and Government, John F. Kennedy School, Harvard University, Cambridge, MA, pp. 1–31

11. Ibid. Page 7.

12. For further discussion of recent U.S. legislative activity on climate change, see this book's Current Developments chapter.

13. Peskin, H.M. 1978. Environmental Policy and the Distribution of Benefits and Costs. Chapter 5 in *Current Issues in U.S. Environmental Policy* Paul R. Portney, ed. Resources for the Future, Johns Hopkins University Press, Baltimore, MD, pp. 144–163.

14. The Byrd-Hagel resolution passed 95–0 and stated that the United States should not sign any agreement that mandates new commitments to limit or reduce greenhouse gas emissions for the Annex I parties, unless the Protocol or other agreements also mandate new specific scheduled commitments to limit or reduce GHG emissions for developing country parties within the same compliance period. (Senate Resolution 98, July 25, 2000).

15. Many members of Congress have been critical of China and India's reluctance to agree to binding GHG reduction targets.

16. Nichols, R.W., and J.H. Ausubel. 1992. *Science and Technology in U.S. International Affairs.* Carnegie Commission on Science, Technology, and Government, USA (January), pp. 21-38; Neuzil, M.R. 1991. *On the Front Burner: How the Greenhouse Effect Entered the Public Arena of the Mass Media.* Thesis (M.A.), University of Minnesota (January), pp. 43–53; Gregory, J., and S. Miller. 1998. *Science in Public: Communication, Culture and Credibility.* Plenum Trade, New York, NY, pp. 13, 34–37.

17. Schelling, T.C. 1992, Some Economics of Global Warming. *The American Economic Review* 82/1:1. pp 1–14 (March).

18. Nye, J.S., P.D. Zelikow, and D.C. King. 1997. The Evolving Scope of Government. In *Why People Don't Trust Government.* Harvard University Press, Cambridge, MA, pp. 36-54.

19. Steltzer, I.M., and P.R. Portney. 1998. *Making Environmental Policy: Two Views.* AEI Press, Washington D.C., pp. 13-20.

20. Sarewitz, D., and R. Pielke, Jr. 2000. Breaking the Global Warming Gridlock. *Atlantic Monthly* (July).

21. Hansen, J.E., Chief of NASA's Goddard Institute for Space Studies. Testimony before Congress June 24, 1988.

22. Hahn, R.W., and R.N. Stavins. 1999. *What has Kyoto Wrought? The Real Architecture of International Tradeable Permits.* BCSIA Discussion Paper, Harvard University, Cambridge, MA (February), p. 18.

23. North American Free Trade Agreement, Chapter 20 in *Institutional Arrangements and Dispute Settlement Procedures,* Subchapter B Dispute Resolution.

24. See Byrd-Hagel resolution.

25. Clinton, William Jefferson, President, and Vice President Albert Gore Jr. October 1993. *The Climate Change Action Plan.* Executive Office of the Vice President, Washington D.C.

26. *The Kyoto Protocol to the United Nations Framework Convention on Climate Change*, December 11, 1997.

27. A Sharper Warning on Warming. *The New York Times*, October 28, 2000, Section A, p. 14.

28. President's Committee of Advisors on Science and Technology. *Federal Energy Research and Development Challenges of the Twenty-First Century.* Executive Office of the President of the United States, November 1997.

+

+

+

Climate Change Mitigation in Japan

Kenichiro Yamaguchi, Naoki Matsuo

Japan occupies an area of roughly 377,000 square kilometers, and has a population of about 126 million. Thus, Japan has nearly half the population of the United States, squeezed into an area about 4 percent its size (the size of Montana). Japan's population density is roughly on a par with densely populated countries like the Netherlands. However, two-thirds of Japan is mountainous terrain, mostly covered with mature secondary forest. Thus, the population density of Japan's habitable areas is very high, and there is little room for additional afforestation.

One of the richest countries in the world, with per capita gross domestic product (GDP) generally above the average for members of the Organization for Economic Cooperation and Development (OECD), Japan is also the world's fourth largest emitter of greenhouse gases (GHGs) after the United States, China, and Russia.

Among developed countries, Japan is one of the least endowed with energy resources. It relies on imports for most of its fuel, raw materials, and food. In 1998, 94 percent of Japan's total primary energy supply was based on imported fuels.[1] Energy diversification (to non-oil sources) and energy efficiency have been pursued for decades due to the paucity of energy resources.

Even this cursory glance gives some insight into the climate policy background of Japan, in that much of what classifies as climate-related policies have already been carried out, and further emissions reduction is a daunting task. The dense population has given rise to a well-developed public transport system. Mature forest cover means little room for additional afforestation. Energy efficiency rates are the highest in the world, and further improvements are hard to achieve.

Status and Characteristics of GHG Emissions

Trends Since 1990

Japan has agreed to a reduction target of 6 percent from its Kyoto baseline.[2] In September 2000, the Environment Agency,[3] the ministry that oversees environmental affairs, released the latest figures for GHG emissions in Japan since 1990 (see Table 1).

Table 1 shows that Japan's emissions grew to 8.7 percent above its Kyoto baseline of 1,272.5 million metric tons carbon dioxide equivalent by 1996, then declined for two years in a row, to about 5 percent above the baseline (or 8 percent above the 1990

The authors and the Pew Center are grateful for the input of Yasuko Kawashima, Hiroki Kudo, and Miranda Schreurs, who reviewed early drafts of the report and provided helpful comments.

Table 1

Annual GHG Emissions in Japan Since 1990

(in million metric tons CO_2-equivalent)

	1990	1991	1992	1993	1994	1995	1996	1997	1998	1999
CO_2	1,124.4	1,147.8	1,162.2	1,144	1,214.1	1,221.1	1,236.9	1,233.9	1,187.6	1,234.8
CH_4	32.3	31.9	31.6	31.5	31.1	30.9	30.2	29.0	28.6	
N_2O	18.1	17.6	17.7	17.6	18.9	19.3	20.3	21.1	19.9	
HFCs	17.6	18.1	19.4	20.9	28.1	29.8	30.0	33.6	31.6	
PFCs	5.7	6.4	6.4	8.7	11.7	15.3	16.2	16.4	17.8	
SF_6	38.2	43.5	47.8	45.4	45.4	52.6	50.2	49.7	50.0	
Total	**1,236.3**	**1,265.2**	**1,285.2**	**1,268.1**	**1,349.4**	**1,369.0**	**1,383.8**	**1,383.7**	**1,335.5**	
% change from Kyoto baseline:										
	−2.8%	−0.6%	1.0%	−0.3%	6.0%	7.6%	8.7%	8.7%	5.0%	9.8% (CO_2)

Sources: Environment Agency, 2000 (emissions data); Institute for Global Environmental Strategies (IGES).[4]

level).[5] However, economic recovery and other factors returned CO_2 emissions almost to the 1996 level in 1999 (about 10 percent above the 1990 level). CO_2 emissions decreased slightly in 1993, followed by a rapid increase in 1994. This was mainly due to an unusually cool summer followed by a very hot one.[6]

Between 1990 and 1998, GHG emissions in the United States, Canada, and Australia grew by about 11, 13, and 17 percent, respectively; while those of the European Union (EU), not including fluorinated gases, decreased until 1993, and have crept back up to just slightly below their 1990 levels since then.

It is noteworthy that Japan's emissions per unit of GDP, a proxy figure for efficiency, decreased for three consecutive years since 1995. This suggests that, while recession was the chief cause of the latest decline in GHG emissions in Japan, other factors seem to have played a role. An example of such factors is increased capacity utilization of nuclear power plants.[7]

Recent trends of emissions per unit of GDP and per capita are shown in Figure 1. Emissions per capita in Japan are roughly on a par with those of the EU in the early 1990s and less than half of those of the United States. However, Japan's energy-related CO_2 emissions per capita surpassed the EU average in the early 1990s, and now stands at about 9 tons of CO_2 per year (T-CO_2/yr). As for emissions per unit of GDP, Japanese figures are lower than those of most OECD countries except for those with large sources of non-fossil energy (e.g., France, Norway) in both real and purchasing power parity terms.

Emissions Profile

One notable characteristic of Japan's emissions is that CO_2 accounts for 90 percent of total GHG emissions. This means that energy-related measures play a dominant role in mitigating climate change in Japan. Sources of non-CO_2 GHGs, such as landfills (CH_4),

Figure 1

GHG Emissions per Unit of GDP and per Capita (1990 - 1998)

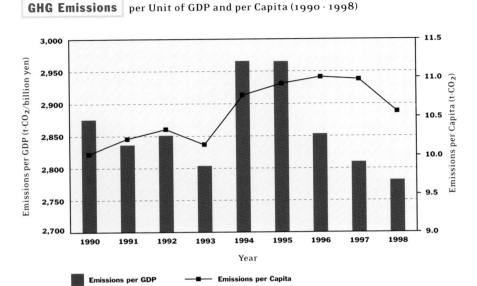

Sources: Environment Agency (GHG Emissions), International Monetary Fund (population),
Prime Minister's Office (GDP).

agriculture (CH_4 and N_2O), cattle farming (CH_4), and the aluminum industry (PFCs),[8] are limited in Japan. This is in contrast to the EU and the United States, where non-CO_2 gases account for nearly 20 percent of GHG emissions and various measures to drastically reduce non-CO_2 emissions may "alleviate" the burden on CO_2. For example, recovery and utilization of landfill methane provides a major opportunity for GHG reduction in many countries, but much of Japan's municipal waste is incinerated due to the lack of space for landfills.

Background and Long-Term Trends

As shown above, CO_2 emissions from fossil fuel consumption is the dominant contributor to Japan's GHG emissions. Thus, addressing the problem of GHG emissions in Japan is virtually synonymous with addressing the energy problem. This section attempts to clarify the characteristics of the energy-related situation in Japan in the context of the long-term historic background.

Japan's energy situation over time can be divided into the following three periods:

- Period up to 1973

- From 1973 to 1986

- From 1986 to present

Period Leading up to the First Oil Crisis (1973)

In the years leading up to the first oil crisis in 1973, GDP, primary energy consumption, and CO_2 emissions grew at almost the same pace as the economy (around 10 percent per year). This indicates that energy conservation and fuel switching to low-carbon energy sources such as hydro and nuclear did not take place. At the time of the first oil crisis in 1973, the energy supply structure of Japan was very vulnerable because the dependence on imported oil in the primary energy supply reached around 77.4 percent.[9]

Decoupling of Energy and Economy (1973 to 1986)

A turning point was reached in 1973, when the first oil crisis struck Japan and wreaked havoc on its society. During the course of two oil crises, Japan began promoting energy conservation and development of alternative energy sources (coal, nuclear, and renewables) in addition to increasing its oil storage, to mitigate the impact of elevated oil prices. As a result, Japan succeeded in decoupling GDP growth and energy consumption growth. The shift from an industry-oriented to a service-oriented economy also helped this transition. Despite the high annual GDP growth (at 3.5 percent per year), domestic primary energy consumption grew only by 0.7 percent per year, and CO_2 emissions decreased 0.3 percent per year during this period. New policies such as the Energy Conservation Law (1979) targeted the supply and demand sectors to reduce oil dependence. Voluntary efforts to conserve energy were taken not only in the industry sector but also in the residential and commercial sectors. These experiences provided opportunities for Japanese industries to develop energy efficient technologies and strengthen their international competitiveness.

Because of its heavy dependence on imported fossil fuels, Japan was one of the countries most affected by the oil crises. Due to its bitter experience during these crises, energy security became the primary objective of Japan's energy policy, and various measures aimed at reducing the dependence on oil imports (especially from the Gulf states) and pursuing an "optimal" energy mix were put in place. This has led to the promotion of diverse energy sources (in particular nuclear) and energy conservation. Promotion of non-fossil energy sources was carried out, using earmarked taxes on imported oil and natural gas (the General Petroleum Excise Tax and the Petroleum Tariff), and a tax on electricity consumption was used for new power plant development (the Electric Power Source Development Promotion Tax).[10]

Facing Difficulty (1986 to present)

The period after 1986 is characterized by two contrasting economic conditions: the period of rapid growth during the "bubble economy"[11] of 1986 to 1991, followed by a period of economic stagnation, which is ongoing. Energy consumption soared in the former, but stagnated in the latter. Both, however, can be characterized by the lack of improvement in overall energy efficiency.

After the sharp decline of oil prices in 1986,[12] GDP and energy consumption resumed a concurrent growth pattern, with both growing by 3 percent per year or more. The bubble economy of the late 1980s boosted this tendency. Since 1986, it has become ever more difficult to decouple GDP and energy consumption, as described below.

- **Industrial sector.** Recently, industrial energy efficiency has peaked and might be declining, mainly because of product diversification. Energy consumption by the industrial sector in 1999 was roughly 7 percent above its 1990 levels, even though industrial output declined by 5 percent between 1990 and 1999.

- **Residential/Commercial/Transportation sectors.** Although per capita consumption in these sectors is low compared to other major OECD countries, it is increasing. In these sectors, energy efficiency of appliances has more or less stagnated, and in some cases is worsening, due to such factors as popular preference for larger-sized goods like automobiles and appliances.

- **Electricity sector.** Average energy efficiency of fossil-based power generation reached 40 percent (at the point of generation) in 1998, and the transmission loss rate in Japan is about 5 percent. This means that the efficiency of electricity generation and transmission is at the highest level in the world, and cannot be significantly improved. However, electricity demand increased

Figure 2

Historic Energy Consumption Trends in Japan, by Source

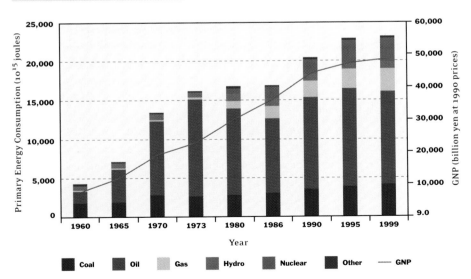

Sources: Institute of Energy Economics Japan, (1960-1995); Resources and Energy Agency, Prime Minister's Office (1999).

Climate Change: Science, Strategies, & Solutions

steadily by 2.6 percent per year for the period 1990–1998 due to a shift in end-use energy toward electricity. Fuel switching in Japan has mainly occurred in this sector by expanding nuclear and gas.

The linkage between GDP and energy consumption continued into the 1990s, when the economy hardly grew. From 1996 to 1999, energy consumption was almost stable, while GDP declined. This implies a decline in energy efficiency. As mentioned before, this has not led to increased CO_2 emissions per GDP, due to factors such as increased use of nuclear power.

Figure 2 illustrates the above arguments. The decoupling of energy and economic growth during the period from 1973 to 1986 is quite noticeable, in that while the economy grew steadily, primary energy supply hardly grew. It is also noteworthy that a substantial change in energy composition (representing a shift toward increased use of non-oil sources) took place during that period.

Future Profiles of Energy and CO_2 Emissions

In this section, we present future profiles of CO_2 emissions in Japan. Table 2 compares the latest forecasts from three different sources with past trends.

The Institute of Energy Economics, Japan (IEEJ), a nonprofit energy research institute principally funded by the energy-intensive industries and the government, released its energy supply and demand forecasts under two scenarios in December 1998. Between 1997 and 2010, the IEEJ expects energy conservation and fuel switching to increase by around 0.5 percent and 0.2 percent per year, respectively, in both scenarios. The Central Research Institute of the Electric Power Industry (CRIEPI), a nonprofit research institute founded by the electric utility industries, released its base-case forecast in 2000, expecting more energy conservation (0.9 percent/yr), and less fuel switching (0.1 percent/yr). The Energy Information Administration (EIA) of the U.S. Department of Energy forecasts lower rates of energy conservation (0.2 percent/yr) and no fuel switching in its *International Energy Outlook 2000*.

Since energy consumption in Japan "re-coupled" with economic growth after 1986, GHG emissions are bound to be dependent on economic growth. Thus, Japan must achieve energy conservation and fuel switching rates high enough to offset GDP growth in order to reduce its CO_2 emissions. This is no easy task despite the low economic growth, since historical trends have shown almost no improvement in overall energy conservation since 1986, and bottom-up efficiency is already at a high level. In other words, much of what is economically feasible was carried out from 1973 to 1986 (see Figure 2), and Japan must implement unprecedented measures to stabilize its emissions once the economy recovers.

Efficiency improvements would be especially needed in sectors such as residential, commercial, and transport. As for fuel switching, a shift to nuclear and gas has been

Table 2

Factor Analysis of CO₂ **Emissions Trends** in Japan

(percent change per year)

		GDP growth	Energy growth	CO₂ growth	Rate of energy conservation	Rate of fuel switching
1965–73		9.3	11.3	11.3	–2.0	0.0
1973–86		3.5	0.7	–0.3	2.8	0.9
1986–99		2.4	2.4	2.0	0.0	0.4
1997–2010 Outlook	IEEJ case 1	1.3	0.9	0.7	0.4	0.2
	IEEJ case 2	1.0	0.5	0.3	0.5	0.2
	CRIEPI base	1.3	0.4	0.3	0.9	0.1
	USDOE/EIA	1.0	0.8	0.8	0.2	0.0

Note: "Rate of energy conservation" is calculated by subtracting GDP growth from energy growth. "Rate of fuel switching" is calculated by subtracting CO₂ growth from energy growth.

promoted in the energy supply sector since the first oil crisis of 1973. As shown in Table 2, this has contributed to a lower rate of CO_2 growth compared to growth in energy consumption during the past 30 years. However, continuation of the trend may prove to be difficult because the siting of new nuclear power plants is facing difficulty. Furthermore, gas stands to face stiffer competition in Japan since it is imported as liquefied natural gas (LNG), which is more expensive than pipeline gas. Ongoing deregulation of the Japanese electricity industry could lead to more emissions since it could bring about participation of independent power producers (IPPs), many of which rely on carbon-rich fuels such as coal and residual fuel oil.

Meeting the Kyoto Target

The Kyoto Protocol was adopted in December 1997 at the third session of the Conference of Parties (COP3) to the United Nations Framework Convention on Climate Change (UNFCCC). The Protocol currently stands as the only major international convention that bears the name of a former Japanese capital. This seems to have had a considerable influence on Japan's position in the international negotiations. It is a product of Japan's domestic energy and environmental situation, as well as the political will to have the Protocol enter into force at an early date and seal Kyoto's place in history.

The Blueprint

Just after COP3, the government produced a framework plan (hereafter referred to as the "blueprint") to achieve compliance with the Kyoto target of 6 percent reduction, as follows:

(a) Reductions from business-as-usual levels to 1990 levels through energy-related measures (the Energy Conservation Law, voluntary commitments by the industries, incentives for efficiency improvement, etc.);

(b) Additional 0.5 percent reductions from the baseline by measures related to N_2O and CH_4;

(c) Additional 2 percent reductions through revolutionary technologies and change in social behavior;

(d) Allowing for a 2 percent increase through increased use of HFCs as CFC substitutes;

(e) Further reductions of 3.7 percent through increased domestic sequestration; and

(f) Acquiring the remaining 1.8 percent in reductions through Kyoto mechanisms.[13]

Since Japan's business-as-usual emissions in 2010 are assumed to be about 20 percent above 1990 levels, this plan would allocate more than 90 percent of necessary reductions to domestic measures.[14] Thus, the targeted "ceiling" on Kyoto mechanisms that Japan set for itself (i.e., 1.8 percent of baseline emissions) was much tougher than quantitative ceilings proposed by the EU at COP6.[15] Under that proposal, Japan would be able to acquire credits of as much as 94 million metric tons of CO_2 (MMT-CO_2) per year,[16] or about 7 percent of its Kyoto baseline.

Since the blueprint was issued, the difficulty in adhering to the framework plan has become increasingly apparent. The main reasons are summarized below.

The Nuclear Issue

Incidents such as the accident at a fuel processing plant in Tokaimura (northeast of Tokyo) in September 1999, during which radioactive substances reached a critical mass, have severely hampered the public acceptance of nuclear power. As a result, plant siting and construction has become difficult. Nuclear power generation planned for the year 2010 has been scaled back from 480 terawatt hours (TWh) per year[17] (20 additional plants) to 392 TWh/yr (13 additional plants or 12.63 gigawatts (GW) new capacity from 2000 to 2009).[18] The decrease would be largely offset by an increase in the use of coal, which, along with nuclear, constitutes the bulk of baseload generation in Japan.[19] This could raise CO_2 emissions in the year 2010 by about 100 MMT-CO_2, or about 7 percent of Japan's total GHG emissions.[20]

Some say that even the plan to build 13 additional plants is too optimistic since only four nuclear power plants are currently under construction and only two more are in preparation for construction. The combined capacity of these six plants is 6.86 GW, slightly above the halfway mark of the revised schedule.

Land Use Change and Forestry

In international negotiations, Japan has proposed a method of accounting for sinks that would entitle it to a reduction equal to 3.7 percent of its 1990 GHG emissions — with 0.3 percent coming from afforestation, reforestation, and deforestation (ARD) activities, and 3.4 percent from additional land use, land use change, and forestry

(LULUCF) activities. Japan publicly announced its 3.7 percent target just after COP3 and has staunchly adhered to it. However, meeting the target through LULUCF activities is becoming increasingly unlikely since the method of calculation proposed by Japan (i.e., the U.N. Food and Agriculture Organization's activity-based accounting method)[21] has not received support from other countries.

Sectoral Issues

In 1990, the industry and electricity sectors accounted for about 45 percent of Japan's energy-related CO_2 emissions. With increased efforts for efficiency improvements coupled with economic stagnation and foreign outsourcing, the industry sector managed to decrease its CO_2 emissions throughout the 1990s. Emissions in other sectors have increased, however, offsetting reductions in the industrial sector. Between 1990 and 1998, emissions from the residential and commercial sectors increased by 9.3 and 16.1 percent, respectively, due to the continued pursuit of amenities such as air-conditioning, as well as to the onset of information technology (many network computers are permanently switched on). Transport-related emissions have risen by 21.1 percent. A growing preference for larger cars and increased automobile ownership have more than offset improvements in fuel efficiency. Furthermore, as discussed before, factors such as electricity deregulation may increase Japan's CO_2 emissions.

Non-CO_2 Gases

The target of achieving a 0.5 percent reduction in total emissions by reducing CH_4 and N_2O emissions amounts to a reduction of about 13 percent for these two gases. Emissions of the two gases decreased by about 4 percent between 1990 and 1998, with an increase in N_2O emissions offset by a decrease in CH_4 emissions. The target of allowing for a 2 percent increase in total emissions by increasing use of HFCs, PFCs, and SF_6 indicates that emissions of these gases can be increased by about 25 percent from their 1995 levels. By 1998, emissions of these gases had increased by about 17 percent. Policies and measures to reduce these gases are under formulation.

Thus, the original blueprint is in serious need of revision due to the possibility of missing the CO_2 target by a considerable margin. The blueprint is being revised by two ministries, which are working somewhat independently and often find themselves at odds with each other: the Ministry of International Trade and Industry (MITI)[22] and the Environment Agency.

Ongoing Activities

Institutional Framework

The Action Plan to Arrest Global Warming, adopted in 1990, established the framework to comply with Article 4.1 (b) of the UNFCCC, which requires parties to formulate and implement measures to mitigate climate change. Though the provisions of the

Climate Change: Science, Strategies, & Solutions

action plan are general, the plan clearly specifies the interim goals of stabilizing CO_2 emissions on a per capita basis (and possibly in absolute terms) and stabilizing CH_4 emissions at 1990 levels by the year 2000. The importance of climate change was also recognized in the Basic Environment Law (1993) and the Basic Environment Plan (1994), and institutional developments have taken place to improve the integration of climate change concerns into sectoral policies. As a result, the Global Warming Prevention Headquarters, led by the prime minister, was established in December 1997 to promote and oversee comprehensive measures.[23] The Guideline on Measures to Prevent Global Warming, which includes principles for achieving the quantified commitment under the Kyoto Protocol (through the blueprint discussed above), was adopted in June 1998. Based on the guideline, a variety of new measures have been introduced or planned.

In April 1999, The Law for the Promotion of Measures to Cope with Global Warming, requiring action plans to be established by central and local governments, entered into force. The law promotes public reporting of plans and the status of their implementation by the central government, local authorities, and businesses. The law promotes actions to be pursued by the national and local governments (formulation of plans, etc.), and recommends voluntary and non-binding activities to be taken up by industry and the general public.

The most important institutional framework can be the Law Concerning the Rational Use of Energy (Energy Conservation Law). The law has served as a basis for conserving energy since the second oil crisis of 1979, and has contributed to enhancing the international competitiveness of Japanese companies and facilitating development of new technologies. In order to meet the Kyoto target, the law was substantially strengthened in June 1998 following the governmental program of "Comprehensive Energy Conservation Measures Toward the Year 2000."

The Energy Conservation Law, as amended, designated two categories of energy-intensive factories. Large factories (those which consume more than 3,000 kiloliters oil-equivalent of fuel or 12 GWh of electricity annually) must send in future plans to the government and promote energy saving with a view to improving energy intensity by 1 percent annually. Mid-sized factories (those that consume more than 1,500 kiloliters oil-equivalent of fuel or 6 GWh of electricity annually) must appoint energy managers and record energy consumption.

Another notable feature of the amendment is the incorporation of a new approach to energy efficiency standards. This so-called "top-runner" approach, which applies to automobiles and several home/office appliances, mandates that new products be more efficient than the most efficient model currently on the market,[24] and companies that fail to achieve the target could face punitive measures, including fines. This is a historic attempt, since most similar laws merely attempt to "pull the laggards up to average performance." The law also provides for measures to improve efficiency in the building sector. Discussions are under way to strengthen the existing framework of the law through broadening the coverage of regulated appliances and strengthening the standards. It should be noted, however, that the law is aimed at improving energy efficiency, but not at conserving energy or switching to renewables.

Nuclear and renewable energy are promoted by the Law Concerning the Promotion of Development and Introduction of Oil Alternatives and other legal instruments for the energy supply sector. A variety of subsidies like low-interest loans and tax exemptions are used to support these purposes.

Voluntary Approach by Industry

In many countries, voluntary approaches by industry through agreements, contracts, or self-declared targets are important constituents of their climate policies. Japan is no exception. The Keidanren (the Japan Federation of Economic Organizations) initiated the Voluntary Action Plan in 1997, declaring that its goal is to reduce overall CO_2 emissions in the targeted industries to 1990 levels by 2010.[25] Contrary to similar plans in some European countries, its targets are non-binding. The action plan, however, is regarded by both government and industry as one of the keys to achieving the blueprint.

The target of the plan is to "control CO_2 emissions from the industrial and energy conversion sectors as a whole at or below the 1990 levels in the year 2010." Specifically, there are four options for setting the quantified target: each industry sector can choose to establish targets using total energy-related CO_2 emissions, total energy consumption, CO_2 intensity, or energy intensity as the benchmark for reductions. The total emissions target mentioned above is calculated by aggregating industry-specific targets after converting to CO_2 emissions.

Although each industry does not have direct responsibility to achieve the target, it is required to contribute indirectly to the achievement of the industry-specific target. The action plan, with the participation of 34 business unions from the industrial and energy sectors, covers 42.6 percent of total CO_2 emissions in Japan and 76.5 percent of emissions from both sectors as of 1990. Collective emissions of the sectors have declined 0.1 percent between 1990 and 1999, from 479.07 MMT-CO_2 to 478.65 MMT-CO_2.

Emerging Industry Initiatives

These new developments, as well as the ongoing deregulation of the energy sector, have prompted the energy industry to launch new initiatives. Energy service companies (ESCOs) are emerging. Utilization of wind energy, which was relatively neglected in Japan's renewable energy portfolio, is rising. In November 2000, the Tokyo Electric Power Company, Japan's largest electricity company, established a company to utilize wind energy. The company also began issuing certificates for renewable energy, which are anticipated to be considered as offsets within the energy saving regulatory framework.

Additional Measures

It is widely recognized that additional measures are necessary to comply with the Protocol target. Since the Japanese government is highly motivated to have the Protocol in force, new policies are under development. One approach currently being considered is to broaden the coverage of targets subject to the Energy Conservation Law. Another is the contentious issue of environmental taxation.

New environment taxes have become a heated topic of discussion among many ministries. This is exacerbated by a comprehensive ministerial reorganization that began in 2001, since various ministries are attempting to safeguard or develop their source of revenue. The Environment Agency is considering a low-rate carbon tax with revenues earmarked to subsidize energy efficiency measures, and the Ministry of Finance is cautious about earmarking the revenues for any specific purpose, anticipating the potential of such revenues to alleviate the mounting budget deficit and/or ease the burden on companies. The Ministry of International Trade and Industry (MITI) has mixed views on such taxes, voicing concern about their potential effects on industry.

Tax reform in the transport sector is particularly contentious. The Ministry of Transport and the Environment Agency have together tried to reform the automobile tax system toward one that favors energy-efficient vehicles, but this effort is strongly opposed by MITI and the Ministry of Construction.[26] The discussions could involve the touchy issue of amending gasoline and diesel fuel excise taxes, most of which are currently earmarked for road construction.

In 2001, concrete measures are expected to be proposed by the government. Environmental tax(es) and domestic emissions trading are the two most contentious issues that remain to be resolved before Japan can prepare legal instruments to ratify the Protocol. New forecasts of long-term energy demand and supply, scheduled for June 2001, will play a key role in discussions leading to ratification since they will, for the first time, include a detailed analysis of the implications of the Kyoto agreement.[27]

These discussions have led to analyses of various policies and measures. It can be said, however, that the effect of a portfolio of policies, measures, and instruments is not well recognized or widely discussed in Japan. Research on the combined effect of taxation, regulation, deregulation, etc., is called for, and some progress is being made.[28]

Kyoto Mechanisms

Japan's original plan was to assign a limited role to the Kyoto mechanisms, but this plan is facing difficulty. Furthermore, it became apparent since COP3 that Japan stands most to gain from the Kyoto mechanisms, since Japan's marginal emissions reduction cost is generally considered to be among the highest in the world.[29] As a result, the Japanese government and the private sector are increasingly looking toward utilization of the Kyoto mechanisms as a cost-effective way to achieve its Kyoto commitments. An overview of current activities is provided below.

Discussions Within the Governmental Committees

Environmental policies taken by the government have traditionally relied on command-and-control measures. Utilization of economic instruments is a relatively new concept, and emissions trading an even newer one (witness Japan's reluctant acceptance of emissions trading at COP3). Discussion on how to utilize the Kyoto mechanisms has been conducted independently by MITI and the Environment Agency since early 1998. Both ministries have established a series of committees consisting of academicians, representatives from industry, consumer organizations, etc. There have been extensive discussions, most notably with respect to definition of project baselines.

JI and CDM Programs

Following the COP1 decision on launching the pilot phase of the joint implementation program (called the "activities implemented jointly" or AIJ program), the AIJ Japan Program was established. So far, 21 AIJ projects are under way, though only five of them are registered by the UNFCCC. Most of them take place in Southeastern Asia and China and are supported by governmental funding. The project with the largest CO_2 reductions is one that improves the energy efficiency of a steel plant in China, offsetting 87,434 metric tons of CO_2 (MT-CO_2) per year.

MITI initiated a program in 1998 to fund the private sector in undertaking feasibility studies on potential joint implementation/CDM projects in the industry sector. In the first year, 37 projects were selected from nine countries. Over half of them concern renovation of old power/heat plants in Russia, with CO_2 emissions reductions reaching up to several million tons per year. None of these projects has been implemented, since cost recovery was deemed difficult. Thus, more rigorous criteria on the financial viability of projects were imposed for subsequent years. The Environment Agency has launched a similar program, mainly on carbon sequestration and renewable energy projects.

Approaches by the Private Sector

So far, private sector activities on implementation of the Kyoto mechanisms in Japan have been slow to develop. The reasons include the following:

- The concept itself is foreign to many corporate managers.

- Public perception toward "trading the right to pollute" is still largely negative.

- Managers face little risk of being criticized for not purchasing inexpensive credits now, as opposed to waiting until a formal Kyoto Protocol regime is in place, when they might have to engage in more costly emissions reduction measures.[30] Thus, there is little incentive to initiate pilot efforts.

- The industry federation Keidanren remains opposed to any form of emissions cap, and is reluctant to enter into discussion with the government.[31]

Despite such adverse conditions, private sector activities to explore the possibilities of the Kyoto mechanisms are emerging. These activities are summarized below:

- *Forestry Activities in Australia.* The planting of eucalyptus and pine trees is carried out by various electricity, trading, and paper manufacturing companies, as well as by Toyota. Some of these projects are designed solely for sequestration purposes; others seek both logging and sequestration opportunities, depending on future regimes and relative prices.

- *Investment in Trading Firms.* Mitsubishi Corporation, a major trading house, bought a stake in Natsource, a major energy/environment broker in the United States. This is to prepare itself for future energy and emissions markets.

- *Carbon Offset Initiative.* Mitsubishi Research Institute, a Japanese research and consulting firm, launched a program in 1999 aimed at identifying and implementing GHG reduction projects by the private sector. Many projects in Eastern Europe, Asia, and Latin America have been identified, and discussions toward implementation are currently under way.

- *Participation in International Schemes.* Six Japanese electricity companies, as well as two trading companies, have joined the World Bank Prototype Carbon Fund. The ESCO fund by the European Bank for Reconstruction and Development has also attracted Japanese participants.

Conclusions

Japan's GHG emissions growth since 1990 has been lower than that of the United States, Canada, and Australia, though it has been higher than the EU average. Much of this lower growth rate can be attributed to economic stagnation, but factors such as increased use of nuclear power also played a role. Energy efficiency improvement, however, has stagnated since the mid-1980s.

Since CO_2 accounts for about 90 percent of Japan's GHG emissions, Japan's climate change policies are focused on energy-related measures. Japan's energy supply/utilization efficiency level, however, remains the highest in the world, and improving it is difficult. Previous analyses suggest that, among major Annex I countries, Japan has the highest marginal GHG emissions reduction cost. Various other factors impede reduction of energy-related CO_2 emissions. Potential for enhancement of carbon sinks is also limited.

The original blueprint for achievement of Japan's Kyoto Protocol commitment was largely dependent on domestic measures. Recent difficulties in nuclear plant siting as well as international negotiations on LULUCF suggest that the blueprint is in serious need of revision. New policy developments such as amendments to the Energy Conservation Law and ongoing developments on environmental taxation could contribute to achieving reduction goals. Electricity deregulation, on the other hand, could potentially negate such achievements. These conflicts and synergies in policy need to be studied. In order to comply with the Kyoto target, however, substantial utilization of the Kyoto mechanisms would be called for.

Since Japanese climate policies were primarily focused on domestic actions, industry response toward the Kyoto mechanisms has been less than enthusiastic. The Japanese government, however, is highly motivated to have the Protocol in force. Japanese positions on various Protocol-related issues have changed considerably since COP3, as a result of external circumstances. Therefore, the industry would do well to take a proactive stance toward utilization of the Kyoto mechanisms, and potentially take part in national policy-making.

Endnotes

1. If nuclear energy is considered as a quasi-indigenous energy source, the ratio declines to 80 percent.

2. "Kyoto baseline" is defined as the aggregated GHG emissions using global warming potentials (GWPs) with appropriate base years; i.e., the baseline of 1,272.5 MMT-CO_2(equivalent)/yr can be calculated from Table 1 by adding 1990 emissions for CO_2, CH_4, and N_2O, to 1995 emissions for HFCs, PFCs, and SF_6.

3. Currently the Ministry of the Environment after a major government reorganization in January 2001.

4. The 1999 data are based on IGES estimates. The MITI estimate — limited to energy-related CO_2 — is 8.9 percent above the 1990 level.

5. The recent decrease has led the Japanese representative at COP5 to proclaim that "GHG emissions in Japan have clearly changed from a pattern of gradual increase to gradual decrease," a view opposed by the environmental NGOs.

6. The bulk of the base-load power generation in Japan is from non-CO_2 emitting sources such as nuclear and hydropower, as well as coal, whereas peak-load generation is supplied chiefly by oil, gas, and pumped storage hydro generation. Hot summers require increased peak-load generation, mainly due to air-conditioning during office hours.

7. Capacity utilization factor, which affects the CO_2 emissions in the electricity generation sector, increased from 72.7 percent in the year 1990 to a record 84.2 percent in the year 1998. It fell to 80.1 percent in the year 1999.

8. CH_4 (methane), HFCs (hydrofluorocarbons), N_2O (nitrous oxide), PFCs (perfluorocarbons), and SF_6 (sulfur hexafluoride).

9. Now the figure has dropped to 52–57 percent, depending on the fluctuation of various factors, as oil plays a role of "buffer" in energy supply. Reducing this rate has been one of the central pillars of energy policy in Japan.

10. It is interesting that these earmarked taxes are uniform without any exemptions or reliefs. In this sense, these taxes are similar to the ideal carbon or energy tax with broad coverage.

11. The "bubble economy" is a combined result of a boom in the economic cycle, a relaxed fiscal policy, and a rapid rise in land prices. This has led to an overheating of the economy, which lasted from 1986 until about 1991. The GDP grew by about 6 percent in 1988.

12. In spite of the sudden increase in oil prices in the year 2000, prices are at pre-oil-crisis levels in real terms in Japan (imported CIF price incorporates the effect of exchange rate). The Plaza agreement of 1985 brought about a continuous appreciation of the Japanese currency, which also helped lower the oil price.

13. The Kyoto mechanisms (which include emissions trading, joint implementation, and the Clean Development Mechanism) are procedures that allow Annex 1 parties to meet their commitments under the Kyoto Protocol based on actions outside their own borders.

14. Using a baseline for Japan of ~1,272 MMT-CO_2 equivalent (CO_2E), Japan's 2010 BAU would be ~1,526 MMT-CO_2E. Reduction from the BAU to the baseline thus requires a reduction of ~254 MMT-CO_2E. ('a' under the Blueprint). Reductions due to (b) and (e) entail reductions of ~6 and ~47 MMT-CO_2E respectively. Since (c) and (d) cancel each other out, the 254, 6 and 47 MMT-CO_2E reductions constitute total

reductions from domestic measures (i.e., ~307 MMT-CO$_2$E). This is ~93 percent of the total 330 MMT-CO$_2$E reductions needed to reduce emissions from the 2010 BAU to the target (~1196 MMT-CO$_2$E)

15. COP6 was held in The Hague in November 2000 (see this book's chapter titled, "Reflections on Climate Talks in The Hague" for more details).

16. One of the methods proposed by the EU limits acquisition of Kyoto mechanisms permits/credits to 50 percent of the difference between the actual annual emissions of any year between 1994 and 2002 multiplied by five, and its assigned amount. Assuming that maximum emissions between that period is 1,384 MMT-CO$_2$ (emissions in the year 1996), the amount will be 94 MMT-CO$_2$. Using the other method (5 percent of average between its Kyoto baseline and its assigned amount), Japan can claim up to 62 MMT-CO$_2$.

17. Electricity supply targets adopted by the Electric Utility Council (June 1998).

18. The Federation of Electric Power Companies, Electricity Supply Plan (March 2000).

19. Another significant source of base-load generation is hydropower, but its additional capacity is limited.

20. Including other factors such as an increase in forecasted electricity demand.

21. Intergovernmental Panel on Climate Change. 2000. *Land Use, Land-Use Change, and Forestry.* Watson, R.T., I.R. Noble, B. Bolin, N.H. Ravindranath, D.J. Verardo, and D.J. Dokken, eds. A Special Report of the IPCC. Cambridge University Press, Cambridge, U.K.

22. Currently, the Ministry of Economy, Trade, and Industry.

23. In addition, the Kyoto Initiative was formulated in December 1997, which consists of strengthening environmental support to developing countries in climate change related projects.

24. This approach well represents the (government-driven) technology-oriented policy of MITI.

25. The definition of "voluntary" action plan differs considerably throughout the world. Keidanren's plan requires reporting to government councils with prescribed formulas to avoid double counting. Unlike similar plans in some European countries, however, it does not mandate formal agreements with the government, nor does it entail punitive measures for failing to achieve the target.

26. Ministry of Transport and Ministry of Construction, with differing views on automobile taxation, are now merged into Ministry of Land, Infrastructure, and Transport.

27. Full liberalization of the electricity retail market is anticipated after a review in 2003.

28. On December 14, 2000, a sub-committee on global warming mitigation under a council established by the Environment Agency released five policy package model options in a public consultation report. These options include: 1) continuing the existing framework, 2) strengthening existing voluntary commitments, 3) environmental taxation, 4) environmental taxation combined with emissions trading, and 5) voluntary emissions trading. It is unclear whether these options provide a basis for further domestic negotiations in the government.

29. See, for example, the report, *International Emissions Trading,* in the Economics chapter of this book.

30. Recognition of this fiduciary responsibility is a significant factor in private sector emissions trading in the United States.

31. This is in contrast to the ongoing industry-government debate at the UK Emissions Trading Group.

Electric Power Futures in Five Developing Countries

William Chandler, ed.

The goal of stabilizing atmospheric concentrations of greenhouse gases stated in the United Nations Framework Convention on Climate Change (UNFCCC) cannot be met without the participation of developing nations.[1] Developing countries will surpass developed country emissions of carbon dioxide (CO_2) in the next quarter-century. But developing countries have been reluctant to accept binding emissions targets, declaring that richer nations caused the build-up of atmospheric greenhouse gases and must be the first to take action to reduce emissions. This position may be changing slowly with the recognition that climate change threatens all nations, and that cooperation in emissions mitigation can provide benefits for development.

In most developing countries, electric power generation accounts for approximately one-third of total carbon dioxide emissions. The power sector is also responsible for about one-third of particulate and up to two-thirds of sulfur dioxide emissions. Power generation is thus a critical area for reducing both local and global pollutants. To address these issues, the Pew Center commissioned studies of five key developing nations and their electric power futures. This paper summarizes that work.[2]

The countries chosen for this research are five fast-growing economies: Argentina, Brazil, China, India, and South Korea, of which only Argentina has expressed willingness to accept emissions targets under the UNFCCC. Argentina and Korea have shown leadership in climate policy, and China has made exceptional progress in reducing the carbon intensity of its economy. Brazil and India produce relatively little carbon dioxide from power generation, the former due to heavy reliance on waterpower, the latter due to very low per capita power consumption. To ensure accuracy and credibility, the Pew Center and Battelle engaged national experts from each of the five nations to study their power futures. Each country team estimated year 2015 emissions for a baseline, or status quo, case, as well as scenarios of carbon emissions in various alternative futures, including efforts to promote market reform, natural gas, energy efficiency, nuclear power, and renewable energy, and to reduce local pollution. The authors estimated the economic cost of each of these options, incorporating environmental externalities into the economic analysis to the extent possible. Special attention was paid to capital investment requirements, policies related to competition,

This report summarizes five case studies — on Argentina, Brazil, China, India, and Korea — previously published by the Pew Center. The author and the Pew Center acknowledge the authors of the original reports, who are listed in endnote 2, and Jeffrey Logan in particular for his contribution to the overall effort.

decentralization, private ownership, and demand-side implications. The conclusions reported here are based on a consensus view of the authors involved.

Why these five countries and not some others? This selection provides a fair cross-section of income, technology, policy, and climate insofar as developing countries are concerned. The economy of each of these nations is growing rapidly, and, in India, the population continues to grow quickly as well. As a result, the carbon emissions from these five nations, which together today amount to about the same as the U.S. total, (see Figure 1), could easily double over the next 15 years. China now ranks first in the world in population and, after the United States, second in the world in energy consumption and emissions of greenhouse gases. It plays a leading role among developing nations in the field of energy and climate policy. Korea, though much smaller, occupies a unique place in energy use and climate change. A member of the Organization for Economic Cooperation and Development since 1996, Korea is similar to developed nations in per capita income and energy use, but is counted among the developing countries by the UNFCCC. India may soon overtake China as the world's most populous nation, but generates only about one-third as much electricity, and its per capita power use is just 4 percent that of the United States. Argentina, a relatively prosperous country, boasts a distinctly market-oriented electricity generating system. Power

Figure 1

Carbon Emissions   Five Developing Nations versus the United States

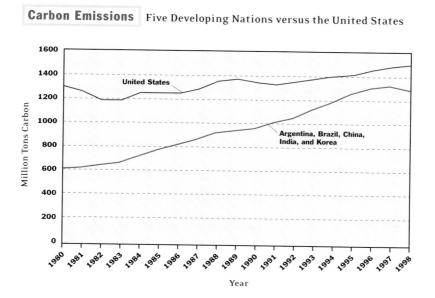

Source: Energy Information Administration. International Energy Database, December 1997: http://www.eia.doe.gov.

sector reforms have progressed further than in most nations, including the United States, and hold important lessons for climate policy. Per capita carbon emissions in Brazil are less than half the world average, largely because of the country's heavy reliance on hydropower, which produces few greenhouse gas (GHG) emissions. Each of these nations is thus uniquely significant for climate change policy.

The Current Context

Argentina, Brazil, China, India, and Korea have in common rapid electric power demand growth. They differ substantially, however, in their choice of fuels to meet supply power. (See Table 1.) China and India rely heavily on coal, Korea on nuclear power, Argentina on gas, and Brazil on falling water (hydropower). These choices are shaped by government policy as much as fuel availability and cost. Korea, for example, pays a premium for nuclear power rather than rely on oil imports. China relies almost exclusively on coal, although many companies would like to develop its gas and coal-bed methane resource for power generation. Moreover, policy plays a role in the rate of power demand growth relative to the overall economy, remarkably so in China. China made a conscious decision two decades ago to improve the energy efficiency of its economy in order to reduce the amount of capital that had to be spent on coal mining and transport and power generation. Consequently, power demand over that period has grown slower than the overall economy — an unusual circumstance for a developing country.

Argentina

Competition in Argentina has favored natural gas over hydropower and nuclear power, thus increasing emissions at the margin, but has also virtually eliminated coal from the market despite its abundance. While competition has lowered the price of electricity, thereby increasing demand, it has done so by reducing inefficiency; that, in turn, has reduced carbon emissions. Privatization and competition in the energy sectors of Argentina and several

Table 1

Electric Power Supply Circa 2000 (shown in terms of percent)					
	Argentina	**Brazil**	**China**	**India**	**Korea**
Coal	1	2	81	64	35
Oil	3	3	1	<1	8
Gas	49	0	1	8	13
Hydro	37	91	17	25	3
Nuclear	11	1	1	3	42
Other*	—	3	—	<1	—

Source: Battelle Memorial Institute.

* "Other" means wind power in India, and bagasse in Brazil.

Climate Change: Science, Strategies, & Solutions

other South American countries are influencing power reform across the continent.

Argentina's electric power sector consumes about 22 percent of that nation's total energy supply. Overall energy demand growth is driven by transportation energy use, which increased by half since 1990. The residential sector grew by more than one-quarter over the same period. Abundant natural gas provides one-third of total energy use and continues to increase market share. Transportation and agriculture still rely on oil, but industry, commercial buildings, and residences have increasingly switched to direct use of natural gas. Argentina also exports oil and natural gas, currently about one-eighth of total production. The country has a relatively strong energy conservation and efficiency program focusing on co-generation of heat and power, energy appliance labeling, and efficient lighting.

The country is now also emerging as a leader in environmental issues. In October 1999, Argentina announced a voluntary effort to restrict greenhouse gas emissions within a range of 2 percent to 10 percent below the projected baseline level during 2008–2012. Argentina became the first developing country to establish such a voluntary target under the UNFCCC. This action could motivate some of the relatively small emitters to take on similar voluntary targets.

While Argentina's power demand is expected to continue to grow over 6 percent each year, growth will not necessarily mean a corresponding increase in emissions. Carbon emissions in particular could be offset by improving energy conversion efficiencies or by promoting low- or non-carbon fuels. But such reductions would most likely occur only as a result of new policies, such as participation in the Clean Development Mechanism (CDM).

Brazil

Brazil generates over 90 percent of its electricity by capturing the energy in falling water. Carbon emissions per capita in Brazil are less than half the world average, in part because of this heavy reliance on hydroelectric power, which produces few greenhouse gas emissions.[3] A significant portion of the country's new power plants, however, will likely use natural gas because hydropower plants are increasingly expensive, controversial, economically risky, and slow to come on-line.

The country's electricity consumption rose at an average annual rate of 7.9 percent from 1970 through 1997, increasing the share of power generation as a part of total energy consumption from 16 to 39 percent.[4] Brazil's economy is also characterized by heavy dependence on electricity-intensive industries. Demand for both primary energy and electricity have grown much faster than the economy, with energy use rising 50 percent faster than the gross domestic product (GDP), and power use growing 170 percent faster. These rates are higher than those typically found in other industrializing countries.[5]

Brazil's economy has suffered bouts of hyperinflation and recession over the past 20 years, and GDP has expanded at an average yearly rate of only 2.1 percent since 1980. Financial panic spread to São Paulo in late 1998 and early 1999 when international investors — fearing that reforms were proceeding too slowly — withdrew capital *en masse*, leading to a collapse of the real, Brazil's national currency.[6] Interest rates and inflation have since stabilized, however, and the economy appears to be rebounding.

Power sector reform in Brazil has been driven by the need to reduce the likelihood of power shortages, lessen the need for state-sponsored investment, and provide incentives for cost reduction. Significant delays have occurred in the transfer of state-owned assets to the private sector due to macroeconomic instability and lack of definition of initial transmission and generation prices. The reform has been relatively successful in privatizing distribution companies, but has failed to attract private investment to expand generation capacity. If this expansion does occur, foreign investors are expected to provide most of the capital.

A foundation of Brazil's new policy is the creation of a competitive wholesale energy market. The new approach will include three semi-competitive business segments: (1) generation companies, which operate in an open marketplace, making both spot and contract sales to large purchasers (distributors and heavy users); (2) transmission companies, which guarantee open access to the grid and operate under a fixed tariff framework, allowing a regulated return on assets; and (3) distribution companies, which have both "free" customers (large users that have permission to purchase electricity from their choice of power providers at a market price) and captive customers (users that purchase from the distributor under regulated tariffs).

China

China ranks second in the world in energy use, but only because of its large population. The nation still uses only about one-third as much electricity as the United States. Per capita electricity consumption remains under 1,000 kilowatt-hours per year — just one-third the world average. (See Table 2.) China relies on coal for power generation much more heavily than most countries. Coal-fired power plants rank as China's largest source of sulfur dioxide, and account for one-third of its carbon emissions. How to meet power demands at least cost — while taking into account environmental impact — is a topic of great concern for decision-makers in the Chinese government and the power industry.

Remarkably, for more than a decade China's demand for power has grown slightly less rapidly than its

Table 2

Electric Power Capacity and Use, 1998

	Million Kilowatts* installed capacity	Kilowatt-Hours per capita consumption
Argentina	221	2,092
Brazil	62	2,027
China	254	920
India	100	424
Korea	44	4,463

Source: Derived from the Energy Information Administration's website at: http://www.eia.doe.gov. U.S. Department of Energy, Washington, D.C., updated July 2000.

* 1 million kilowatts = 1 gigawatt.

Box 1

A Guide to Linear Programming
for Power Sector Analysis

Analysts use linear programming (LP) models to optimize combinations of inputs whose values are valid only over specific ranges. For example, power planners and electric utilities use LP models to determine the types of power plants required to meet least-cost power demand over time while meeting limitations in pollution emissions, energy sources, and manufacturing capacity. Models can help planners analyze alternatives, but non-quantitative factors must also be considered when designing real-life systems.

Researchers use two classes of models to analyze energy systems. LP models are often called "bottom-up" models because they contain detailed information about technology and costs. They have rich engineering detail and rely on user input to simulate broader economic conditions. "Top-down" models, on the other hand, begin from a higher level of economic reality by simulating the interaction of supply and demand in the main sectors of an economy. While top-down models have less detailed information about energy technologies and costs, they capture the reality of consumer behavior better than bottom-up models. Some models, like MARKAL-MACRO, try to integrate the economic reality of top-down models with the engineering detail of bottom-up models.

Researchers at Battelle created a generic LP model which each of the country teams in this study modified to analyze least-cost power options according to the conditions in their specific countries. The model can choose among 17 different types of power plants (coal, petroleum, natural gas, nuclear, hydroelectric, and renewable) to meet power demand. The model divides the country into as many as five regions to capture the

variation in energy availability, fuel cost, and environmental limitations. Simulation begins with a base year (1995) and then determines the amount of new capacity from each type of power plant needed to meet demand over 5-year intervals.

After analysts enter technology and cost characteristics of the power plant options, the model calculates the levelized, or lifecycle, costs of power generation. Levelized cost analysis accounts for all the costs of building, fueling, operating, and controlling pollution from power systems and spreads them out over the economic life. In this way, the costs of delivering power to users from nuclear plants (with high construction and low fuel costs) can be compared directly with the costs of providing power from combined-cycle plants (low construction costs and high fuel costs). Analysts also enter the regional power demand over time. These values are calculated separately according to estimates of economic growth and power demand intensity.

The actual linear program will then find the minimum cost combination of power plants needed to meet the demand. Additional constraints can include emissions caps on pollutants such as sulfur dioxide, manufacturing limitations for power generation equipment such as nuclear reactors, energy supply limitations such as hydropower capacity, and transmission line characteristics that limit the amount of power that can be sent from one region to another. For a given time period, the LP model will select the least-cost power source available and continue to use that technology until a constraint prevents its use. LP models need expert input to define when constraints are needed to simulate reality.

Table 3

Cost and Performance of Selected Technologies

Assumed for 2005

	Investment $/kW	O&M Cost $/kW/year	Efficiency percent	Lead Time years
Small Steam Turbine (natural gas)	900	18	35	4
Large Steam Turbine (natural gas)	850	17	35	4
Pulverized Coal	600–1,100	20	35	4
Small Gas Turbine	450–540	10	20	1
Mid-Sized Gas Turbine	380	10	30	1
Large Gas Turbine	310	12	34	1
Small Combined-Cycle	450	20	45	2
Large Combined-Cycle	400–795	18	53	2
Nuclear	1,700–1,800	30	34	5
Small Hydro	840–1,000	12	—	5
Large Hydro	1,300–1,570	15	—	7
Wind	950–1,000	13	—	1.5
Photovoltaic	4,800	7	8	1
IGCC	1,025–1,600	—	—	—

Sources: Bariloche Foundation, 1999; Energy Information Administration, 2000.

generation technologies for capital, fuel, operations, and associated environmental costs, and converted these to costs per kilowatt-hour. Third, the analysts tested alternative policies for their impact on average generation costs and especially for changes in greenhouse gas emissions relative to the present and to the baseline.

The results indicate increased or reduced economic cost compared to the baseline, along with changes in technology, power plant capacity, utilization, and emissions.

Five power generation technologies were included in the analysis: coal-fired power plants with scrubbers, oil-fired combined-cycle units, gas-fired combined-cycle units, nuclear power plants, and hydropower plants. (See Tables 3 and 4 for a summary of capital and fuel cost assumptions.) The modeling also included integrated gasification combined-cycle (IGCC) plants, none of which has yet been built in any of these nations.

Various renewable technologies such as wind, biomass, and geothermal technologies were included in the modeling as options, but in the time period considered, they did not compete with the other technologies. The analysis did consider how much costs would have to decline for renewable and other advanced energy technologies (like fuel cells) before becoming competitive. The modeling was not an attempt to forecast power futures or even the necessary power capacity. It simply served as a tool to compare the impact of different policy options on technology choices and on environmental quality.

carbon elimination. Least-cost modeling simulated these scenarios through changes in emissions fees and caps, costs for advanced technologies, demand side efficiency, and clean energy supplies. Specifically, the scenarios produced the following results.

Baseline. This scenario assumes that institutional reform such as privatization and increased competition among generators is successfully implemented over the coming decade. The installed capacity grows from 56 gigawatts in 1995 to 94 gigawatts in 2015, an increase of 68 percent. Natural gas plants increase from essentially zero to 11 percent of installed capacity over the period of analysis. Energy efficiency and cogeneration play important roles in limiting an even greater reliance on fossil fuel power generation. The total cost of meeting demand is $183 billion, which includes capital, fuel, and operation and maintenance costs. Carbon dioxide emissions rise over four-fold from 3.4 million tons of carbon in 1995 to 14.5 million tons in 2015. However, the intensity of CO_2 in Brazil remains low, even in 2015, as hydropower still accounts for 74 percent of total generation. Sulfur dioxide and particulate emissions grow proportionately with power generation, while nitrogen oxides increase five-fold to reflect the greater use of natural gas in power generation turbines.

Advanced Technology. The advanced technology scenario simulates capital cost reductions for power plant equipment due to technological progress driven by government incentives. Environmental costs are also at least partially accounted for in the least-cost analysis by including some of the external costs of emissions, hydropower construction, and nuclear decommissioning that are normally ignored. Wind power increases from zero to almost 2 percent of total installed capacity by 2015 due to the environmental fees imposed on fossil-fuel use. The total cost of this scenario is $181 billion, 1.6 percent less than the baseline, mainly due to the cheaper costs of building and operating combined-cycle power plants in the later years. This figure does not include the research, development, and deployment costs needed to improve technologies. Carbon dioxide emissions drop slightly from the baseline, reaching 13 million tons of carbon in 2015. Sulfur dioxide emissions decline by approximately 50 percent due to the elimination of diesel generators after 2005.

Local Environmental Control. In this scenario, renewable energy policies and higher environmental externalities influence the technologies employed. The environmental costs of pollution are assessed at a higher value than in the technology scenario, and cost reductions for cleaner, advanced technologies are also assumed. Hydropower plays a larger role in this scenario, rising to over 87 percent of total installed capacity. The environmental and social impacts of expanding hydroelectric power production this much are difficult to estimate, but could be significant. Biomass capacity rises from 2 percent in the 2015 baseline case to 5 percent here. The cost of this scenario is $179 billion. Carbon dioxide emissions drop from 3.4 million tons of carbon in 1995 to 2.6 million tons in 2015. Sulfur dioxide emissions decline substantially, while particulate emissions increase due to the growth in biomass combustion for power generation.

Carbon Elimination. In the carbon elimination scenario, Brazil installs electric power generation technologies that produce no net carbon dioxide emissions and only minor impacts on watersheds and landscapes. Installed capacity in 2015 reaches 97 gigawatts, and hydropower continues to account for over 80 percent of installed capacity. Renewable energy accounts for all but 3 percent of Brazilian power generation in 2015, with biomass accounting for over 16 percent. The total cost of this scenario is $208 billion, 14 percent above the baseline case. Carbon emissions all but cease from the power sector, and sulfur dioxide emissions drop, but particulate emissions rise five-fold due to the heavy reliance on biomass.

Brazilian power demand will continue to increase over the next two decades regardless of the country's current economic difficulties. Reforms under way in the power sector, however, will greatly influence how power demand is met and the emissions that result. Hydropower will continue to dominate through 2015, but its share will most likely decrease.

Carbon emissions more than quadruple in the baseline scenario to 14.5 million tons, but remain extremely low in absolute terms, equivalent to the emissions from 10 large coal-fired power plants. Biomass and wind power might play a larger role in Brazil's power future if the government focuses on developing advanced technologies and accounts for at least some of the costs to the environment. Coal-based technologies are not competitive with other forms of power generation, allowing Brazil to largely avoid the tradeoff between improving the quality of the local environment and reducing global greenhouse gas emissions.

In the local environmental control and carbon elimination scenarios, there is a strong interdependence between electricity generation based on sugar-cane bagasse and ethyl alcohol production for automotive use. By accounting for the environmental impacts of local pollutants or restricting power generation options to those with no carbon dioxide emissions, sugar-cane bagasse becomes feasible, making it the power generation option that is most widely used in both scenarios after hydropower. This indicates that Brazil has the potential to service the electricity market without carbon emissions if the market or the international community can support the 14 percent higher costs.

In all four scenarios, energy efficiency and cogeneration play an important role in the least-cost power solution. Saving electricity through increased efficiency offsets the need for new supply and has enormous potential in Brazil's industrial sector. Efficiency also reduces the environmental burden associated with electricity production and transmission (most likely via natural gas combined-cycle plants) without compromising the quality of the services end users demand.

Carbon dioxide emissions from Brazil's power sector will remain low in absolute terms over the next two decades. Brazil appears able to play a unique role within the context of the UNFCCC by fostering economic growth that does not sacrifice local or global environmental quality. Achieving clean development would serve as a powerful example for other developing countries.

China

Baseline case. Power generating capacity and consumption are expected to nearly triple by 2015 from their values in 1995, requiring some $449 billion in total costs.[17] Coal would provide 85 percent of power, and coal use for power generation alone would reach 1 billion tons per year. Emissions of sulfur dioxide and carbon from the power sector reach roughly 20 million tons and one-half billion tons per year, respectively. This scenario assumes that the current environmental policy remains the same, which is increasingly unlikely.

Sulfur emissions controls. Annual sulfur dioxide emissions from the power sector could be cut to 12.5 million tons by 2015, a 40 percent reduction from the baseline level, by imposing regulatory caps on emissions or fees ranging from $360–$960 per ton of sulfur released. Total costs using the sulfur fees would rise by 4 percent. Sulfur control policies would reduce total coal use very little, but would greatly increase coal washing and flue gas desulfurization. These options cost less in China than alternatives such as nuclear power, hydropower, and advanced coal technologies that reduce sulfur emissions by a comparable amount. Achieving sulfur reductions would also require stricter regulatory enforcement. However, greenhouse gas emissions would change little as a result of stricter sulfur dioxide emissions control.

Carbon control. This scenario tested the effect of reducing carbon emissions in the power sector by 10 percent, or 50 million tons per year, by 2015. The study simulates these reductions by assuming the construction of new, less carbon-intensive power plants; it does not consider alternatives to lower emissions in existing plants. A 10 percent reduction from the baseline would add $20 billion to total costs by 2015, an increase of about 4 percent. Greater reliance on washed coal, hydropower, and nuclear power, and fuel switching to natural gas would be the cheapest ways of reducing emissions. Moderate carbon taxes were also tested in this analysis, but they were not found to be particularly effective in encouraging fuel switching. Only very high taxes (over $75 per ton of carbon) produced significant emissions reductions.

Natural gas. China currently uses very little natural gas for power generation. The government will need to establish new policies and reforms to increase the availability of natural gas. This scenario simulates the impact of policies to boost gas use in the power sector. Increased availability of low-cost natural gas in the power sector (combined with improved turbine efficiency and a $300 fee per ton of sulfur dioxide emissions) could cut carbon and sulfur dioxide emissions by about 14 and 35 percent, respectively, from the baseline. Natural gas power in this scenario is cheaper than coal-fired power only along the coastal regions (where coal is relatively expensive), but gas would need to be available for $3 per gigajoule. The power sector would consume approximately 65 billion cubic meters of gas, accounting for roughly half of China's total gas demand in 2015.

Clean coal. A set of scenarios tested the effect of reducing the cost of advanced coal technologies such as integrated gasification combined-cycle (IGCC) or pressurized fluidized bed combustion (PFBC) to help them capture additional market share relative to the baseline. A 40 percent reduction in capital costs for IGCC and PFBC, combined with a mid-level sulfur dioxide emissions fee of $300 per metric ton, would reduce carbon dioxide and sulfur dioxide emissions by 9 and 75 percent, respectively. However, approximately $140 billion in additional investment (perhaps through international cooperation on technology transfer and clean development) would be required to subsidize the cost of building these plants.

Efficiency. This scenario tested the effect of reducing electric power use by 10 percent compared to the baseline. Such a reduction would lower carbon and sulfur dioxide emissions by 13 percent by 2015 and save $55 billion in investment and fuel costs by postponing the need for 52 gigawatts of coal-fired generation capacity. The analysis did not consider the required policies or costs to lower power demand.

These scenarios revealed two important findings. First, policy options exist to reduce carbon emissions substantially in the Chinese power sector at relatively low incremental cost. Emissions reductions equal to more than 10 percent, compared to projected baseline emissions in 2015, can be achieved for less than 5 percent of the total cost of power. Continued improvement in demand-side efficiency is a particularly attractive option to lower carbon emissions. Second, not all of these reductions will be achieved for reasons that are in China's own interest, such as reducing sulfur dioxide emissions.

The Chinese team concluded that without a strong environmental policy, China's electric power mix would continue to rely heavily on coal. When policy measures such as fuel availability, technical performance, and full cost accounting are considered, however, the mix of electric power generation technologies (if not necessarily the fuels) changes significantly. Cooperation with other countries would be required to achieve more dramatic results.

India

The Indian case study projected a baseline of current trends, then tested the effects market reform, efficient technology, local environmental pollution controls, and sustainable development strategies could have on India's power sector and economy. The goal of this analysis was to present policy options based on the relative differences between scenarios and other qualitative information about India's power sector.

Baseline. Power capacity in 2015 will grow to two-and-a-half times the 1995 level. Coal technologies will continue to account for the largest share of new additions to capacity, but will decline from 62 percent in 2000 to about 55 percent in 2015. Clean-coal and natural gas technologies will play larger roles, while hydropower will decline slightly. Capital requirements will total $151 billion from 2000 to 2015, more than all other scenarios except a high-growth case. Sulfur dioxide and carbon emissions both more than double from their 1995 levels to 5.9 million and 217 million tons, respectively, by 2015.

Growth Scenarios. Two additional alternative baselines were run as sensitivity tests for India: a high- and a low-growth case. The high-growth scenario, which assumes 7 percent economic growth, requires $218 billion in investments by 2015 — more than any other scenario. Surprisingly, high growth cuts sulfur emissions significantly — to 3.3 million tons in 2015 — due to the availability of sufficient funds. This is less than half the baseline level. Carbon emissions reach 225 million tons by 2015, about 2.5 times the 1995 level, though only 4 percent more than the base-case total. In a low-growth scenario, capital requirements amount to just $117 billion per year by 2015, about 45 percent lower than baseline requirements. Ironically, this scenario produces the second highest level of sulfur dioxide — more than 4 million tons in 2015, or 50 percent more than the market reform case. Carbon emissions in 2015, reach 145 million tons, about one-third lower than baseline emissions, but almost 60 percent higher than 1995 emissions levels.

Market Reforms. This scenario assumes that India will further liberalize domestic policies and become more receptive to foreign direct investment. The scenario results in only minor changes to the power supply mix compared to the baseline. However, less capacity overall is needed due to greater energy efficiency and utilization of existing capacity, thus reducing capital requirements by 11 percent. Carbon emissions decline 7 percent from the baseline by 2015, falling to 203 million tons per year. Sulfur dioxide emissions fall by 56 percent by 2015, to 2.6 million tons, due to increased use of pollution control equipment with conventional coal technologies.

Efficient Technology. This scenario simulates the impact of cost reductions in advanced power generation and end-use technologies. Investment requirements in this scenario decline 20 percent below the baseline due to the higher capacity and efficiency factors of advanced technologies. Advanced technologies in the end-use sector reduce total installed capacity from 246 gigawatts to 207 gigawatts, while the power mix shifts from coal to gas and hydroelectric. Carbon emissions total 168 million tons in 2015, about 23 percent lower than the baseline. Sulfur emissions fall by 60 percent in 2015, to 2.3 million tons.

Local Environment Control. This scenario simulates stricter control of nitrogen and sulfur oxides and particulates. Capacity additions closely resemble the base case in this scenario, but capital costs increase by 5 percent due to more expensive control technologies. The marginal cost of electricity is 3 percent higher than the baseline and higher than in most other scenarios. Compared to the base case, fitting coal technologies with sulfur control equipment cuts sulfur emissions by 40 percent in 2015. Carbon emissions reach 218 million tons by 2015, similar to base case emissions, primarily because there is little substitution of gas for coal.

Sustainable Development. The country team simulated a combination of progressive policy options in this scenario, including decentralized management, environmental conservation, efficiency and renewable energy promotion, and regional cooperation. Requirements for capacity additions in this scenario fall 22 percent below the baseline, but still require $117 billion of cumulative investment between 2000 and 2015. Ironically, sulfur emissions are not as low as in the high-growth case, but do fall dramatically to about 2.3 million tons. This scenario produces the lowest carbon emissions in 2015 — about 141 million tons — which is still 60 percent higher than emissions today.

The Indian case study concludes that near-exclusive government ownership, a supply-side orientation, and tariff distortions led the Indian power sector on an unsustainable path before 1990. The analysis yields three additional insights. First, strict control of local pollutants may not necessarily lead to reduced carbon emissions. A least-cost option might simply promote early penetration of pollution control equipment suited to conventional coal power technologies. Second, high economic growth does not have to lead to excessive coal-fired emissions; instead, stricter emissions control and greater financial resources might cause a shift to cleaner fuels and more energy-efficient coal technologies. Third, the share of natural gas increases in all scenarios, indicating that enhancing the gas supply is a vital energy policy measure.

A useful hedging strategy would be to keep the energy and technology mix flexible. Gas is a robust option that meets multiple objectives like low emissions and peak load requirements and multiple constraints like low investment costs and short construction time. Another option is to gain experience with emerging renewable technologies. It would also be useful to integrate regional grids into the national grid to enhance the efficiency and reliability of the power supply. Nepal and Bhutan own hydro resources while Bangladesh and Myanmar possess gas. Cooperation with these neighboring countries will help diversify the capacity mix, reduce costs, and improve environmental performance.

India could reduce electricity losses in the power grid by updating technology and management practices. Furthermore, improved grid reliability could lower the need for captive power generation (often inefficient diesel generators) at industrial sites. The costs required to accomplish this would be offset by the savings in building and operating new power plants and the associated reduction of harmful emissions.

The Indian power sector has substantial potential to reduce its carbon emissions from baseline projections. Short-run policies such as promoting clean technologies and reducing energy demand are likely to curb local pollution substantially and reduce carbon emissions by one-quarter by 2015, or a total of 600 million tons between 2000 and 2015. However, measures that control sulfur and nitrogen oxide pollutants — such as coal washing, sulfur scrubbing, and integrated gasification combined-cycle plants — have little impact on carbon emissions. Climate change mitigation policies for the Indian power sector, therefore, will have to be crafted for their own sake.

Korea

The Korean team evaluated trends in that country's power sector to estimate the size and cost of alternative energy strategies and their effect on the nation's greenhouse gas emissions. A baseline case represents the status quo, which itself incorporates a strong measure of liberalization of the power sector over the coming decade. Against that base case, the analysis tested a scenario of rapid reform and restructuring in the power sector — a move that would culminate quickly in a competitive supply market. That scenario would have private suppliers make the power supply choices that are now made by the Korean government and its state-owned monopoly. The authors also tested a scenario in which Korea's high energy intensity gradually falls to the level of Japan's by 2015. This case assumes that economic activity will shift from heavy to light industry and services, and that energy efficiency will play an even stronger role in Korea than it does now. A fourth scenario tests a case in which natural gas supplies increase and prices decline due to greater availability of imported pipeline gas. In a fifth scenario, environmental externalities were monetized and incorporated into the price of electricity supplies. The sixth scenario analyzes the ability of nuclear power to compete with other options based on a sensitivity test of capital costs. The final case analyzed carbon dioxide control measures. The authors also estimated the cost reductions necessary for advanced power generation technologies such as fuel cells and wind energy to be competitive with fossil fuels.

Baseline. Korea's electricity consumption will likely double from the 1995 level by the year 2015. In the baseline scenario, consumption of LNG would triple, while coal use would grow even more. The least-cost power capacity mix for the baseline in 2015 includes coal, LNG, nuclear, oil, and hydropower at 35, 25, 19, 10, and 6 percent, respectively. But, in terms of electrical output, coal and nuclear account for 41 and 27 percent, respectively, with LNG and oil supplying 17 and 11 percent, respectively.

Korean coal resources are limited, expensive, and unlikely to be the fuel of choice among cost-sensitive, market-oriented producers. Imported coal is the cheapest source of power, however, and in a pure market would likely be the cheapest option. Coal-fired plants using imported coal are likely to form the second largest source of power generation in Korea over the next decade unless environmental policy alters the economics.

Restructuring. Restructuring would shift the mix of power plant capacity from coal to petrol and gas-fired plants. Total costs would increase due to greater demand for less expensive power, but by less than 0.5 percent. Carbon dioxide emissions would decline by 9 percent, and sulfur emissions would drop more than 60,000 tons (24 percent) compared to the baseline case. Nuclear power is unlikely to be adopted by private power developers because of high capital costs, siting delays, and the scale of investment required. Reform would also reduce power demand growth rates by lowering subsidies for energy-intensive large industries and for power production itself. Private power developers would prefer the efficiency, low capital costs, and flexibility of small-scale power plants fueled by LNG, which produces less than half as much carbon per unit of delivered electricity as coal. Reforms may enable Korea to improve economic efficiency and satisfy local environmental concerns, and may help mitigate global climate change.

Energy Efficiency. Carbon and sulfur emissions would fall by 21 and 25 percent, respectively, if Korea were to utilize cost-effective energy efficiency options. Greater use of co-generation, district heating, and performance contracting, combined with the elimination of subsidies to energy intensive heavy industry, could help reduce Korea's energy use per unit of GDP to that of Japan by 2015. The country would save almost $8 billion in power plant construction and operation costs, and significantly reduce energy imports.

Natural Gas Policy. Liberalizing natural gas imports would make gas-fired power generation the most cost-competitive option. Such a scenario requires meeting 36 percent of power capacity with natural gas while bringing costs down slightly from the baseline by reducing import tariffs, which might allow gas to be imported from Russia. Installed capacity of gas-fired units would increase to 30 gigawatts by 2015. Sulfur dioxide and carbon dioxide emissions would drop by 64,000 tons (25 percent) and 5.5 million tons (11 percent), respectively.

Environmental Protection. Including the shadow environmental externality costs in the planning of electricity significantly alters the power mix in 2015 and makes coal more expensive than natural gas or nuclear. While total costs would rise about $2.3 billion over the baseline scenario, sulfur and carbon emissions would decline by 59 and 28 percent, respectively. Wind power would become competitive in this scenario by 2010.

Nuclear Power Promotion. Free of greenhouse gas emissions, nuclear power is the largest source of electric power in Korea and is likely to remain so for at least another decade. This is true despite the fact that nuclear power is no cheaper for the end user than coal, oil, or gas-fired power. Capital costs range from three to four times that of gas- and oil-fired systems, offsetting the significantly lower fuel costs that nuclear power enjoys. The dominance of nuclear power is a policy choice of the central government, which regulates the power sector closely, and reflects its concern for energy security and air pollution control. As a result, Korea may be paying a premium for power in order to enhance its security. A scenario modeled for this report simulating an expanded role for nuclear power indicates that total energy costs would rise 6 percent, although carbon and sulfur dioxide emissions would fall significantly.

Nuclear power becomes the cheapest alternative only if capital costs fall to about $1,200 per kilowatt, about one-third less than the cost today. If nuclear costs were reduced as a result of research and development, carbon dioxide and sulfur dioxide emissions would decline sharply by 23 million tons (46 percent) and 157,000 tons (61 percent), respectively. Total installed nuclear capacity in this case would reach 29 gigawatts in 2015. However, when more likely capital costs are assumed, nuclear power is not competitive.

Carbon Control. This scenario estimated the effects of carbon control on the power sector. Two means of control were assessed, including a policy decision to lower carbon emissions by 10 percent from the baseline by 2015. As an alternative, the size of a carbon tax to achieve a similar effect was estimated.

The least-cost way to reduce carbon emissions by 10 percent from the 2015 baseline would be to switch from coal and petroleum to gas-fired plants. The model also indicates that a $20 tax per ton of carbon emissions on all new generation would be enough to change the least-cost power mixture significantly. Different results would have been achieved if the tax were applied to existing plants as well. The tax is low because initial costs of switching between coal and oil- and gas-fired plants were not large. If Korea were interested in selling carbon permits internationally, fuel switching from coal to natural gas would be a simple and cost-effective way (although energy security would be another issue to consider).

The carbon tax — which makes the levelized cost of coal and petrol plants higher than gas plants — would result in much lower coal use and much greater use of LNG. In the model, this is fundamentally different from the carbon cap — which is a physical constraint on the amount of carbon that can be released. The cap is less costly since no external tax is applied, but results in less mitigation of carbon and sulfur dioxide. Neither case assumes that Korea will use carbon trading to lower emissions.

Advanced Technology Options. The Korean team also provided a sensitivity analysis demonstrating how far costs will need to decline before other advanced and renewable energy technologies are able to compete with fossil fuel plants. Wind turbines currently cost about $1,100 per kilowatt installed. Capital costs would need to decline by another 30 percent to $775 per kilowatt before the levelized cost of wind could match that of combined-cycle turbines. Korea does not appear to have significant wind resources located in prime locations, but extensive resource assessments have not yet been completed. Wind costs will likely continue to decline due to advances in technology. This resource may be able to contribute to the country's energy needs and security by 2015 if quality wind sites are found.

The capital cost of integrated gasification combined-cycle systems (coal gasification coupled with gas-turbine power generation) would likewise need to be reduced from their current level of an estimated $1,700 per kilowatt to $1,000 per kilowatt. This result is based on an assumption that sufficient coal is available at $30 per ton. For comparison, capital costs in the other model scenarios are assumed to decline to $1,400 per kilowatt, and power plant efficiency rises from the current value of 43 percent to 46 percent by 2015 without any significant action.

Fuel cells would need to be substantially less expensive to compete in the power sector. If the capital costs of fuel cells can be reduced to $800 per kilowatt from approximately $3,000 per kilowatt now, and if their efficiency can be pushed to 75 percent, fuel cells would be competitive. However, fuel cells are highly reliable compared with internal combustion engines — a factor that should be considered in any direct comparison.

If Korean power generation becomes competitive and if reductions in sulfur, nitrogen, and particulate emissions are imposed, LNG will probably be the fuel of choice. The capital cost for LNG-fired power plants is low and efficiency very high — a combination that would beat all the competition. Petroleum-fired, combined-cycle power plants may also be used, partly to diversify energy sources. These are also relatively low in greenhouse gas emissions. Modeling results show that a restructured power sector would reduce carbon emissions by approximately 10 percent by 2015 and slightly lower costs per unit of electricity generated. Hydropower and wind energy are limited and expensive in Korea, and thus noncompetitive in the near term. However, these resources may be effective tools for providing environmental protection and energy security in the future.

In summary, Korea could boost economic performance, improve environmental quality, and ensure greater energy security by accelerating energy efficiency efforts. To accomplish this, the country will need to reduce subsidies to heavy industry and support even greater development of demand-side management, co-generation, district heating, and energy service companies. Korea could improve least-cost power planning by considering the full economic and environmental impacts of electricity generation options. Reducing the taxes and duties on LNG imports could make combined-cycle power plants more competitive. Importing pipeline natural gas from Russia would further lower the cost of power from combined-cycle plants. Korea's economy and environment could benefit from advanced technologies such as fuel cells and wind power if research and development is accelerated and capital costs decline as a result.

Conclusions

In virtually every one of the more than 25 scenarios run by the country study teams, emissions increase dramatically. (See Table 5.) Total power sector emissions in the year 1999 are roughly 350 million tons for the five countries. Year 2015 emissions levels are projected in the baselines for the five countries to reach more than twice that, or 770 million tons. The total of even the lowest emissions scenario for each of the five countries would put power-sector emissions in 2015 for the five countries at 575 million tons, an increase of two-thirds. Emissions per capita and electric power consumption in these developing countries will nevertheless remain low compared to the current annual U.S. levels of about 5.8 tons and 12,500 kWh, respectively.

An examination of what works in the scenarios to reduce emissions without significantly driving up costs leads to five conclusions. First, markets, competition, and globalization generally reduce greenhouse gas emissions. This result flows from the fact that natural gas is the most attractive fuel in many countries, and that natural gas contains only about half as much carbon as coal, which is often subsidized. Hydropower and wind energy are expensive in most countries and noncompetitive in the near term. Nuclear power is usually more expensive than coal, oil, or gas-fired power. Capital costs range from three to four times that of natural gas and oil-fired systems, offsetting the significantly lower fuel costs that nuclear power enjoys.

Second, addressing local environmental air pollution problems increases the cost of power little, but dramatically reduces the power sector's emissions of these pollutants.

Economic Analysis of Global Climate Change Policy: A Primer

Robert N. Stavins

What is the point of conducting a detailed analysis of the economics of global climate change policy, some might ask. Surely — the thinking might go — the performance of the economy is largely independent of the quality of the environment, and policy choices regarding environmental quality should be made without attention to economic considerations; moreover, economics cannot shed much light on ways to solve environmental problems.

Quite to the contrary, there are numerous bi-directional linkages between economic performance and environmental quality. Economic considerations can help inform policy decisions regarding environmental protection, and economics provides powerful analytical methods for investigating environmental problems, and hence can provide valuable insights about those problems' potential solutions. The reasons for all of this are essentially two-fold: first, the causes of environmental degradation, at least in market economies, are fundamentally economic in nature; and second, the consequences of environmental problems have important economic dimensions.

Global climate change, perhaps even more than other environmental problems, can be addressed successfully only with a solid understanding of its economic dimensions. First, the fundamental cause of anthropogenic emissions of greenhouse gases, almost by definition, is economic: excessive emissions are an example of an externality, a well-understood category of market failure, where markets left to their own devices tend not to produce social efficiency.[1] Second, economic analysis is clearly necessary to estimate the costs that will be incurred when and if nations take action to reduce the risk of global climate change. Third, because of the large costs that will be involved in any serious climate change strategy, there is considerable interest in economic-incentive or market-based policy instruments that can reduce the costs of addressing the problem. And fourth, turning to the other side of the ledger, the biophysical consequences of global climate change can be evaluated with economic methods in order to identify the benefits, or avoided damages, of global climate policy action.

This chapter consists of six reports, five of which examine specific aspects of the economics of global climate policy. This introduction develops the analytical framework for carrying out economic analyses of policies intended to address the threat of global

---

The Pew Center and the author appreciate the input of Larry Goulder and Michael Scott, who reviewed this paper and provided helpful comments.

climate change.[2] Along the way, analytical issues particularly germane to climate policy analysis are highlighted,[3] and in a concluding section, the relationships between these issues and the subsequent reports in this chapter are described.

Three broad economic questions are raised by the challenge of addressing the threat of global climate change:[4] what will be the benefits of reducing the risk of global climate change; what will be the respective costs; and how can this information about the benefits and costs of alternative policy regimes be assimilated in ways that are useful to decision-makers?

The Benefits of Global Climate Change Policy

Economics is fundamentally anthropocentric; if an environmental change matters to any person — now or in the future — then it should, in principle, show up in an economic assessment. And environmental changes do matter to people in a wide variety of ways. The economic concept of environmental benefits is considerably broader than most non-economists seem to think.[5] From an economic perspective, the environment can be viewed as a form of natural asset that provides service flows used by people in the production of goods and services, such as agricultural output, human health, recreation, and more amorphous goods such as quality of life. This is analogous to the manner in which real physical capital assets (for example, factories and equipment) provide service flows used in manufacturing. As with real physical capital, a deterioration in the natural environment (as a productive asset) reduces the flow of services the environment is capable of providing.

Note that ecological benefits are very much part of the picture. Here, it is important to distinguish between ecosystem functions (for example, photosynthesis) and the environmental services produced by ecosystems that are valued by humans, since it is only the latter that are potential benefits in an economic framework (Freeman 1997). The range of these services is great, including obvious environmental products such as food and fiber, and services such as flood protection, but also including the quality of recreational experiences, the aesthetics of the landscape, and such desires (for whatever reasons) as the protection of marine mammals. The economic benefits of global climate change policies range from direct and specific impacts, such as those on agricultural yields and prices, to less direct and more general effects on biodiversity.

Protecting the environment usually involves active employment of capital, labor, and other scarce resources. Using these resources to protect the environment means that they are not available to be used for other purposes. The economic concept of the "value" of environmental goods and services is couched in terms of society's willingness to make trade-offs between competing uses of limited resources, and in terms of the sum of individuals' willingness to make these trade-offs.[6] Economists' tools of valuation were originally developed in a more limited context — one in which policy changes mostly caused changes in individuals' incomes and/or prices faced in the market. Over the last 30 years, however, these ideas have been extended to accommodate changes in the quality of goods, to public goods that are shared by individuals, and to other non-market services such as environmental quality and human health.[7]

The economist's task of estimating the benefits or loss of benefits resulting from a policy intervention is easiest when the benefits and costs are revealed explicitly through prices in established markets. When it comes to measuring environmental impacts, however, placing a value on benefits is more difficult, and requires indirect methods. With markets, consumers' decisions about how much of a good to purchase at different prices reveal useful information regarding the benefits consumers anticipate receiving from various items. With non-market environmental goods, it is necessary to infer this willingness to trade off other goods or monetary amounts for additional quantities of environmental services by using other techniques. Environmental economists have developed a repertoire of techniques that fall broadly into two categories: indirect measurement and direct questioning. Both sets of valuation methods are relevant for assessing the anticipated benefits of global climate change policies.

Economists prefer to measure trade-offs by observing the actual decisions of consumers in real markets, using so-called revealed preference methods. Sometimes the researcher can observe relationships that exist between the non-marketed (environmental) good and a good that has a market price. Thus, individuals' decisions to avert or mitigate the consequences of environmental deterioration can shed light on how people value other types of changes in environmental quality (averting behavior estimates). In other cases, individuals reveal their preferences for environmental goods in the housing market (hedonic property value methods), or for related health risks in labor markets (hedonic wage methods). In still other cases, individuals reveal their demand for recreational amenities through their decisions to travel to specific locations (Hotelling-Clawson-Knetsch and related methods). These various estimation techniques are well established for measuring the conceptual trade-offs that are the basis of environmental valuation. However, they are applicable only in limited cases.

In many other situations, it is simply not possible to observe behavior that reveals people's valuations of changes in environmental goods and services. This is particularly true when the value is a passive or non-use value. For example, an individual may value a change in an environmental good because she wants to preserve the option of consuming it in the future (option value) or because she desires to preserve the good for her heirs (bequest value). Other people may envision no current or future use by themselves or their heirs, but still wish to protect the good because they believe it should be protected or because they derive satisfaction from simply knowing it exists (existence value). With no standard market trade-offs to observe, economists must resort to surveys in which they construct hypothetical markets, employing stated preference, as opposed to revealed preference methods. In the best known stated preference method, commonly known as contingent valuation, survey respondents are presented with scenarios that require them to trade off, hypothetically, something for a change in the environmental good or service in question.

Although great uncertainty exists regarding the magnitude (and, in some cases, even the direction) of regional climate impacts, global climate change is anticipated to have a variety of impacts that will affect human welfare, including: changes in resource productivity (for example, in some cases, lower agricultural yields, and scarcer water resources); damages to human-built environments (including coastal flooding due to sea-level rise); human-health impacts (such as increased incidence of tropical diseases in more temperate climates); and damages to various ecosystems.[8] The uncertainties surrounding these various physical impacts are very great, and those uncertainties are compounded by imprecise estimates of respective economic consequences.

Whereas impacts on marketed goods and services (such as agricultural output) can be estimated with some reasonable degree of precision, monetary estimates for nonmarketed goods are notoriously imprecise. Furthermore, existing economic estimates in both categories come from industrialized nations, particularly the United States. Much less is known about anticipated economic damages in developing countries, which is especially troubling because they are particularly vulnerable to the impacts of global climate change.[9]

The Costs of Global Climate Change Policy

The task of estimating the costs of global climate change policies may seem straightforward, compared with the conceptual problems and empirical difficulties associated with estimating the benefits of such policies. In a relative sense, this is true. But as one moves toward developing more precise and reliable cost estimates, significant conceptual and empirical issues arise. More attention has been given by economists to analyzing the costs than the benefits of global climate policy action (largely because of existing uncertainties regarding regional biophysical impacts of climate change). Hence, my treatment of the cost side of the ledger is proportionately more extensive.

The economist's notion of cost, or more precisely, opportunity cost, is linked with — but distinct from — everyday usage of the word. Opportunity cost is an indication of what must be sacrificed in order to obtain something. In the environmental context, it is a measure of the value of whatever must be sacrificed to prevent or reduce the chances of an environmental impact. These costs typically do not coincide with monetary outlays — the accountant's measure of costs. This may be because out-of-pocket costs fail to capture all of the explicit and implicit costs that are incurred, or it may be because the prices of the resources required to produce environmental quality may themselves provide inaccurate indications of the opportunity costs of those resources.

Hence, the costs of global climate policies are the forgone social benefits due to employing scarce resources for global climate policy purposes, instead of putting these resources to their next best use.[10]

A taxonomy of environmental costs can be developed, beginning with the most obvious and moving toward the least direct (Jaffe, Peterson, Portney, and Stavins 1995). First, many policy-makers and much of the general public would identify the on-budget costs to government of administering (monitoring and enforcing) environmental laws and regulations as the cost of environmental regulation. This meets the economist's notion of (opportunity) cost, since administering environmental rules involves the employment of resources (labor and capital) that could otherwise be used elsewhere. But economic analysts would also include as costs the capital and operating expenditures associated with regulatory compliance. Indeed, these typically represent a substantial portion of the overall costs of regulation, although a considerable share of compliance costs for some regulations fall on governments rather than private firms.[11] Additional direct costs include legal and other transaction costs, the effects of refocused management attention, and the possibility of disrupted production.

Next, there are what have sometimes been called "negative costs" of environmental regulation, including the beneficial productivity impacts of a cleaner environment and the potential innovation-stimulating effects of regulation.[12] General equilibrium or multi-market effects associated with discouraged investment[13] and retarded innovation constitute another important layer of costs, as do the transition costs of real-world economies responding over time to regulatory changes.

In the give-and-take of policy debates, abatement costs of proposed regulations have sometimes been over-estimated (Harrington, Morgenstern, and Nelson 2000; Hammitt 2000). This may partly be due to the adversarial nature of the policy process, but it is also a natural consequence of employing short-term cost analyses that do not take into account potential future cost savings due to technological change, some of which may be a consequence of the regulatory regime.

Although the task of estimating the costs of environmental protection efforts might be somewhat more straightforward than that of estimating environmental protection benefits, costs seldom can be estimated with great precision, and producing high-quality cost estimates requires careful analysis. Conceptually, there are four steps required to appraise empirically the cost of an environmental-protection measure. First, it is necessary to identify the specific policy instrument that is associated with the measure. For example, is a conventional instrument, such as a technology standard, or a market-based instrument, such as an emissions charge, to be employed? This can be important because the same target, such as a given reduction in carbon dioxide (CO_2) emissions, may be achieved at very different total costs with different policy instruments. The second conceptual step is identifying the specific actions that sources will take to comply with the statute or regulation, as implemented with the given policy instrument. Some of these actions may involve the adoption of a new piece of equipment, but others may involve a change in process. Third, it is necessary to identify the true cost of each action, which requires more than assessing required monetary outlays. Fourth, it is often necessary to aggregate these costs across society and over the relevant time frame.

In the case of climate change, the opportunity cost of taking action may include: direct outlays for control (for example, the incremental cost of employing natural gas rather than more carbon-intensive coal for energy generation); partial equilibrium costs to both producers and consumers (for example, accelerated depreciation of fixed capital); and general equilibrium costs that arise in related markets as prices adjust (Hourcade et al. 1996). In this last regard, it is important to keep in mind that the ultimate consequences of a given environmental policy initiative depend on interactions between the new policy and existing regulations or tax policies. In particular, additional costs can arise from interactions between climate policies and pre-existing distortions in the economy, such as those due to taxes on labor (Goulder 1995).

The baselines, or anticipated business-as-usual paths, utilized for climate policy cost (and benefit) analyses are very important. Indeed, a striking finding from a wide range of integrated assessment models (which layer economic models upon underlying scientific models of climate change relationships) is that differences in welfare impacts[14] across plausible baseline assumptions are greater than the welfare impacts attributable to climate policy itself (Goulder 2000). These baselines are built upon various assumed time paths of future economic growth, encompassing overall rates of growth plus relevant sectoral changes, and a particularly important aspect of alternative baselines is the assumed rate and direction of technological change.

The cost of achieving any given global climate target depends critically upon the "physical scope" of policy action. Does the policy being analyzed affect only emissions, for example, of CO_2 by encouraging fuel switching? Or does the policy also provide mechanisms for: increased biological uptake of carbon through carbon sequestration, presumably through changes in land use (Sedjo, Sampson, and Wisniewski 1997; Stavins 1999); carbon management, that is, removal and storage of CO_2 in the deep ocean or depleted oil and gas reservoirs (Parson and Keith 1998); and/or geo-engineering, such as various means of increasing the earth's reflectivity (National Academy of Sciences 1992)? More broadly still, do the cost estimates allow for adaptation policies, which in many cases may be less costly than "equivalent" measures that work through emissions reduction, sequestration, management, or geoengineering (Pielke 1998; Kane and Shogren 2000)? Finally, does the policy being assessed focus exclusively on CO_2 or is a larger set of greenhouse gases being targeted? This is a crucial question, since broader targets enhance flexibility, and, in some cases, can reduce costs of achieving a given climate goal substantially (Hansen et al. 2000).

Just as the allowed physical scope of policy response will affect the costs of achieving any given climate target, the policy instrument chosen to affect change will have profound impacts on costs, both in the short term and the long term. On the domestic front, the portfolio of potential policy instruments includes conventional technology and uniform performance standards (so-called command-and-control approaches), as well as the newer breed of economic-incentive or market-based policy instruments, such as taxes, tradeable permit systems, and various information policies (Stavins 1997).[15] And at the international level, the set of instruments that have been subjected to analysis include international taxes, harmonized domestic taxes, international tradeable permits, joint implementation, and the Clean Development Mechanism of the Kyoto Protocol (Fisher et al. 1996). These two sets of climate policy instruments — domestic and international — should not be thought of as functioning independently of one another. Indeed, the relative cost-effectiveness of what may be one of the most promising mechanisms, the international tradeable permit system recognized by Article 17 of the Kyoto Protocol, will depend greatly upon the particular set of domestic policy instruments adopted by participating nations (Hahn and Stavins 1999).

The outcome of any cost comparison among greenhouse policy instruments also depends upon the sophistication of the underlying analytical models. With many environmental problems, relatively simple analytical models can be employed for comparing policy instruments, since it is reasonable to utilize static (short-term) cost-effectiveness as a criterion for comparison. But the long-term nature of global climate change and related policies means that it is important to employ a dynamic (long-term) cost-effectiveness criterion for comparisons. In this context, the intertemporal flexibility provided by some policy instruments, such as banking and borrowing in a tradeable permit system, can turn out to be very significant (Manne and Richels 1997).

More importantly, the very long time horizons typically employed in global climate policy analysis mean that it is essential to allow for the effects of alternative policy instruments on the rate and direction of relevant (cost-reducing) technological change (Jaffe, Newell, and Stavins 1999, 2000). Three stages of technological change (Schumpeter 1939) can be analyzed: invention, the development of a new product or process (Popp 1999); innovation or commercialization, the bringing to market of a new product or process (Newell, Jaffe, and Stavins 1999); and diffusion, the gradual adoption of new products and processes by firms and individuals (Hassett and Metcalf 1995; Jaffe and Stavins 1995). Most large-scale analyses of global climate policy have not allowed for technological improvements in response to economic stimuli, but this is beginning to change (Goulder and Schneider 1999; Nordhaus 1999; Goulder and Mathai 2000).[16]

Since the compliance costs associated with most climate policies are initially incurred by private firms, it is important to analyze correctly the behavioral response of such firms to various policy regimes. Most economic analyses treat firms as atomistic profit-maximizing or cost-minimizing units. This is satisfactory for many purposes, but it can lead to distorted estimates of the costs brought about by some policies. For example, one potentially important cause of the mixed performance of implemented market-based instruments is that many firms are simply not well equipped internally to make the decisions necessary to fully utilize these instruments. Since market-based instruments have been used on a limited basis only, and firms are not certain that these instruments will be a lasting component on the regulatory landscape, most companies have not reorganized their internal structure to fully exploit the cost savings these instruments offer. Rather, most firms continue to have organizations that are experienced in minimizing the costs of complying with command-and-control regulations, not in making the strategic decisions allowed by market-based instruments (Hockenstein, Stavins, and Whitehead 1997).

The focus of environmental, health, and safety departments in private firms has been primarily on problem avoidance and risk management, rather than on the creation of opportunities made possible by market-based instruments. This focus has developed because of the strict rules companies have faced under command-and-control regulation, in response to which companies have built skills and developed processes that comply with regulations, but do not help them benefit competitively from environmental decisions (Reinhardt 2000). Absent significant changes in structure and personnel, the full potential of market-based instruments will not be realized. Economic models may thereby underestimate the relative costs of employing such instruments to achieve global climate targets.

Finally, the costs of achieving any given global climate target, indeed the very feasibility of achieving such targets, will depend upon the nature of respective international agreements and the institutions that exist to support those agreements. This is an area where economic analysis (along with political science and legal scholarship) can also contribute. A principal issue is the architecture of such agreements and the breadth of the coalitions that are parties to them (Jacoby, Prinn, and Schmalensee 1998; Schelling 1998). From an economic perspective, a fundamental challenge is the necessity of overcoming the strong incentives for free riding that exist with a global commons problem (Carraro and Siniscalco 1993; Barrett 1994; Cooper 1998). More specifically, there is a pressing need to design international policy instruments that can provide incentives over time for more nations — in particular, developing countries — to join the coalition and take on binding targets or other responsibilities (Manne and Richels 1995; Rose, Stevens, Edmonds, and Wise 1998; Frankel 1999; Bohm and Carlén 2000).

Assimilating Benefit and Cost Information

The next analytical challenge, after the benefits and costs of proposed global climate change policies have been assessed, is to assimilate this information in ways that are useful for decision-makers. Two major categories of analysis are required: one is to provide an overall characterization of a policy in terms of its likely benefits and costs (aggregate analysis); and another is to describe the distribution of those benefits and costs across relevant populations, defined, for example, by geographic location, economic sector, income level, or time period (distributional analysis).

Aggregate Analysis

It seems reasonable to ask whether the gains (to the gainers) outweigh the losses (to the losers) of some public policy, and thus determine, on net, whether society as a whole is made better or worse off as a result of that policy. Benefit-cost analysis is the standard technique used to carry out this comparison of the favorable effects of risk reductions (the benefits) with the adverse consequences (the costs). A policy that achieves maximum aggregate net benefits is said to be an efficient one. Although efficiency is surely an important criterion for sound policy analysis, most economists think of benefit-cost analysis as no more than a tool to assist in decision-making. Virtually all would agree, however, that the information in a well-done benefit-cost analysis can be of great value in helping to make decisions about risk reduction policies (Arrow et al. 1996).

Time is a critical and prominent dimension of global climate change policy. First, greenhouse gases accumulate in the atmosphere over very long periods (up to hundreds of years), because of their very slow natural decay rates. Second, changes in the capital stock that are made in response to the threat of climate change have long lives: for example, 50 to 70 years for electricity generators, and 60 to 100 years for residential buildings (Jaffe, Newell, and Stavins 1999). Third, technological change is a long-term phenomenon that has great bearing on global climate change and policies to address it. For all three reasons, benefit-cost analyses of global climate policies must involve the dimension of time, and over very long intervals, at that.

When adding the value of net benefits over time, it is essential to recognize that people are not indifferent to receiving a given economic benefit (or paying a given economic cost) today as opposed to 10 or 20 years from now. For this reason, all future net benefits are typically discounted (expressed in terms equivalent to the time-value of today's net benefits); that is, the present value of net benefits in each year is computed before aggregating net benefits over time.[17]

While the concept of discounting has a sound rationale, it can lead to conclusions that many people, including economists, find unpalatable. For long-run policy problems, such as global climate change, little weight will be given in an analysis to the long-term benefits of taking action, compared with the up-front costs of those actions. This conundrum has stimulated an active area of research (Portney and Weyant 1999), as economists have considered how best to address the apparent dilemma. One avenue of this research has suggested a theoretical basis for employing lower discount rates for longer run analyses (Weitzman 1999). Considerations of time can thus have profound effects on aggregate analysis of the benefits and costs of alternative climate policies.[18]

In addition to time, uncertainty is a prominent feature of global climate change, on both the benefit and the cost side of the ledger. In effect, the risks of premature or unnecessary actions need to be compared with the risks of failing to take actions that subsequently prove to be warranted (Goulder 2000). Because of this, many economic analyses have indicated that climate change may best be addressed through sequential decision-making, with policies being modified over time as new information becomes available and uncertainties are reduced. Because such new information is potentially of great value, flexible policies that adapt to new information have very significant advantages over more rigid policy mechanisms.

The significant uncertainties associated with global climate change interact with the intertemporal nature of the problem to yield another important dimension — irreversibility (Kolstad and Toman 2000). It is well known that when uncertainty is combined with long-lived impacts (economic, if not physical, irreversibility), there is a value (called quasi-option value by economists) in delaying those impacts until more information is available (Hanemann 1989). This value should, in principle, be included in the calculation of benefits and costs. In the global climate context, the irreversibilities include both the accumulation of greenhouse gases in the atmosphere and the accumulation of capital investments that cannot easily be reversed. These two effects push a stochastic benefit-cost analysis of global climate policy in opposite directions. Which is dominant? Although it has been argued that the second effect is more important (Kolstad 1996), it is ultimately an empirical question (Ulph and Ulph 1997; Narain and Fisher 2000).

Distributional Analysis

This discussion of benefits and costs, as well as the way the two are compared, has glossed over an important point, and one that is exceptionally important in the context of global climate change policy. Specifically, benefit-cost analysis is silent about the distributional implications of policy measures. Although considerable thought has been given by economists over the years to the possibility of using weights to incorporate distributional considerations into determinations of efficiency, there is no consensus, nor likely to be one, on what those weights ought to be. It seems reasonable, instead, to estimate benefits and costs, and also provide as much information as possible to decision-makers about gainers and losers.

Assessments of national, intra-national, and intergenerational distributions of the benefits and costs of alternative policy regimes are necessary for the identification of equitable climate strategies. A number of criteria merit consideration (Goulder 2000). First, the criterion of responsibility would suggest that — other things equal — nations that are most responsible for the accumulation of greenhouse gases in the atmosphere should take on the greatest burden for containing the problem. Second, the criterion of ability to pay is premised on the notion that wealthier nations possess greater capacity to respond to the problem. Third, the criterion of the distribution of benefits suggests that nations, which stand to benefit most from action taken ought to take on greater shares of the cost burden. The first two considerations suggest that industrialized nations should bear the principal burden for dealing with the prospect of climate change, while the third consideration favors action by developing countries.

Climate Change: Science, Strategies, & Solutions

Because of the long time horizon of global climate policy analysis, important issues of intergenerational equity arise. But it should be noted that the use of discounting in benefit-cost analysis has ambiguous effects. For example, some have called for not discounting future costs and benefits at all when time-horizons are very great. At first, this might seem to be a course of action that would favor future generations. In an important sense, however, it does not. If, by using a zero or very low discount rate, we adopt policies that do not pay off until the distant future, we are favoring climate policy action over other policies for which a standard (higher) discount rate is used. As a result, we may pass up opportunities to employ other, non-climate policies that could benefit future generations. Thus, it is not clear that we make future generations better off by using a low rate of discount. More broadly, Schelling (1998) has highlighted the trade-off that may exist between policies to address intergenerational equity and those that address (current) distributional equity: by taking actions to protect future generations (who presumably will be better off than current ones), we reduce the resources available to help today's poor in developing countries.

The Path Ahead

Global climate change — perhaps even more than other environmental problems — can be addressed successfully only with a solid understanding of its economic dimensions. A substantial body of economic literature can be brought to bear on the three broad questions that are raised by the threat of global climate change: what will be the benefits of global climate policies; what will be their costs; and how can this information about alternative policies be assimilated in ways that are ultimately most useful for decision makers? Although the existing literature from decades of economic analysis is helpful in addressing these questions, the truth is that global climate change policy — because of the magnitude of its anticipated benefits and costs, its great time horizons, massive uncertainties, and physical and economic irreversibilities — presents unprecedented challenges to economic research, as it does to the other social and natural sciences.

The five reports that follow this one stake out a number of the frontiers of that research, addressing in turn five particularly timely and important aspects of the overall puzzle. First, in an overview of economic models, John Weyant provides a much-needed user's guide to the large-scale integrated assessment models that continue to be central to much of the research and many of the policy debates on global climate change. Weyant reviews the structure, data, and findings of 14 of the most prominent large-scale economic models, and explains how the models differ, how they do not, and how their results relate to one another.

One of the key determinants identified by Weyant — the role played in respective models by technological change — becomes the sole focus of the second report, "Technological Change and its Effects on Mitigation Costs," by Jae Edmonds, Joseph Roop, and Michael Scott. These authors emphasize that understanding the way technologies evolve and penetrate the market is essential to understanding methods of addressing global climate change. Their focus, then, is on the ways in which technological change is captured by climate change policy modelers, with particular attention to two idealized approaches: top-down and bottom-up. Their conclusion is consistent with Weyant's, namely that in order to understand the implications of large-scale economic models of the climate change problem, it is essential to understand first the assumptions that have been made regarding the path of technological progress.

Much of the analysis of technological change in the global climate policy context has focused directly on products and processes related to the generation and use of energy, because of the prominence of fossil-fuel combustion and consequent CO_2 emissions as a major contributor to the accumulation of greenhouse gases in the atmosphere. The area of technological change that has been most dramatic over the past decade, however, has surely been information technology. What effects, if any, will the emergence of the related "new economy" have on the costs of achieving global climate targets?

This question is the topic of the third report, "The New Economy: Implications for Climate Change Policy," by Everett Ehrlich and Anthony Brunello. Drawing upon both economic theory and empirical evidence, they first argue that with the marginal cost of processing information falling, there will be substitution of information and information-related activities for energy in the production of goods and services. Second, they find that this will lead to more specialization and greater outsourcing of energy service management. Third, empirical analysis suggests that consequent changes in the capital stock will mean that policies intended to reduce CO_2 emissions will have more benign effects on the economy than otherwise. Their overall conclusion is that the information revolution will cause the economic costs of climate change policies to be lower than previously thought.

If there is one lesson that has been learned from the 30 years of environmental policy experience that began with the first Earth Day, it is that flexible environmental policies cost less than more rigid ones. In particular, market-based instruments, such as taxes and tradeable permit systems, can enable governments to achieve their environmental targets at lower aggregate costs than via conventional, command-and-control approaches. Given the magnitude of the global climate change problem, the potential for cost savings with such instruments is enormous, and this is documented in the fourth report, "International Emissions Trading," by Jae Edmonds, Michael Scott, Joseph Roop, and Christopher MacCracken.

With some simple, but powerful numerical examples, and with a review of the results from eight models of carbon trading, the authors are able to document the degree to which international greenhouse gas emissions trading would lower overall mitigation costs. They highlight the fact that the cost savings will increase if greater flexibility is provided in trading mechanisms, such as by allowing trade among the major greenhouse gases, across types of sources, and over time. But the authors also note that the full economic potential of these trading regimes will be reached only if crucial issues of program design and institutional structure are successfully addressed.

Finally, as I emphasized above, because climate policy compliance costs are initially incurred by private firms, it is essential to analyze the behavioral response of such firms to various policy regimes. Nearly all economic analyses treat firms as atomistic profit-maximizing or cost-minimizing units, but firms are vastly more complex. Although such simplifying assumptions are satisfactory for many purposes, it is useful to go inside the black box of the firm to understand private industry's approach to the climate problem and potential solutions. This is precisely the purpose of the final report in this chapter, "A Business Manager's Approach to Climate Change," by Forest Reinhardt and Kimberly Packard.

The authors address a key question: how can managers reconcile the goals of improving both shareholder value and environmental performance? This is a question that has generated considerable debate in the past, but that debate has all too often been dominated by extreme and misleading views: on the one hand, by wishful thinkers who see "win-win opportunities" even where there are severe tradeoffs between environmental and private financial goals; and, on the other hand, by ideologues who portray all environmental regulations as crippling for business. Packard and Reinhardt make sense of this confused and confusing debate, and thus describe a more sensible path for business managers in the face of real concerns about global climate change and the new policies that such concerns may bring forth.

Overall, the five reports that follow provide instructive examples of how economics can offer powerful analytical methods for investigating the problem of global climate change, and how economic analysis can thereby provide valuable insights about potential solutions to this very challenging problem.

Endnotes

1. Looked at somewhat differently, but still well within the framework of conventional economics, the problem is that the atmosphere is treated as "common property," and hence a freely-available receptacle for waste products.

2. A generic, but more detailed treatment of the basic analytical framework can be found in U.S. Environmental Protection Agency (1998).

3. For a more extensive treatment of some of these issues, see Kolstad and Toman (2000).

4. Although the pace of economic research on global climate change has accelerated greatly in the past decade, the earliest work appeared more than 25 years ago. See, for example, Nordhaus (1977, 1982).

5. For a summary of myths that non-economists seem to have regarding economics, and a set of responses thereto, see Fullerton and Stavins (1998).

6. Reference is typically made to "willingness-to-pay" for environmental improvement or "willingness-to-accept" compensation for environmental degradation.

7. For a comprehensive treatment of the theory and methods of environmental benefit estimation, see Freeman (1993).

8. In assessing these economic damages, economists recognize that humans typically adapt to risk — to some degree — in order to lower their anticipated losses.

9. A review of the likely economic damages of global climate change is provided by Pearce et al. (1996).

10. Costs and benefits are thus two sides of the same coin. Environmental benefits are created by taking some environmental policy action, while other benefits are thereby foregone. Hence, the cost of an environmental-protection measure may be defined as the gross decrease in benefits (consumer and producer surpluses) associated with the measure and with any price and/or income changes that may result (Cropper and Oates 1992).

11. One example in the United States is the (federal) regulation of contaminants in drinking water, the cost of which is borne primarily by municipal governments.

12. The notion that environmental regulation can foster economic growth is a controversial one among economists. For a debate on this proposition, see Porter and van der Linde (1995); and Palmer, Oates, and Portney (1995). It is also important to recognize that good economic analysis can be used (and has been used) to identify circumstances where policies involve real "negative opportunity costs," such as policies that increase energy efficiency by reducing distortionary energy subsidies. In these cases, economic analysis can identify true "win-win" policy options.

13. For example, if a firm chooses to close a plant because of a new regulation (rather than installing expensive control equipment), this would be counted as zero cost in narrow compliance-cost estimates, but it is obviously a real cost.

14. This is measured by net benefits: the difference between benefits and costs.

15. For a comprehensive review of worldwide experiences with market-based instruments for environmental protection, see Stavins (2000).

16. As mentioned earlier, climate policy instruments can impose additional costs through their interaction with pre-existing distortionary taxes. This raises another issue for cost comparisons since some policy instruments, such as taxes and auctioned permits, generate revenues, which can be used by governments to reduce pre-existing taxes, thereby reducing what the overall costs of the policy would otherwise be (Goulder 1995).

17. What rate should be used to carry out this discounting? There is extensive literature in economics that addresses this question. A comprehensive summary was provided by Lind (1982), and a more recent exploration was organized by Portney and Weyant (1999).

18. The time dimension is also crucial, of course, in one type of distributional analysis, namely intertemporal distribution, as is discussed below.

References

Arrow, K., M. Cropper, G. Eads, R. Hahn, L. Lave, R. Noll, P. Portney, M. Russell, R. Schmalensee, K. Smith, and R. Stavins. 1996. Is There a Role for Benefit-Cost Analysis in Environmental, Health, and Safety Regulation? *Science,* April 12.

Barrett, S. 1994. Self-Enforcing International Environmental Agreements. *Oxford Economic Papers* 46: 878–894.

Bohm, P., and B. Carlén. 2000. *Cost-Effective Approaches to Attracting Low-Income Countries to International Emissions Trading: Theory and Experiments.* Working Paper, Stockholm University, Sweden.

Carraro, C., and D. Siniscalco. 1993. Strategies for the International Protection of the Environment. *Journal of Public Economics* 52(3): 309–328.

Cooper, R. 1998. Toward a Real Global Warming Treaty. *Foreign Affairs* 77, March/April, pp. 66–79.

Cropper, M.L., and W.E. Oates. 1992. Environmental Economics: A Survey. *Journal of Economic Literature* 30: 675-740.

Fisher, B., S. Barrett, P. Bohm, M. Kuroda, J. Mubazi, A. Shah, and R. Stavins. 1996. Policy Instruments to Combat Climate Change. *Climate Change 1995: Economic and Social Dimensions of Climate Change.* J.P. Bruce, H. Lee, and E.F. Haites, eds. Intergovernmental Panel on Climate Change, Working Group III. Cambridge University Press, Cambridge, UK, pp. 397–439.

Frankel, J.A. 1999. *Greenhouse Gas Emissions.* Policy Brief No. 52, The Brookings Institution, Washington, D.C. (June).

Freeman, A.M., III. 1993. *The Measurement of Environmental and Resource Values: Theory and Methods.* Resources for the Future, Washington, D.C.

Freeman, A.M., III. 1997. On Valuing the Services and Functions of Ecosystems. *Ecosystem Function and Human Activities: Reconciling Economics and Ecology.* D.R. Simpson and N.L. Christensen, Jr., eds. Chapman and Hall, New York, NY.

Fullerton, D., and R. Stavins. 1998. How Economists See the Environment. *Nature* 395: 433–434 (October 1).

Goulder, L.H. 1995. Effects of Carbon Taxes in an Economy with Prior Tax Distortions: An Intertemporal General Equilibrium Analysis. *Journal of Environmental Economics and Management* 29: 271–297.

Goulder, L.H. 2000. *Central Themes and Research Results in Global Climate Change Policy.* Paper prepared for the Intergovernmental Panel on Climate Change, Working Group III, Third Assessment Report. Mimeo, Stanford, California (May).

Goulder, L.H., and K. Mathai. 2000. Optimal CO_2 Abatement in the Presence of Induced Technological Change. *Journal of Environmental Economics and Management* 39(1): 1–38.

Goulder, L.H., and S.H. Schneider. 1999. Induced Technological Change and the Attractiveness of CO_2 Abatement Policies. *Resource and Energy Economics* 21(3–4): 211–253.

Hahn, R.W., and R.N. Stavins. 1999. *What Has the Kyoto Protocol Wrought? The Real Architecture of International Tradeable Permit Markets.* The AEI Press, Washington, D.C.

Hammitt, J.K. 2000. Are the Costs of Proposed Environmental Regulations Overestimated? Evidence from the CFC Phaseout. *Environmental and Resource Economics* 16: 281–301.

Hanemann, W.M. 1989. Information and the Concept of Option Value. *Journal of Environmental Economics and Management* 16: 23–37.

Hansen, J., M. Sato, R. Ruedy, A. Lacis, and V. Oinas. 2000. Global Warming in the Twenty-First Century: An Alternative Scenario. *Proceedings of the National Academy of Sciences* (August 15).

Harrington, W., R. Morgenstern, and P. Nelson. 2000. On the Accuracy of Regulatory Cost Estimates. *Journal of Policy Analysis and Management* 19(2).

Hassett, K.A., and G.E. Metcalf. 1995. Energy Tax Credits and Residential Conservation Investment: Evidence from Panel Data. *Journal of Public Economics* 57: 201–217.

Hockenstein, J.B., R.N. Stavins, and B.W. Whitehead. 1997. Creating the Next Generation of Market-Based Environmental Tools. *Environment* 39(4): 12–20, 30–33.

Hourcade, J.C., K. Halsnæs, M. Jaccard, W.D. Montgomery, R. Richels, J. Robinson, P.R. Shukla, P. Sturm, W. Chandler, O. Davidson, J. Edmonds, D. Finon, K. Hogan, F. Krause, A. Kolesov, E. Rovere, P. Nastari, A. Pegov, K. Richards, L. Schrattenholzer, R. Shackleton, Y. Sokona, A. Tudini, and J.P. Weyant. 1996. A Review of Mitigation Cost Studies. *Climate Change 1995: Economic and Social Dimensions of Climate Change.* J.P. Bruce, H. Lee, and E.F. Haites, eds. Report of Working Group III to the Second Assessment Report of the Intergovernmental Panel on Climate Change. Cambridge University Press, New York, NY, pp. 263–306.

Jacoby, H.D., R. Prinn, and R. Schmalensee. 1998. Kyoto's Unfinished Business. *Foreign Affairs* 77, July/August, pp. 54–66.

Jaffe, A.B., R.G. Newell, and R.N. Stavins. 1999. Energy-Efficient Technologies and Climate Change Policies: Issues and Evidence. *Climate Issue Brief No. 19.* Resources for the Future, Washington, D.C. (December).

Jaffe, A.B., R.G. Newell, and R.N. Stavins. 2000. Technological Change and the Environment. *Handbook of Environmental Economics.* K.G. Mäler and J. Vincent, eds. Elsevier Science, Amsterdam, forthcoming.

Jaffe, A.B., S.R. Peterson, P.R. Portney, and R.N. Stavins. 1995. Environmental Regulation and the Competitiveness of U.S. Manufacturing: What Does the Evidence Tell Us? *Journal of Economic Literature* 33: 132–163.

Jaffe, A.B., and R.N. Stavins. 1995. Dynamic Incentives of Environmental Regulation: The Effects of Alternative Policy Instruments on Technology Diffusion. *Journal of Environmental Economics and Management* 29: S43–S63.

Kane, S., and J. Shogren. 2000. Adaptation and Mitigation in Climate Change Policy. *Climatic Change,* forthcoming.

Kolstad, C.D. 1996. Fundamental Irreversibilities in Stock Externalities. *Journal of Political Economy* 60: 221–233.

Kolstad, C.D., and M.A. Toman. 2000. The Economics of Climate Policy. In *The Handbook of Environmental Economics.* K.G. Mäler and J. Vincent, eds. North-Holland/Elsevier Science, Amsterdam.

Lind, R.C., ed. 1982. *Discounting for Time and Risk in Energy Policy.* Johns Hopkins University Press, Baltimore, MD.

Manne, A.S., and R. Richels. 1995. The Greenhouse Debate: Economic Efficiency, Burden Sharing, and Hedging Strategies. *The Energy Journal* 16(4): 1–37.

Manne, A.S., and R. Richels. 1997. *On Stabilizing CO_2 Concentrations — Cost-Effective Reduction Strategies.* Working Paper, Stanford University, Stanford, CA.

Narain, U., and A. Fisher. 2000. *Irreversibility, Uncertainty, and Catastrophic Global Warming.* Giannini Foundation Working Paper 843, Dept. of Agricultural and Resource Economics, University of California, Berkeley.

National Academy of Sciences. 1992. *Policy Implications of Greenhouse Warming.* Committee on Science, Engineering, and Public Policy, Panel on Policy Implications of Greenhouse Warming. National Academy Press, Washington, D.C.

Newell, R.G., A.B. Jaffe, and R.N. Stavins. 1999. The Induced Innovation Hypothesis and Energy-Saving Technological Change. *Quarterly Journal of Economics* 114: 941–975.

Nordhaus, W.D. 1977. Economic Growth and Climate: The Case of Carbon Dioxide. *American Economic Review* (May).

Nordhaus, W.D. 1982. How Fast Should We Graze the Global Commons? *American Economic Review* 72: 242–246.

Nordhaus, W.D. 1999. *Modeling Induced Innovation in Climate-Change Policy.* Paper prepared for IIASA-Yale Conference on Induced Innovation (June).

Palmer, K., W.E. Oates, and P.R. Portney. 1995. Tightening Environmental Standards: The Benefit-Cost or the No-Cost Paradigm? *Journal of Economic Perspectives* 9(4): 119–132.

Parson, E.A., and D.W. Keith. 1998. Fossil Fuels without CO_2 Emissions: Progress, Prospects, and Policy Implications. *Science* 282(5391): 1053–1054 (November 6).

Pearce, D.W., R.K. Cline, S. Pachauri, S. Fankhauser, R.S.J. Tol, and P. Vellinga. 1996. The Social Cost of Climate Change: Greenhouse Damage and the Benefits of Control. *Climate Change 1995: Economic and Social Dimensions of Climate Change.* J.P. Bruce, H. Lee, and E.F. Haites, eds. Report of Working Group III to the Second Assessment Report of the Intergovernmental Panel on Climate Change. Cambridge University Press, New York, NY.

Pielke, R.A., Jr. 1998. Rethinking the Role of Adaptation in Climate Policy. *Global Environmental Change* 8(2): 159–170.

Popp, D. 1999. *Induced Innovation and Energy Prices.* Dept. of Economics Working Paper No. 1999-4, University of Kansas, Lawrence, KS.

Porter, M.E., and C. van der Linde. 1995. Toward a New Conception of the Environment-Competitiveness Relationship. *Journal of Economic Perspectives* 9(4): 97–118.

Portney, P.R., and J.P. Weyant, eds. 1999. *Discounting and Intergenerational Equity.* Resources for the Future, Washington, D.C.

Reinhardt, F.L. 2000. *Down to Earth: Applying Business Principles to Environmental Management.* Harvard Business School Press, Boston, MA.

Rose, A., B. Stevens, J. Edmonds, and M. Wise. 1998. International Equity and Differentiation in Global Warming Policy: An Application to Tradeable Emission Permits. *Environmental and Resource Economics* 12: 25–51.

Schelling, T.C. 1998. The Cost of Combating Global Warming. *Foreign Affairs* 76(6): 8–14.

Schumpeter, J.A. 1939. *Business Cycles, Volumes I and II.* McGraw-Hill, New York, NY.

Sedjo, R.A., N. Sampson, and J. Wisniewski, eds. 1997. *Economics of Carbon Sequestration in Forestry.* CRC Press, New York, NY.

Stavins, R.N. 1997. Policy Instruments for Climate Change: How Can National Governments Address a Global Problem? *The University of Chicago Legal Forum,* pp. 293–329.

Stavins, R.N. 1999. The Costs of Carbon Sequestration: A Revealed-Preference Approach. *American Economic Review* 89(4): 994–1009 (September).

Stavins, R.N. 2000. Experience with Market-Based Environmental Policy Instruments. *Handbook of Environmental Economics.* K.G. Mäler and J. Vincent, eds. Elsevier Science, Amsterdam, forthcoming.

Ulph, A., and D. Ulph. 1997. Global Warming, Irreversibility, and Learning. *Economic Journal* 107: 636–650.

U.S. Environmental Protection Agency. 1998. Benefit/Cost Analysis for Integrated Risk Decisions. Prepared by the Environmental Economics Advisory Committee, Science Advisory Board, as Chapter 4 of *Integrated Environmental Decision-Making in the Twenty-First Century.* Washington, D.C.

Weitzman, M.L. 1999. Just Keep Discounting, But.... *Discounting and Intergenerational Equity.* P.R. Portney and J.P. Weyant, eds. Resources for the Future, Washington, D.C.

Economic Models:
How They Work & Why Their Results Differ

John P. Weyant

What are the potential costs of cutting greenhouse gas (GHG) emissions? Can such reductions be achieved without sacrificing economic growth? Interest groups active in the climate change debate believe the stakes are high. Some fear the environmental and socioeconomic costs of climate change itself. Others are more fearful of the economic consequences of trying to avoid climate change. This debate is, to a large extent, played out through economic analysis of climate change policy. Hundreds of these analyses have been published over the past decade, and this pace is likely to continue. Large computer models are used to perform economic analyses.

These analyses are rich and extensive, but widely divergent in their results. Many predict the domestic costs of complying with the Kyoto Protocol. Some groups have concluded that the United States can reduce its carbon emissions to significantly below its Kyoto target, with net economic savings. Others have predicted rather significant costs to the U.S. economy.

Through these economic analyses, people translate their expectations into concrete assumptions about the future. The set of assumptions that describe what happens in the future if nothing is done to control GHG emissions is known as the "base case" (or as the "baseline" or "business-as-usual" case). The base case may embody optimism or pessimism about the economy, about GHG emissions, about the changes in climate that will occur as a result of these emissions, and about what will happen to the environment as a result of this climate change. The higher the base case emissions, the more emissions must be reduced to achieve a particular target, and therefore the higher the control costs. The greater the base case climate impacts, the greater the benefits of controlling emissions.

Two other assumptions drive projections of what will happen if society does control GHGs. Will new, low-cost, low-emitting technologies become available? To what extent will consumers and producers be able to meet their needs through substituting lower-emitting products and services for higher-emitting ones rather than sacrificing satisfaction of some of those needs? Economic analyses may embody optimism or pessimism on either of these fronts.

This paper is a condensed version of a report published in July 2000 by the Pew Center: An Introduction to the Economics of Climate Change Policy. The Pew Center and the author wish to acknowledge input from Dick Goettle, Larry Goulder, Henry Jacoby, Bill Nordhaus, Rich Richels, and Bob Shackelton. We also wish to acknowledge the use of material from a background paper prepared by Duncan Austin, Gwen Parker, and Robert Repetto of the World Resources Institute.

A fourth assumption is related to how society goes about requiring GHG control, i.e., what policies the government will put in place. Will the policies be flexible, allowing targets to be met at significantly lower costs? For example, one key aspect of the policy regime is the extent to which emissions trading is allowed. Another key aspect is the inclusiveness of the policy. Will carbon-absorbing activities, such as tree-planting, count as an offset to carbon emissions? Will all GHG emissions count, and will inter-gas trading be permitted?

Finally, most quantitative analyses of control costs do not include the environmental benefits of reducing GHGs. Including these benefits could lead to net costs of control that are significantly lower than the gross costs. In cost-benefit analyses, the optimal level of emissions control just balances the cost of the last increment of emissions control with the environmental benefits it produces.

Two of the five key assumptions — substitution and innovation — are structural features of the economic models used to make emissions projections. The other three are external factors or assumptions. The results summarized in this paper illustrate the importance of these five determinants and the large role played by the external factors or assumptions. Cost projections for a given set of assumptions can vary by a factor of two or four across models because of differences in the models' representation of substitution and innovation processes. However, for an individual model, differences in assumptions about the baseline, policy regime, and emissions reduction benefits can easily lead to cost estimates that differ by a factor of 10 or more. In understanding how these five determinants influence cost projections, decision-makers will be better equipped to evaluate the likely economic impacts of climate change mitigation. For example, among the 12 models and dozens of model runs reviewed for this paper, the base case forecasts range from a 20 percent to a 75 percent increase in carbon emissions by 2010.

One rough measure of economic costs is the carbon price — the amount of money one would have to pay to reduce emissions by a ton of carbon. Among the model results reviewed here, carbon price forecasts for meeting the emissions targets of the Kyoto Protocol ranged from less than $20 per ton to over $400 per ton. (In this paper "ton" means "metric ton.") These differences may result from using different models, as well as from using the same model with different input assumptions.

The economic consulting firm Wharton Econometric Forecasting Associates (WEFA, 1998) has projected a carbon price of $360 per ton, which reflects pessimism about the development of new technologies and the ability of businesses to plan ahead and begin responding early. It also assumes relatively inflexible government policies. In contrast, the President's Council of Economic Advisers published an "official" analysis in 1998 in which the carbon price under Kyoto would be quite low — on the order of $20 per ton — largely because it assumed that the United States would purchase most of its emissions reductions overseas through international emissions trading.

In many cases the assumptions that drive economic models are readily apparent; in other cases they are difficult to tease out because they are embedded in detailed aspects of the model's structure. The goal of this paper is to demystify what is driving these model results, thereby enabling the reader to participate fully in one of the most important debates of our time.

Five Determinants of Climate Change Cost Estimates

Five major determinants of GHG mitigation cost and benefit projections are discussed in this paper. Understanding how different forecasters deal with these determinants can go a long way toward understanding how individual estimates differ from one another. The five major determinants are:

- Projections for base case GHG emissions and climate damages;
- The climate policy regime considered (especially the degree of flexibility allowed);
- The representation of substitution possibilities by producers and consumers;
- How the rate and processes of technological change are incorporated in the analysis; and
- The characterization of the benefits of GHG emissions reductions in the study.

In addition to understanding the determinants themselves, it is important to understand how a given analysis treats uncertainties relating to these determinants.

Projections of Base Case Emissions and Climate Damages

Projecting the costs associated with reducing GHG emissions starts with a projection of GHG emissions over time, assuming no new climate policies. This is often an important and under-appreciated determinant of the results. The higher the base case emissions projection, the more GHG emissions must be reduced to achieve a specified emissions target. If a base case is higher, though, there may be more opportunities for cheap GHG mitigation due to a slow rate of technological progress assumed in the base case.

The base case emissions[1] and climate impact scenarios, against which the costs and benefits of GHG mitigation policies are assessed, are largely the product of assumptions that are external to the analysis. Each GHG mitigation cost analysis relies on input assumptions in three areas:

- **Population and economic activity.** Very few of the researchers analyzing GHG emissions reductions make their own projections of population and economic growth.[2] Rather, they rely on projections from organizations such as the World Bank and the United Nations.

- **Energy resource availability and prices.** The prices of fossil fuels — oil, natural gas, and coal — are important because producers and consumers generally need to substitute away from these fuels when carbon emissions are restricted.

- **Technology availability and costs.** These tend to be critical determinants of energy use in both the base case and control scenarios. Most analysts use a combination of statistical analysis of historical data on the demand for individual fuels and process analysis of individual technologies in use or under development in order to represent trends in energy technologies.

Projections of the benefits of reductions in GHG emissions are also highly dependent on the base case scenario employed. The greater the base case damages (i.e., the damages that would occur in the absence of any new climate policies), the greater the benefits of a specific emissions target. The magnitude of the benefits from emissions reductions depends not only on the base case level of impacts but also on where they occur, what sectors are being considered, and how well the affected populations can cope with any changes that occur. The task of projecting base case climate change impacts is particularly challenging because: (1) most assessments project that serious impacts resulting from climate change will not begin for several decades; and (2) most of the impacts are projected to occur in developing countries where future conditions are highly uncertain. How well developing countries can cope with future climate change will depend largely on their rate of economic development.

In a recent study organized by the Energy Modeling Forum (EMF) at Stanford University (called EMF 16), each modeling team was asked to prepare its own base case projection of carbon emissions in each world region.[3] Base case carbon emissions projections from this study (as well as a few other studies) for the United States are shown in Figure 1.

Projections of future carbon emissions vary widely. Even by 2010 there are significant differences. These differences are the result of different assumptions. Projections for the year 2010 range from 1,576 to 1,853 million metric tons of carbon (MMT-C), with a median of about 1,800 MMT-C. For the year 2020, they range from 1,674 to 2,244 MMT-C, with a median of about 1,950 MMT-C.

The Climate Policy Regime Considered

The policy regime considered is a crucial source of differences in cost and benefit projections. Once one constructs a base case scenario, one must specify the types of policies that nations may use to satisfy their GHG emissions obligations.

The Kyoto Protocol explicitly includes flexibility in determining which GHGs can be reduced, where they can be reduced, and when they can be reduced. If the parties to the Protocol can avoid restricting these flexibility mechanisms, and can surmount definitional and other obstacles to their implementation, the mechanisms can reduce the economic impacts of achieving the Kyoto targets by a factor of ten or more.

An important flexibility mechanism in the Kyoto Protocol is that nations have agreed to consider six principal GHGs: carbon dioxide (CO_2), methane (CH_4), nitrous oxide (N_2O), sulfur hexafluoride (SF_6), perfluorocarbons (PFCs), and hydrofluorocarbons (HFCs), and to allow inter-gas trading.[4] These GHGs differ both in their heat-trapping

Climate Change: Science, Strategies, & Solutions

capacity and in the length of time they remain in the atmosphere. CO_2 is the most significant contributor to global climate change among global GHG emissions. The Protocol also specifies that taking CO_2 out of the atmosphere (i.e., using CO_2 "sinks") through afforestation and reforestation, as well as through avoiding deforestation, counts toward each country's emissions reduction commitment.

The Protocol also allows participants to average emissions over a five-year period (2008–2012, also known as the first budget period) in satisfying their emissions reduction requirements. Averaging allows corporations and households to shift their reductions in time to reduce the economic impact of the required emissions reduction. A number of studies indicate that this emissions averaging can be quite effective in cutting the cost of reducing cumulative emissions, and that an even longer averaging period could be very advantageous.[5]

Figure 1

Base Case Carbon Emissions Projections for the U.S.

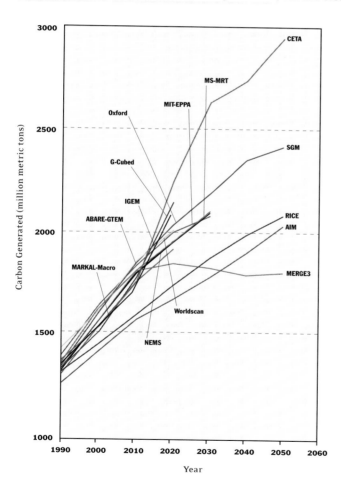

Emissions Trading

Another flexibility mechanism included in the Kyoto Protocol is international emissions trading. There are benefits from international emissions trading because the costs of reducing emissions vary greatly from country to country.[6] If the cost of emissions reductions in any country is higher than it is in any other country, it is advantageous to both countries for the higher cost country to buy emissions "rights" from the lower cost country at a price that is between the two cost levels.[7] If one aggregates all regions participating in the trading system together, one can compute the supply and demand for emissions rights at any price, as well as the equilibrium emissions rights price that balances the available supply with the amount demanded.

Figure 2 illustrates that the broader the scope of trading, the greater the benefits. The figure shows carbon price results for the United States for four alternative trading regimes. A recent EMF study (Weyant and Hill, 1999) started with four relatively simple alternative interpretations of the trading provisions in the Protocol in order to get a rough idea of what is at stake in determining the rules governing the trading regime: (1) No Trading of international emissions rights, (2) Full Annex I (developed countries or Annex B[8]) Trading of emissions rights, (3) the Double Bubble, which considers separate trading blocks for the European Union (EU) and for the rest of the Annex I countries, and (4) Full Global Trading of emissions rights, with the non-Annex I countries constrained to their base case emissions. Despite some limitations, a number of general conclusions can be drawn from the model results.

Figure 2

Year 2010 **Carbon Tax Comparison** for the U.S.

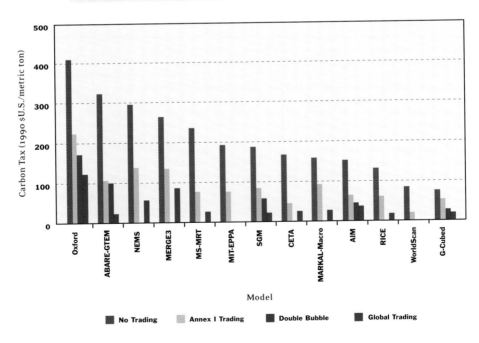

The potential advantages of expanding the scope of the trading regime are evident. Moving from the No Trading to the Annex I Trading case lowers the carbon price required in the four regions by a factor of two. (This is because almost all of the models project that a significant amount of carbon emissions rights will be available from Russia, an Annex I country.) The advantages of Global Trading relative to Annex I Trading are also significant. They result primarily because non-Annex I countries can reduce emissions more inexpensively relative to their unconstrained allocation of emissions rights than can the Annex I countries relative to their much more tightly constrained Kyoto allocation.

Domestic Policy

Although the focus of this paper is the global analysis of a global regime, assumptions about domestic policies within international models can make a difference. The Kyoto Protocol allows each national government to choose the policies it will use to comply with the Kyoto targets. Thus, different governments may allow emitters different degrees of flexibility. For example, governments may establish prescriptive regulations. Or governments may utilize price, either directly or indirectly. Governments may impose emissions taxes to motivate GHG reductions; or they may cap the total amount of emissions, distribute or sell emissions "allowances," and let the market determine the price and distribution of these allowances. In either case, fossil fuel combustion and other GHG-generating activities become more expensive. How producers and consumers respond to these price increases is described below.

Governments also have flexibility as to how they utilize any tax revenue or proceeds from allowance sales. Theoretical work indicates that the costs of carbon taxes can be significantly reduced if tax revenues are used to finance cuts in income tax rates. Assumptions about such so-called "revenue-recycling" can also be important.

Representation of Substitution Possibilities by Producers and Consumers

It is important to understand that flexibility in substitution of less GHG intensive activities is not a policy choice. It is a characteristic of the economy and depends ultimately on choices made by individual consumers and firms. Producers adjust to price increases — including policy-induced price increases — by substituting inputs (i.e., switching to inputs that generate fewer GHG emissions in manufacturing any particular product), and by substituting outputs (i.e., producing different products that require fewer GHG emissions to make). Different substitution responses are possible, depending on the time frame. In the short term, consumers and producers might turn their thermostats down; in the intermediate term, they might buy more efficient heating systems; and in the long term, they might take advantage of completely new technologies.

The extent to which inputs can be shifted depends on the availability and cost of appropriate technologies as well as the turnover rate of capital equipment and infrastructure. These two factors, as well as consumer preferences, determine an industry's ability to produce and sell alternative mixes of products. Increases in the costs both of fossil fuels, and of products that depend on fossil fuel combustion, will reduce consumers' real incomes. Consumers will simultaneously decide: (1) the extent to which they wish to adjust their mix of purchases towards less carbon-intensive products, and (2) how to adjust their mix of work and leisure time to compensate for the reduction in their real income.

Two Approaches to Representing Substitution Possibilities

There are two basic approaches to modeling substitution possibilities:

- Aggregate (longer-term models that characterize general, or aggregate, technological trends); and

- Technology-by-technology (short-term models intended to shed light on precise technology choices by specifying many separate technologies).

The choice of approach depends, in part, on the time frame under consideration and the level of technological detail desired. Many models blend the two approaches. Some models employ a technology-by-technology approach for the energy sector, but use an aggregate approach for the rest of the economy. Some models employ an aggregated approach for new capital investment, yet utilize a technology-by-technology approach for each vintage (i.e., all equipment of a particular age), once it has been installed. Similarly, a model may employ an aggregated approach for conventional fuels, yet stipulate discrete technologies for a particular non-carbon fuel.

Models Employing the Aggregate Approach

Four characteristics of these models are important in analyzing the time horizon for meeting the Kyoto targets (Jacoby and Wing, 1999):

- **The time interval over which a model solves its equations.** If a model uses a ten-year time interval, this limits its ability to be used in analyzing phenomena occurring within a decade, such as the consequences of accepting a 2008–2012 Kyoto target after the year 2000.

- **The level of detail about capital stock and how goods are produced.** This determines the degree of substitution that is possible within the model's structure.

- **The specification of economic foresight.** Models that assume perfect foresight allow emissions targets to be met at lower costs because investment decisions are made in the full certainty that emissions limits will be set and achieved. Models that assume some degree of myopia generate higher costs because investors must scramble to alter the capital stock as the target period approaches, prematurely scrapping existing capital (e.g., coal-fired power stations) and quickly investing in less carbon-intensive alternatives. In practice, investors do not have perfect foresight, nor do they suffer from complete myopia. While there is inevitable uncertainty regarding future economic conditions, policy-makers can reduce uncertainties by making credible commitments to meet targets or to initiate market-based policies.

- **How models capture the aging of capital.** In evaluating short-term adjustment to climate policies, it may be unrealistic to assume that the current capital stock can be transformed into more efficient and less carbon-intensive alternatives. However, for analysis of the long run, after fuel prices have settled at a new equilibrium level relative to other goods and services, that assumption makes more sense. In this post-adjustment phase, the inherited capital stock will be increasingly fuel-efficient and competitive under prevailing conditions, because those conditions will more closely match the conditions in place at the time the investments were made.

Model Results

Model results clearly demonstrate that the more convinced investors are that emissions targets will become binding, the less costly the transition to lower carbon emissions. The more lead time corporations and individuals have, the more flexibility they have in matching investments to the new policy regime.[9]

Unfortunately, it is very difficult to compare how models treat substitution since definitions, points of measurement, and level of aggregation of parameters differ greatly from model to model. However, Figures 3a and 3b are a valuable starting point in the process of understanding differences in model results by plotting the projected carbon price against the percentage reduction in carbon emissions for each of the trading regimes considered. This yields an approximate marginal carbon emissions reductions cost curve for each model for each region in each year.

The plots show how difficult it will be to make the required adjustments. The steeper the marginal cost curve, the higher the carbon price required to achieve a given percentage reduction in base case emissions. The steepness of these curves depends on the base case emissions projected by the model, the magnitude of the substitution and demand elasticities embedded in it, and the way capital stock turnover and energy demand adjustments are represented. Elasticities measure the responsiveness of the demand for a product to the price of the product. All three factors work together, so that models with higher base case emissions lead to higher adjustment costs. If the elasticities are high and the adjustment dynamics rapid, adjustment costs are lower.

Most of the results shown in this paper come from comparisons across models. Future studies looking at the sensitivity of substitution parameters within models rather than among models would be a very useful exercise in understanding why model cost estimates differ.

Marginal Cost of U.S. **Carbon Emissions Reductions in 2010** with No Emissions Trading Under Kyoto Scenarios

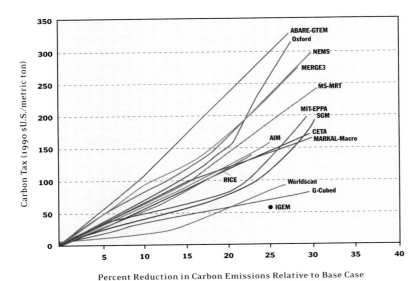

Percent Reduction in Carbon Emissions Relative to Base Case

Note: Only one data point was available for the IGEM; it was run only for a scenario that considered stabilizing U.S. carbon emissions at 1990 levels through 2000 prior to the formulation of the Kyoto Protocol.

Technological Change

New technologies and products can lower carbon emissions, thereby reducing the need for emissions reductions and/or making them less expensive. Such technologies might become available over time as a result of research and development (R&D) expenditures or cumulative experience. The emergence of these new technologies might be related to energy price increases, the base case trend of all other prices, or simply the passage of time.

Technological change is sure to be one of the dominant solutions to the problem of global climate change over the next century. However, there are large uncertainties about the most cost-effective way to accelerate the invention of brand new technologies, innovations in existing technologies, and the diffusion of new technologies. Each of these works on different time scales, and therefore different policies may be required to accelerate them.

There are opportunities for synergy between public support and private investment in new technology development. It would be worthwhile to find an appropriate combination of government interventions and private sector incentives that encourage

Figure 3b

Marginal Cost of U.S. **Carbon Emissions Reductions in 2020**
with No Emissions Trading Under Kyoto Scenarios

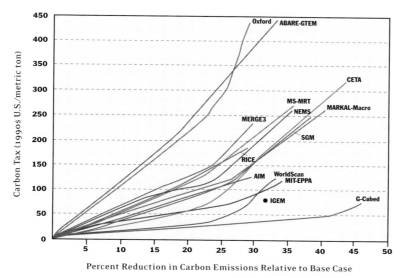

Percent Reduction in Carbon Emissions Relative to Base Case

Note: Only one data point was available for the IGEM; it was run only for a scenario that con-
sidered stabilizing U.S. carbon emissions at 1990 levels through 2000 prior to the formula-
tion of the Kyoto Protocol.

innovation. Thus far, however, most of the policy debate on the influence of technological change on climate change policy has focused not on technology policy options,
but rather on how restrictions on GHG emissions affect the cost of GHG reductions over
time. This effect has been labeled "induced technological change" (ITC).

Induced Technological Change

ITC occurs when the cost of a product — say, carbon-based fuels — rises and, in
response, firms develop new processes or products that use less of the now-costlier
item. In the case of carbon, these new processes or products would lower the cost of
mitigation. ITC has to do with price-induced behavior — i.e., what private firms do
specifically in response to higher prices, as opposed to what firms do in trying to
become more competitive through investing in R&D, or in response to government spon-
sorship of R&D or other direct government technology policies. The potential for ITC to
substantially lower, and perhaps even eliminate, the costs of CO_2 abatement policies
depends on whether technological change can be induced at no cost, or at some cost.

The relationship between price changes and technological change is extremely
complicated. There is an emerging consensus that the effects of ITC will be modest in

the short run, but much more significant in the longer term. Thus it will probably only have modest impacts on the 10-year emissions reduction cost projections presented here. ITC will have more of an impact on the 20-year projections and perhaps a dominant impact on cost projections for the middle to late century.

Empirical Modeling and Analysis of ITC

Some ITC models are based on empirical (i.e., actual) observations of past responses to energy price and policy changes. Empirical modeling of ITC may be valuable for short- to medium-term projections, or for estimating the short- to medium-term cost of policies on the economy. Empirical models may also be valuable in comparing or calibrating short-term projections from other types of ITC models.

A number of researchers have started to do empirical work on the factors influencing particular energy innovations. A recent empirical study of the energy sector by Newell et al. (1996) found that (1) technological characteristics may advance at different rates, with significant influence from "endogenous" factors resulting mainly from changes in energy prices, and (2) a significant component of technical advance is exogenous in that it is independent of changes in energy prices. Endogenous technological change is the term modelers use to describe how models simulate the process of induced technological change. Endogenous means internal to the model (i.e., it is represented by an equation inside the model). In ITC models, it usually means price-induced. However other things besides price can induce technological change — e.g., federal research and development expenditures or the process of learning-by-doing, which is discussed below. "Exogenous," on the other hand, can mean external to the model (i.e., an assumption made from outside the model), or independent of price, or both.

Autonomous Energy Efficiency Improvement

In contrast to the ITC models, many models include exogenous technical change. Many of the models use a single rate that the modeler specifies — the autonomous energy efficiency improvement (AEEI) — that lowers overall energy use per unit of output over time, independent of any changes in energy prices. Many modelers specify the AEEI as a percentage of gross domestic product (GDP) growth, so that the value changes over time. Although the definition of the AEEI varies from model to model, in all models it implicitly represents the general effect of technological progress. In some models it also represents one or both of two additional trends: (1) changes in the structure of the economy, resulting in a shift in the relative contribution of energy-intensive industry output to total economic output; and (2) an improvement in energy efficiency over time, reflecting the gradual removal of market barriers to efficient energy technologies.

Although the AEEI approach allows for energy improvements over time, it is limited in two respects. First, the AEEI approach ignores price-induced technological progress (i.e., ITC). In reality, higher prices do spur greater innovation and more rapid diffusion of energy-saving technologies. Second, it is not clear what an appropriate rate for AEEI should be.

Models employing the AEEI approach show that ongoing evolutionary technological change can have a major impact on base case emissions, which in turn affect GHG mitigation projections. Based on detailed process-engineering models and historical trends, modelers cluster around an AEEI of 1 percent per year. This would translate into a 22 percent improvement over 20 years.

More sophisticated specifications (often used in conjunction with an AEEI parameter) attempt to paint a more detailed picture of technological change by incorporating some degree of price sensitivity, distinguishing different sectors, and assessing changes to specific technologies.

Learning-by-Doing

In practice, much technological advances coming from learning-by-doing (LBD) — the incremental improvement of processes through small modifications and adjustments. It is not until a technology is actually used that important lessons are learned that can be applied to its subsequent development. LBD is an integral part of the innovation process. Observation of past technological innovations show that initial installations are quite expensive, but that costs drop significantly the more the technology is used, and the more lessons are learned from using it. This type of learning may be the result of either exogenous or endogenous technological change.

Characterization of Benefits

The motivation for policies to reduce emissions of GHGs is the reduction in climate change impacts. Some analyses focus exclusively on mitigation costs, showing both the likely range of costs under different policy regimes and the sensitivity of the cost estimates to key model inputs and parameters. These studies often start with emissions targets proposed during the international negotiation process. These studies, sometimes called "cost-effectiveness" analyses, do not estimate the benefits of the reduction in climate impacts, or any other accompanying benefits from the emissions reductions.

Other analysts, however, have projected base case climate impacts, and the change in those impacts resulting from climate policies (Watson et al., 1996). Those analyses that do address climate impacts vary widely in terms of what is included, and what is excluded. Table 1 shows the impact categories that have been considered. Aggregate impact/benefit studies differ as to whether they include only "market" impacts or both "market" and "non-market" impacts. It is much more difficult to place an economic value on non-market impacts.

The benefits of GHG emissions reductions, in terms of avoiding climate impacts for individual sectors, are difficult to assess, although considerable progress has been made (see Science and Impacts chapter). Valuation and aggregation across categories is difficult and controversial. In addition, climate change and impacts are more directly related to atmospheric concentrations of GHGs than to emissions, and GHGs can stay in the atmosphere for a hundred years or more. Thus, an assessment of the benefits of emissions reduction requires a long-run projection of the difference between

Table 1

Environmental Impacts in Economic Models

Environmental Impact	Description
Agriculture	Impacts on the level of productivity of different crops and on farmers' choice of crops to grow.
Forestry	Impacts on the level of productivity of commercial forests.
Sea-Level Rise	Impacts of rising sea levels on coastal development.
Ecosystems	Impacts on ecosystem function and vegetation patterns.
Human Health	Impacts on the incidence of vector- and water-borne diseases and heat and cold stresses.
Wildlife	Impacts on animal life, species diversification, etc.
Biodiversity	Impacts on plant and animal species diversity.
Fisheries	Impacts on commercial fisheries.
Amenity Values	Values individuals place on opportunities such as participation in various recreational activities. (Climate change could affect some forms of recreation either positively or negatively.)

climate impacts with and without controls. The difference must be aggregated over time with some sort of inter-temporal discounting. The assessment and the discounting also need to account for the risk that conditions could turn out to be much worse than expected in the future.

The current range of estimates for the direct benefits of reducing GHG emissions is from $5 to $125 per ton (1990 U.S. dollars) (Bruce et al., 1996). The wide range of estimates reflects variations in model assumptions, as well as a high sensitivity to the choice of a discount rate. Although simulations based on a social discount rate of 5 percent tend to be in the $5 to $12 range, assuming a rate of 2 percent or less can lead to estimates that are higher by a factor of ten. There are three key points to keep in mind when interpreting these numbers:

- The range of benefits projections depends critically on assumptions about both base case impacts and the ability of people and institutions to adapt to these impacts;

- Many analysts and policy-makers believe that costs ought to be weighed against disaggregated impacts that are left in physical (and, therefore, more tangible) and not monetary form; and

- Most analysts now recognize the much greater relative importance of low-probability, but high-consequence extreme events, as opposed to more gradual, linear changes, in our vulnerability to climate change. However, they have only just begun to study them.

Projections of ancillary (i.e., incidental) benefits range from $0 to $20 per metric ton of carbon (1990 U.S. dollars). These projections depend heavily on precisely where the emissions reductions occur. This is because: (1) most of the ancillary benefits are the result of incidental reductions in other air pollutants (e.g., sulfur dioxide, nitrogen oxides, volatile organic compounds, particulates, etc.); and (2) those air pollution benefits depend on both the prevailing meteorology and where people live relative to an emitting plant site or freeway system.

In summary, users of economic analyses should either: (1) focus on cost-effectiveness, taking emissions or concentration targets from other analysts or policy developers, or (2) factor in reductions in physical or monetary impacts and weigh them against mitigation costs. Above all, it is essential to keep the benefits of climate change policies transparent and separate from the costs, both in doing the analysis and in communicating the results.

Conclusions

Analysts have produced a wide range of projections of the costs and benefits of reducing GHG emissions. Understandably, policy-makers want to know what's behind these projections and why they often differ. This paper attempts to simplify the task of understanding differences in projections by focusing on five key areas in which differences in model configuration and in input assumptions drive differences in model results.

Two key determinants of costs and benefits are the base case emissions projection against which emissions reductions are compared, and the policy regime considered. These two factors primarily reflect differences in input assumptions. The model results summarized in this paper demonstrate that the higher the base case emissions, the greater the economic impacts of achieving a specific emissions target. It is also shown that key elements of the policy regime, like the extent to which international emissions trading is permitted, can have a profound effect on the economic impacts of emissions reduction. In general the more flexibility permitted in where, when, and which GHG reductions may be used to satisfy a commitment, the smaller the economic impacts. It also matters greatly how revenues raised through carbon taxation are reused, especially if certain uses are politically feasible.

A third determinant — the extent to which the model accounts for the benefits of emissions reductions often comes from external sources or is omitted (as in a cost effectiveness analysis). Sometimes, however, a cost-benefit analysis is performed or a benefits estimate is subtracted from the gross cost estimate to get a net benefits estimate.

These three external sources of differences in projections account for most, but not all, of the range of cost projections. The residual differences can be traced to how each model's structure accounts for two other key factors — the rate and extent to which available inputs and products can be substituted for one another, and the rate of improvement in the substitution possibilities themselves over time (i.e., technological change).

This paper has not attempted to ascribe the costs and benefits of GHG emissions reductions projected in any particular study to differences in input assumptions. Given the large number of detailed assumptions made in each modeling analysis of GHG policies, this is not feasible. Also, it could possibly be misleading if the wrong set of detailed inputs were selected and incorrectly interpreted. This paper has instead described the major input assumptions and features of the models to consider in interpreting and comparing the available model-based projections of the costs and benefits of GHG reductions.

Endnotes

1. In order to obtain such a base case emissions projection, one must estimate each of the following — population, economic output per person, energy per unit of economic output, and carbon per unit of energy — and then multiply them together. The projection of each of these factors is, in turn, either assumed or inferred from projections of other underlying factors. This is sometimes referred to as the Kaya identity since it was first observed by Yoichi Kaya (1989) of Tokyo/Keio University.

2. See, for example, Bruce et al. (1996).

3. Of the 13 global models included in EMF 16, 11 had an explicit U.S. region. In this report the authors have omitted the other two models — FUND and GRAPE — and added results from three U.S.-only models: IGEM, NEMS, and Markal-Macro.

4. Thus, after converting to "CO_2 equivalents," base year emissions are increased, and whether or not the inclusion of the "other" gases increases or decreases mitigation costs depends on whether the cost of decreases in CO_2 are larger or smaller than the cost of equivalent reductions in the other GHGs. Early studies of the use of this flexibility mechanism in meeting the objectives of the Kyoto Protocol indicates that there may be some cost savings associated with moving from CO_2-only reduction to reductions in all GHGs (Reilly et al., 1999 and Hayhoe et al., 1999). However there are two critical implementation issues associated with this potentially valuable flexibility mechanism: (1) a lack of consensus on the appropriate relative global warming potential of different gases, and (2) if the institutions are not in place to assign credit for reductions in the non-CO_2 gases, then the adoption of a multi-gas approach may actually increase costs by putting more pressure on CO_2 abatement.

5. See, for example, Manne and Richels (1999), Nordhaus and Boyer (1999), and Peck and Teisberg (1999).

6. See the report in this book on International Emissions Trading by Edmonds et al.

7. This is just another example of the gains from trade (Srinivasan and Bhagwati, 1983), albeit for a good that is not now traded.

8. Annex B refers to industrialized countries that are trying to return their GHG emissions to 1990 levels by the year 2000 per Article 4.2 of the Kyoto Protocol.

9. This observation is reflected in the comparison between Figures 3a and 3b where the marginal cost curve for emissions reductions in 2020 is seen to be flatter than that for 2010 for almost all of the models because there is more lead time to work with in reducing emissions by 2020 rather than 2010.

References

Bruce, J.P., H. Lee, and E.F. Haites. 1996. *Climate Change 1995. Volume 3: Economic and Social Dimensions of Climate Change.* Cambridge University Press. Cambridge, UK.

Hayhoe, K., A. Jain, H. Pitcher, C. MacCracken, M. Gibbs, D. Wuebbles, R. Harvey, and D. Kruger. 1999. Costs of Multigreenhouse Gas Reduction Targets for the U.S.A. *Science* 286: 905–906 (October 29).

Jacoby, H.D., and I.S. Wing. 1999. Adjustment Time, Capital Malleability and Policy Cost. J.P. Weyant, ed. In The Costs of the Kyoto Protocol: A Multi-Model Evaluation. Special Issue of *The Energy Journal.*

Manne, A.S., and R.G. Richels. 1999. The Kyoto Protocol: A Cost-Effective Strategy for Meeting Environmental Objectives? J.P. Weyant, ed. In The Costs of the Kyoto Protocol: A Multi-Model Evaluation. Special Issue of *The Energy Journal.*

Newell, R., A. Jaffe, and R. Stavins. 1996. *Environmental Policy And Technological Change: The Effect of Economic Incentives And Direct Regulation On Energy-Saving Innovation.* Working Paper, Harvard University, John F. Kennedy School of Government.

Nordhaus, W.D., and J.G. Boyer. 1999. Requiem for Kyoto: An Economic Analysis of the Kyoto Protocol. J.P. Weyant, ed. The Costs of the Kyoto Protocol: A Multi-Model Evaluation. Special Issue of *The Energy Journal.*

Peck, S.C., and T.J. Teisberg. 1999. *CO_2 Emissions Control Agreements: Incentives for Regional Participation.* Working Paper, Electric Power Research Institute.

Reilly, J., R. Prinn, J. Harnisch, J. Fitzmaurice, H. Jacoby, D. Kicklighter, J. Melillo, P. Stone, A. Sokolov, and C. Wang. 1999. Multi-Gas Assessment of the Kyoto Protocol. *Nature* 401: 549-555 (7 October).

Srinivasan, T.N., and J.N. Bhagwati. 1983. *Lectures on International Trade.* MIT Press, Cambridge, MA.

Watson, R.T, M.Z. Zinyowera, and R.H. Moss, 1996. *Climate Change 1995. Volume 2: Impacts, Adaptations, and Mitigation of Climate Change: Scientific-Technical Analyses.* Cambridge University Press, Cambridge, UK.

Weyant, J.P., and J.N. Hill. 1999. Introduction and Overview to Special Issue. J.P. Weyant, ed. In The Costs of the Kyoto Protocol: A Multi-Model Evaluation. Special Issue of *The Energy Journal.*

Wharton Econometric Forecasting Associates (WEFA). 1998. *Global Warming: The High Cost of the Kyoto Protocol, National and State Impacts.* Eddystone, PA.

Technological Change
& its Effects on Mitigation Costs

Jae Edmonds, Joseph M. Roop, Michael Scott

Stabilizing the global concentration of greenhouse gases (GHGs) in the atmosphere presents one of the toughest challenges for humanity in the twenty-first century. This objective will be pursued while most of the world's peoples seek to attain higher standards of living. These goals will be extraordinarily difficult to reconcile with current technologies. Fortunately, technology does not stand still.

The world today depends primarily on fossil fuels to supply its energy needs. These fuels lead to the release of more than six billion metric tons of carbon per year along with significant quantities of other GHGs. Restricting these emissions means developing and adopting new technologies that either do not use fossil fuels, prevent carbon from entering the atmosphere, or remove carbon from the atmosphere.

Irrespective of the climate change problem, most analysts believe that low-emitting GHG technologies such as nuclear, solar, wind, biomass, hydro, and conservation will continue to improve and achieve larger market shares in the future. But even if substantial energy improvements lie in the future, it will take time for technologies to penetrate the market. It has taken on average a century for the global market share of every major energy technology — from wood to coal to oil — to rise from 1 percent to 50 percent of global consumption.

Thus, understanding the way technology evolves and penetrates the market place is essential to understanding how to address climate change. Economists are in agreement that technological change can dramatically ease the transition to a sustainable climate. No matter how costly one believes this transition may be, technological progress makes it cheaper; no matter how urgent one believes it to be, technological progress allows for longer lead times.

As with all future gazing, one's understanding of future technology becomes murkier the further into the future one attempts to look. Nonetheless, in order to understand predictions regarding the effects of climate change or policies to address it, one must clearly understand the way technological change occurs and the way that process is represented in computer models used to analyze the problem. This paper examines those issues.

This paper is a condensed and updated version of a report published in September 2000 by the Pew Center: Technology and the Economics of Climate Change Policy. The Pew Center and the authors appreciate the valuable insights of several reviewers of this paper, including Nebojsa Nakićenović, Ian Parry, and Alan Sanstad.

What is Technological Change?

Technological change is a major driver in how a country's economic output increases. This happens for several reasons. If a country's population can be productively employed, its output grows as its population grows. If the amount of physical capital (machinery, transportation infrastructure, etc.) associated with that population grows faster than the number of workers, output per worker increases and that causes the economy to grow. The economy will also grow as human capital (i.e., educated and trained individuals) grows. Finally, the economy also grows as a result of improved types, quality, and use of physical and human capital stock. This is called technological change.

Carbon emissions from this growth will depend upon the amounts and types of energy sources and energy technologies that are used, and other consequences of growth such as land clearing. Energy demand will depend on the efficiency with which energy is used and the responsiveness of the use of energy to changes in its price. The carbon emissions consequences of meeting this energy demand will depend on the timing and availability of low-carbon forms of energy and energy technologies. Technological change will include development of backstop technologies (i.e., technologies that substitute for carbon-emitting technologies), will improve the efficiency with which energy is used, and will increase the efficiency with which energy can be discovered and extracted. In myriad ways, technological change will affect both the way a country grows and how its emissions evolve over time.

Figure 1

Global Market Penetration of Fuels 1850-1994

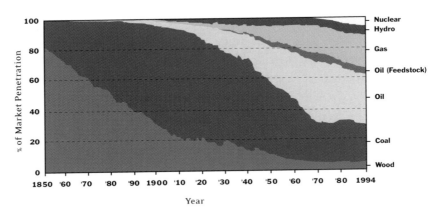

Source: Grübler, 1998.

Figure 2

Market Penetration of Transportation Technologies

in France (1844-1994)

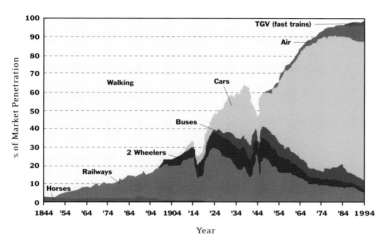

Source: Grübler, 1998.

The Institute of International Applied Systems Analysis (IIASA) of Austria has found that it takes between 50 and 100 years for an energy form to increase its global market share from 1 percent to 50 percent.[1] Figure 1 shows the percentage of market penetration in the world's energy market for seven fuels from 1850–1994. Oil, for example, took 90 years to grow from 1 percent to 40 percent of the energy market and has still not reached the 70 percent market share that coal reached in 1913. Natural gas has captured an increasing share of the market, rising from a 1 percent to a 25 percent share of the global energy system over the past 85 years. Despite the increased growth in oil and natural gas, other fuels such as hydro and nuclear power have also grown, while coal and wood still account for nearly 25 percent of the global energy market.

Technological change also has a profound effect on how people live over both the short and long term. A recent short-term example is the introduction of the personal computer. Since 1980, the computing power of desktop computers has increased by a factor of 100, while their cost has been reduced by a factor of four, adjusted for inflation. These developments, in turn, have opened up entirely new uses for computers and have changed the way that many people work and live.

Over long periods of time, technological change affects communications, transportation, and how businesses function. One hundred years ago, the telephone, the automobile, and the incandescent light bulb were in their infancy. For example, as seen in Figure 2, prior to 1850, passenger transport in France was dominated by walking. It was not until 1940 that walking was overtaken by the combined market penetration of rail, buses, and cars. The market penetration of the internal combustion engine led to the personal automobile eclipsing all other modes combined during the 1960s, and then stagnating in the past two decades due to ascension of air travel and TGV (fast trains).

Technological change alters the economy over time, affects the rate of economic growth, and helps determine carbon emissions. Increasing knowledge will affect future technological change, but it is not known exactly how research and development (R&D) directs technological change. Clearly technological change could have a large impact on the cost of climate change mitigation. The major issues are how technological change affects the economy and how technological change affects carbon emissions. These issues provide a backdrop to how technological change is treated in climate change models.

How Technological Change Affects the Economy

Technology and the economy are inseparable. In the 1950s, Robert Solow of the Massachusetts Institute of Technology developed a theoretical argument for the importance of technological change as the prime long-term determinant of continued increases in the standard of living. Simon Kuznets empirically codified this result in the 1970s. Both men were awarded Nobel Prizes for their work.[2] These and other studies have shown that growth can occur as a result of four basic causes. These are:

- **Capital deepening.** If the stock (amount) of physical capital (i.e., equipment and production facilities) grows at a rate faster than the growth of the working population (that is, capital per worker increases), output per labor hour increases. The gain from foregoing consumption today to invest in capital for greater output tomorrow is what motivates investment. Technological change occurs when the capital put into place is more efficient than the older capital it replaces. Capital grows and output per labor hour increases, perhaps enhanced by improvements in labor quality, which is also an element of technological change.

- **More efficient allocation of resources.** This can be the result of improved division of labor or may occur as a result of specialization and trade. One of the best-known examples of improved division of labor is the assembly line pioneered by Henry Ford. Controlling carbon emissions is an example of an activity that can be done more efficiently through trade, since emissions reductions have different unit costs in different places.

- **Scale effects.** Doubling inputs more than doubles output. One reason is that increasing scale spreads fixed costs over more units. For example, the unit costs of producing one vehicle are lower using technology that produces a million cars per year than using technology that produces only a thousand cars per year.

- **Increases in knowledge.** These can be fundamental insights into how the world works or simply better ways of organizing the production process to increase output. Much of the observed technological change comes from advances in knowledge that lead to gradual improvements in efficiency. These increases in knowledge are the long-term effects of learning-by-doing (LBD) and gradual improvements in technology.

Beyond these gradual changes, however, lie the dramatic shifts in knowledge and technique that can so profoundly influence the structure of economic systems in the future. Mokyr refers to these dramatic changes as "macroinventions" and argues that without them, improvements through LBD would be subject to the same law of diminishing returns as all other economic activity.[3] It is the creation of entirely new technologies — the atmospheric engine of Newcomen (the breakthrough that allowed the development of the first successful steam engine), the converter of Bessemer (that allowed steel to be produced cheaply for the first time), and, more recently, the transistor — that allows progress to continue unabated. These new technologies typically start by providing a specialized service at a high cost in a specialized application. As experience is gained, the technology improves and its costs decline. The decline in cost enables the technology to migrate to a wider set of applications, where further experience leads to further modifications and improvements, and stimulates the development of complementary technologies. But these macroinventions cannot be easily anticipated and they usually take a long time to have an impact on the overall economy.

Whether gradual or "macro" in scale, technological change that results from increases in knowledge is an especially important source of economic growth. To the economist, that portion of our growth that cannot be explained by increases in inputs is considered technological change. Any improvement in the quality of inputs, other measurable indicators of economic efficiency (such as improved inventory control or "just-in-time" delivery of supplies), and things that are not measurable are all part of technological change. Studies of the effect of this unexplained portion of technological change suggest that it accounts for about half of all economic growth.[4]

The perspective of the economist provides an alternative to the perspective of the technologist or engineer. The engineer considers specific technological change to be more understandable — i.e., it is any change in equipment or technique that allows quicker, better, and cheaper ways of getting work done. The economist, on the other hand, would call such changes capital deepening or increases in human capital. So the economist knows what causes technological change, but not what it is; the engineer knows what it is in specific instances, but not what causes it. This difference in perspective is important in determining how technological change is included in economic models.

Technological change usually has its effect on growth over a period of time. The process that transforms inventions into improvements in output is characterized by the sequence: invention-innovation-diffusion. The diffusion of technologies that are economically superior is a gradual process, the more "macro" the invention, the longer it takes to penetrate the market.[5]

How Technological Change Affects Carbon Emissions

There are four different ways in which technological change can affect carbon emissions. All four may be seen as beneficial in some respects, but they don't always reduce GHG emissions.

- Technological change can make carbon-based fuels cheaper (e.g., through improvements in the efficiency of fossil fuel extraction). Part of the reason for the current concern about GHG emissions is that technological change has increased reliance on carbon-based fuels while making them available at modest cost.[6]

- Technological change can also increase the overall rate of growth of the economy through improvements in labor productivity. As with changes in the efficiency of fossil fuel extraction, this too would tend to increase emissions, unless there were concomitant improvements in energy efficiency.[7]

- Technological change can increase the rate of improvement in alternatives to carbon-emitting energy technologies. Non-carbon emitting technologies, such as biomass, wind, and hydro, were the dominant energy forms before the age of fossil fuels. New versions of these technologies, plus technologies that would allow the capture and sequestration of carbon from fossil fuels, could provide the technological mechanisms for controlling carbon accumulation in the atmosphere.

- Technological change can increase the rate of improvement in the efficiency with which carbon-based fuels are used. Improvement in the efficiency of energy use can occur as a result of price changes or technological change, or a combination of both. The efficiency of energy use will improve over time as new investments combine with new knowledge to provide the capital that uses energy more efficiently. These improvements may be stimulated by price increases in energy, such as the United States and most of the world experienced in both the early and late 1970s.

These last two influences are the major routes by which technological change reduces carbon emissions.[8]

An Unknown Technological Future

While much is known about past technological change, much less is known about future technological change. These uncertainties include: where inventions will come from; what inventions will become successful; what any given dollar of R&D will return; how much learning will occur; how quickly a particular product or process will diffuse into wider use; or where the next big breakthrough will come. There is no evidence in the literature that any single technology will provide society with the ability to control the cost of emissions mitigation.

Still, centuries of experience suggest that a suite of new and improved technologies will become available over time, both through incremental improvements and through "macroinventions." Incremental technological improvement and LBD affect existing technologies, and take advantage of existing infrastructure and supporting technological systems. Macroinventions yield entirely new ways of doing things, open new markets, and stimulate the creation and dissemination of new infrastructure and complementary technologies. The large-scale introduction of low-cost carbon capture and sequestration technologies would constitute a macroinvention.

Figure 3 illustrates how the electricity costs for five technologies in the European Union have declined as the level of installed capacity has increased. The figure shows how technologies such as wind, solar photovoltaics (PV), and biomass have much steeper "learning curves" (the relationship between the cost of output and cumulative experience) than advanced fossil fuel technologies such as natural gas and coal, giving the impression that their costs could soon be equal. However, both coal (gasified clean coal) and natural gas technologies continue to have an absolute cost advantage, although costs for electricity generated by wind have been nearly equal to the costs of electricity generated by coal and gas since 1995.[9]

With all the ways that technological change can affect the path of growth and resource use, which is most amenable to human intervention? What can be done to stimulate innovation and technological change? According to *The Economist,* in a recent article discussing a collection of papers on the chemical industry and economic growth:

Figure 3

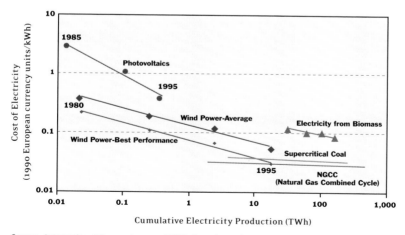

Electric Technologies in the European Union 1980-1995

Source: International Energy Agency, 2000. Experience Curves for Energy Technology Policy, OECD/IEA, p. 21.

"What matters most? What is the main thing governments must do to spur economic growth? Ah, well, that remains a mystery."[10]

While analysts cannot completely solve this mystery, it is possible to shed some light on what might be most effective in reducing the costs of GHG emissions mitigation via technological change. Technological advances in energy efficiency are believed to be extremely important in reducing the future cost of GHG emissions mitigation. The literature is unanimous on this point. It is hard to overestimate the importance of developing and commercializing new and improved energy technologies over the course of this century. The value of future improvements in GHG related technologies, relative to the present set of technologies, has been estimated to be in the trillions of dollars. Models are used to come up with such estimates. How technological advances are integrated into economic models of GHG emissions is the topic of the next section.

Modeling Technological Change

Current understanding of technological progress and its relationship to environmental goals comes from integrated economic, energy, and environmental system models, which are mathematical and often computer-based representations of how the economy and environment interact. Models that explain how technological change relates to climate change can be organized into two idealized approaches — "top-down" and "bottom-up" models (even though many models contain elements drawn from more than one approach), and there are sometimes important distinctions within these groups. Because it is difficult to predict the future of technology, all models addressing this issue rely on some sort of assumption regarding the future course of technological progress.

Bottom-up models are based on engineering, and come to the problem of emissions limitations from the perspective of the cost and performance of specific emissions-reducing equipment and practices (such as energy-efficient space heating, lighting, or motors). They generally begin by assuming that a set of advanced technologies either does or will exist, with predetermined cost and efficiency characteristics. They then compare alternative scenarios of how the world might evolve, depending upon how widely the assumed technologies are used. Thus, their depiction of climate change policy will depend strongly on the assumptions they make regarding the characteristics of the new technologies as well as their rates of market penetration.

Very importantly, because of the narrow emphasis on the comparative cost and performance of individual technologies, these models usually do not reflect many other aspects of the economy's response to climate change. Nor do these models respond to climate change policy, such as broader price-induced changes in energy demand, or the way households work or save. They are poorly suited, therefore, to estimating the societal economic cost of climate change or related policies. Instead, their strength lies in forecasting the near-term impacts of specific advances and in illuminating the economic value of possible technological improvements.

Top-down models, in contrast, are generally broad economic models; they depict the way the economy and environment interact in the aggregate. They reproduce the history of technological change in its broad outline but do not necessarily provide details about design, costs, or performance of specific technologies. Rather, they start with a set of initial conditions based on the current state of the economy, and then extrapolate from past experience to look at the future implications of major economic and technological forces.

As they contain little or no explicit technological detail, many "top-down" models represent technological change in terms of a single societal rate at which energy efficiency will continually improve in the future, a rate usually based on observed values in the past. This exogenous depiction (so named because its description of technology comes from outside the model), therefore, requires the modeler to make an assumption regarding the value of this ongoing, "autonomous," improvement.

Other top-down models have attempted to replace this "exogenous" assumption with a more detailed representation of the very process by which technology is created and adopted by firms in the economy. This approach starts by assuming that the amount of innovative effort in the economy is a direct function of current and anticipated economic conditions, and is called endogenous technological change, because technological change is projected within the model.

What remains unresolved in all of these approaches is the precise relationship between economic stimuli such as research and development (R&D) expenditures, energy prices, taxes, and subsidies, and the direction and rate of technological change, and the subsequent effect on societal cost. The various types of models discussed here have represented these phenomena in different ways. In bottom-up models, the rate of technological change depends on how different the alternatives are to the present technology as well as how quickly they substitute for one another. For top-down models, the rate of change is determined by assumptions about the "autonomous energy efficiency improvement" (AEEI) or elasticities (responsiveness to price and other variables). In top-down models employing endogenous technological change, technological change is often represented as a function of past production, the amount of past R&D, or the extent of energy price changes. Thus, even these endogenous models operate on the basis of an assumption — one regarding what determines technological effort and how that effort translates into progress.

So, regardless of their different structures, all models of technological progress rely on some fundamental assumption regarding future technology, an assumption that plays a central role in determining model results.

Model structure, of course, is important, because different model structures will treat economic phenomena differently. Consider, for example, how changes in the price of energy will affect the rate of technological change. When energy prices increase, the costs of production increase for nearly all goods and services, but more for those goods and services that require larger amounts or more expensive kinds of energy. As these costs flow through the economy, both producers and consumers of these goods and services will search for alternatives to their use. This search leads to innovation and technological change that reduces the need for energy.

This phenomenon is captured in models in a variety of ways, depending on their structure. Bottom-up models will capture the effects of energy price increases by improving the cost advantage of assumed new technologies that will then penetrate the economy more rapidly. Top-down, exogenous models capture the effect of higher energy prices by allowing firms to substitute capital or labor for energy, but generally will not change the underlying "autonomous" rate of technological change. Top-down, endogenous models will capture the effect of rising energy prices directly by having technological change accelerate when prices increase, using whatever causal relationship the modeler chooses. Table 1 lists the various models cited in this paper and their technological change characteristics.[11]

How Technological Change Affects GHG Mitigation Cost Estimates

Numerous international and regional studies have examined how climate change objectives could be achieved. Some compare the results of a variety of models. Others have used individual models to examine mitigation costs and have reported these results in journals, conference proceedings, and other venues. A review of this vast literature[12] leads to the following conclusions:

- All models show that additional technological change in low-carbon energy or energy efficiency will lower mitigation costs.

- Assumptions drive results.

- GHG mitigation costs are partially determined by the form of the technological change.

All models show that additional technological change in low-carbon energy or energy efficiency will lower mitigation costs. Technological change that expands low-and non-carbon technology options and increases energy efficiency lowers mitigation costs,[13] according to a large number of studies using a variety of methods. One method is to assume that specified technologies become available over time at specified costs (these are known as "backstop technologies"), and to explore the implications for mitigation costs. Another method is to adjust the overall rate of exogenous technological change. A third method is to explore the effect of the market penetration rates of the available technologies. A fourth method is to "endogenize" technical change using one or more of the three methods (price effects, explicit R&D, or LBD). A number of studies use a combination of methods. Examples of studies applying these methods are described below.

Table 1

Technological Change Characteristics of Models Cited

Model	Model Category	Technology Change (TC) Characteristics	Reference
CETA Carbon Emissions Trajectory Assessment	Top-down	Exogenous TC Technology Snapshot	Peck and Teisburg (1992)
CRTM Canadian Recursive Trade Model	Top-down	Exogenous TC	Rutherford (1992)
DGEM Dynamic General Equilibrium Model	Top-down	Exogenous TC	Jorgenson and Wilcoxen (1990)
DIAM Dynamic Integrated Assessment Model	Bottom-up	Technology Snapshot Endogenous TC	Grubb et al. (1995)
DICE Dynamic Integrated Climate & Economy Model	Top-down	Exogenous TC	Nordhaus (1994)
ERB Edmonds, Reilly, Barnes Model	Top-down	Exogenous TC Technology Snapshot	Barnes et al. (1992)
ETA-Macro	Top-down	Exogenous TC Technology Snapshot	Manne (1981)
FOSSIL-2	Bottom-up	Technology Snapshot	The AES Corporation and Energy and Environmental Analysis, Inc. (1991)
Global 2100	Top-down	Exogenous TC Technology Snapshot	Manne and Richels (1999)
Goulder and Colleagues	Top-down	Endogeneous TC	Goulder and Schneider (1999); Goulder and Mathai (1998)
Grübler and Gritsevskii	Bottom-up	Technology Snapshot Endogenous TC	Grübler and Gritsevskii (1997)
ICAM-3 Integrated Climate Assessment Model	Bottom-up	Technology Snapshot Endogenous TC	Dowlatabadi (1998)
MARKAL-Macro	Bottom-up	Technology Snapshot Endogenous	Manne and Wene (1994)
MESSAGE	Bottom-up	Technology Snapshot Endogenous TC	Messner (1997)
OECD-GREEN General Equilibrium Environmental Model	Top-down	Exogenous TC	Burniaux et al. (1990)
SGM Second Generation Model	Top-down	Exogenous TC Technology Snapshot	Edmonds et al. (1995)

- *Varying assumptions about backstop technologies.* The Energy Modeling Forum (EMF-12) studied an accelerated technology scenario using four different models: ERB,[14] Fossil-2,[15] Global 2100,[16] and CRTM.[17] All four of the models showed dramatically lower mitigation costs. EMF-14[18] also found that accelerated technology scenarios lowered mitigation costs. Hourcade et al.[19] compared assumptions about the availability of specific technologies from a number of studies of European Union countries. The article concluded that: (1) emissions reduction costs decrease over time, simply because more technologies become available; and (2) the magnitude of this effect depends on the characteristics of the assumed technologies. In this study, for the Netherlands, the researchers assumed the future availability of fuel cells and hydrogen-fired equipment and therefore forecast large cost reductions over time.

- ***Varying the overall rates of exogenous technological change.*** The Organization for Economic Cooperation and Development (OECD) conducted an (accelerated technology) experiment similar to the EMF-12, but allowed models to adjust the rates of exogenous technological change as well as the technology back-stops. The OECD study compared GREEN[20] to some of the other models used by EMF. OECD found that in the period up to 2010, technological change assumptions made little difference, but that beyond 2010, higher rates of technological change significantly lowered mitigation costs. This is because technological change takes time to have an effect.

- ***Varying assumptions about market penetration rates.*** A study by MacCracken et al.[21] found that the costs in a given scenario were highly dependent on the associated advanced technology penetration rates — the lower the rates, the closer the costs to the reference case that achieved the same environmental objective using existing technology. The model results further showed that, under an advanced technology scenario, there are large benefits of moderating the rate of near-term control, allowing new technologies more time to penetrate the market. In fact, moderating the rate of near-term controls allows one to achieve both a more stringent target and a lower cost.

- ***Incorporating endogenous technological change.*** There are three ways that endogenous technological change is introduced into economic models of climate change. The first is to alter the rate of efficiency improvement in energy use by converting the AEEI to a variable determined by the model (rather than an assumption of the model) — i.e., to "endogenize" it. One way to do this is to make energy efficiency improvement a function of energy prices. A second approach introduces an explicit R&D activity that leads to new technologies that affect the efficiency of energy use. The third way is to allow improvement in energy efficiency (or abatement activities) as experience is gained in the production of energy-intensive goods. This modeling of LBD usually allows energy efficiency to improve based on cumulative production: how much is produced over a span of time.

Numerous models have attempted to endogenize technical change using one or more of the three methods (price effects, explicit R&D, or LBD). Grubb et al.[22] used a model with two forms of endogenous technological change: R&D investment and LBD. They found that much greater abatement takes place when endogenous technological change is taken into account than when it is not.[23] Dowlatabadi and Oravetz[24] modeled endogenous technological change through price effects. They found that price-induced efficiency improvement significantly lowered emissions for the United States and reduced the cost of limiting carbon dioxide concentrations. Dowlatabadi later added

LBD to the model and found that it, too, lowered mitigation costs. Goulder and Mathai[25] incorporated LBD and R&D, and found that both lowered costs. However, unlike LBD, R&D is not "free," so R&D costs had to be included in the overall cost of mitigation.

Goulder and Schneider[26] introduced two types of R&D activity[27] — "sector-specific" and "spillover." In their model, sector-specific knowledge is derived from R&D expenditures paid by that industry or sector, and only benefits that sector. R&D that has economy-wide benefit ("spillover") can be financed by government subsidies. Subsidies to R&D had either positive or negative consequences. Industry-specific R&D subsidies led to over-investment in R&D, but R&D subsidies of knowledge spillovers reduced the costs of achieving emissions reductions.

A number of bottom-up modelers have integrated endogenous technological change with technology snapshot models in interesting ways. For example, Sabine Messner[28] found that LBD changes the optimal technology mix considerably (i.e., advanced coal, new nuclear, and solar technologies become dominant) and lowers investment costs (by nearly 20 percent in 2050). Grübler and Gritsevski[29] used an optimizing bottom-up model to explore delays in the adoption of a technology as learning occurs, or as uncertainty arises. They found that LBD makes new technologies penetrate the market more rapidly, but that uncertainty delays penetration.

Assumptions drive results. While models may take many different approaches to representing technological change, they all tell a remarkably consistent story over the long term. For example, studies employing "bottom-up" models usually produce lower estimates of the costs of climate change policy. Bottom-up modelers frequently interpret differences between economic potential and observed market penetration as evidence of inefficient "barriers" that can be overcome by appropriate action. Top-down modelers commonly emphasize the role of markets and prices in deciding whether alternative technologies will be used in production. It is more common for "bottom-up" than "top-down" modelers to make these assumptions, but these assumptions are not themselves part of the structure of the models. The models themselves also make some difference. Bottom-up models are more likely than top-down models to show substantial, immediate penetration of new technologies. This is because bottom-up models assume that individual technologies compete on the basis of cost and performance, and that a more efficient process will capture a significant part of the market.

A recent Stanford Energy Modeling Forum (EMF) study demonstrated that the span of results from "top-down" models was sufficiently wide to encompass all of the bottom-up results. A related EMF study revealed that where it was possible to standardize assumptions for discount rates, capital-stock turnover rates, ancillary environmental benefits, and engineering descriptions of the technologies, any observed gap between top-down and bottom-up studies largely evaporated. Thus, the type of model proves to be a secondary consideration in determining the nature of its results: the assumptions that enter these models drive the process. For example, one study compared the cost of stabilizing the concentration of carbon dioxide at twice the pre-industrial level under two scenarios: one assuming only currently available technologies, and one assuming technologies that forecasters expect to be available by the end of this century. The difference in cost was more than ten times — attributable to differences in technology adoption, performance, and cost alone. Alternatively, recent estimates of the discounted present value of reducing the cost of solar photovoltaic power by one cent per kilowatt hour range from tens of billions of dollars to hundreds of billions of dollars, depending on the CO_2 target and what other technology alternatives are assumed to be available. In short, the way a model represents the climate change problem over a period of decades, if not generations, cannot be understood until the assumptions it has made regarding technological progress have been made transparent.

GHG mitigation costs are partially determined by the form of the technological change. There is virtually no cost to some forms of technological change. Both learning-by-doing (LBD) and autonomous energy efficiency improvement (AEEI) fall into the category of low- or no-cost technological change. In these cases, the simple progress of time or the normal accumulation of experience is all that is required for technological progress to occur. With AEEI, just the passage of time reduces energy input requirements. With LBD, the simple accumulation of production experience (volume of production) lowers the cost and resources required for production. These models specify that no additional resources have to be sacrificed to achieve these changes, so society does not have to give up anything.

On the other hand, both R&D and technology replacement entail considerable societal costs that may partially offset the benefits of technology innovation and diffusion. R&D requires expensive facilities and personnel. The R&D investment may pay off in terms of future technological breakthroughs, but it may not. Technology replacement can be costly because it may involve the premature retirement of useful equipment. For example, the owner of a factory might have made an investment in energy equipment that he reasonably expected to last 10 years. If carbon control policies were implemented, he might have to replace that equipment with new and more efficient technology. If the policies were imposed less than 10 years after he made his investment, he would lose money on that investment. Thus the form of technological change that the model assumes will affect the costs of innovation and diffusion, which, in turn, will affect mitigation costs.

Challenges for Modelers and Policy-Makers

Modeling the rate of technological change is in its infancy. While relatively simple models can be built to illustrate the effects of inducing technological change through R&D expenditures, through LBD, and through price, these models fall far short of the complexity of the real world. Moreover, endogenous technological change

Climate Change: Science, Strategies, & Solutions

models substitute assumptions about what factors affect the rate of technological change for assumptions about technological change itself. While these models provide some insights, they do not fully explain the process of technological change, nor do they relieve the modeler of having to make assumptions about the process.

While economic models can explore what might happen, they cannot forecast what will happen. How these events unfold in the future will depend not only on how technology affects mitigation costs, but also on what decision-makers do in response to the threat of global warming: from R&D investments, to the timing of policy, to international climate change negotiations. Economic models can help to inform those actions. A better understanding of what the models say should lead to better decisions, and reduced costs of GHG emissions mitigation.

The value to a policy-maker of using economic models of climate change to assess the cost of mitigating GHG emissions will depend on the specific question being asked and the design of the model used. If the question is how specific technologies may reduce emissions or mitigation costs, then the model must be capable of allowing that technology to affect the outcome of the model simulation. A general understanding of the model and how technological change is treated in the model will serve the policy-maker well. The questions below provide insight into the value of the model for answering specific policy questions.

What determines the rate of technological change in models? In bottom-up models, the rate of technological change depends on how different the snapshots are, and how quickly they substitute for one another. In endogenous technological change versions of these models, the rate of technological change also depends on LBD. For top-down models with exogenous technological change, the rate of change is determined by assumption about the AEEI or elasticities. For top-down models in which technological change is endogenous, the rate of change depends on the rate of past production, the amount of past R&D, or the extent of energy price changes.

How do energy prices affect technological change? When energy prices increase, the costs of production increase for nearly all goods and services. Costs increase more for those goods and services that require larger amounts or more expensive kinds of energy. As these costs flow through the economy, both producers and consumers will search for alternatives to these goods and services. This search leads to innovation and technological change that reduces the need for energy. This phenomenon is captured in models in a variety of ways. Endogenous models of technological change can capture this directly by having technological change accelerate when prices increase. In technology snapshot models, energy price increases may improve the cost advantage of new technologies so they penetrate faster. Top-down models capture the effect of higher energy prices through the elasticity of substitution between energy and other inputs.

How does R&D affect technological change? R&D affects technological change in three basic ways. First, it contributes to knowledge generally, and leads to future developments that may (or may not) directly improve the efficiency of energy use. Second, R&D can lead directly to changes in the efficiency of energy-using equipment (for example, the research on high-efficiency lighting done by the U.S. Department of Energy). Third, R&D can lead to improvements in the efficiency of extraction of fossil fuels and thus encourage, through lower prices, further consumption of fossil fuels. Indirectly, contributions to knowledge are integrated into production techniques that lower the costs of production and may reduce energy use. The complex chain of events that leads from R&D expenditures to increases in knowledge and hence to improvements in energy efficiency is important, but not well understood, nor the topic of this paper. In models of endogenous technological change, the models assume that R&D at one point in time will lead to improved energy efficiency at a later time.

There are two promising approaches to developing a better understanding of the role of technology in addressing the climate change issue. The first is to combine the best features of the top-down and bottom-up models in a single modeling framework — introducing better engineering representations into a consistent, general, energy-economic setting. And the second is to continue to pursue the development of fully endogenous models of technological change. Both approaches can help estimate the costs of policy as well as identify and rank technology opportunities.

The current state-of-the-art may leave analysts unable to predict the nature, rate, and direction of technological change without resorting to important assumptions. But it also makes clear that one cannot fully understand the climate change problem without understanding technological progress, and that without dramatic technological developments, the road to climate stabilization will be an arduous one.

Endnotes & References

1. Grübler, A. 1998. *Technology and Global Change*. Cambridge University Press, Cambridge, United Kingdom.

2. Simon Kuznets received the Bank of Sweden Prize in Economic Sciences in Memory of Alfred Nobel in 1971, partly on the basis of *Economic Growth of Nations*, Harvard University Press, 1971. Robert Solow received the same prize in 1987 for his contributions to the theory of economic growth.

3. Mokyr, J. 1990. *Twenty-Five Centuries of Technological Change: A Historical Survey*. Harwood Academic Publishers, New York, p. 103.

4. Denison, E.F. 1994. *Accounting for United States Economic Growth, 1929–1969*. The Brookings Institution, Washington, D.C., Table 9.7.

5. One of the first economic studies to document this market penetration is by Griliches. The Griliches citation and others are given in Jaffe, A.B., and R.N. Stavins, 1994. The Energy-Efficiency Gap: What Does It Mean? *Energy Policy* 22(10): 804–810.

6. While this improvement has been dramatic (see, for example, M.N. Fagan. 1995. Resource Depletion and Technological Change: Effects on U.S. Crude Oil Finding Costs from 1977 to 1994. *Energy Journal* 18(4): 91-105, who suggests that the rate of technological change in refining has increased over time, reaching 15 percent per year in cost reductions in 1994 and is important to carbon emissions, there is very little mention of this type of technological change in the literature. Accordingly, there is no further emphasis on this point.

7. Labor productivity directly affects economic growth, in both models and reality. The rate of productivity growth for the economy as a whole can be associated with improvements in inputs (e.g., a more educated labor force), or changes in the ratio of one input to another, (e.g., capital deepening). The work of Denison decomposes the sources of productivity change into about 20 different such improvements (Denison, 1974.) As used in

climate change models, this component of technological change usually shows up as changes in labor productivity (that is, output per hour of labor), and is determined or assumed. Improvements in labor productivity may cause the economy to grow more rapidly, and increase the use of carbon-based fuels, so the likely effect of an increase in labor productivity would be an increase in carbon emissions. Because the focus of this paper is on how models depict technological change that reduces carbon emissions, there is little further emphasis on this point.

8. There is a fifth way that models depict the rate of technological change that is indirect and therefore not highlighted in the literature. That is through the responsiveness of changes in demand to energy prices. As energy price increases become embedded in both goods used to produce other goods and final consumption goods, firms and consumers will economize on these now more expensive goods and services. These shifts in both final consumption and production will alter the structure of the economy over time, and bring about technological change that would otherwise not have occurred. These shifts are additional to the two mechanisms discussed in the text below.

9. The wind and PV cost numbers do not include back-up systems to cover down periods. While this is not a problem for marginal deployment, it becomes an important issue when significant displacement of conventional capacity is addressed.

10. *The Economist,* March 6, 1999, p. 72. This study originally appeared in Arora, A., R. Landau, and N. Rosenberg, eds. 1999. Chemicals and Long-Term Economic Growth. *Economic Focus.* Wiley Interscience, New York, NY.

11. The references for the table are as follows:

Barns, D.W., J.A. Edmonds, and J.M. Reilly. 1992. *The Use of the Edmonds-Reilly Models to Model Energy Related Greenhouse Gas Emissions.* OECD, Economics Dept., Working Paper No. 113, Paris. The Edmonds-Reilly model is found in Edmonds, J.A., and J.M. Reilly. 1985. *Global Energy: Assessing the Future.* Oxford University Press, New York.

Burniaux, J.M., J.P. Martin, G. Nicoletti, and J.O. Martins. 1990. *The Costs of Policies to Reduce Global Emissions of CO_2: Initial Simulation Results with GREEN.* OECD, Dept. of Economic Statistics, Resource Allocation Division. Working Paper No. 103. OECD/GD (91), Paris.

Dowlatabadi, H. 1998. Sensitivity of Climate Change Mitigation Estimates to Assumptions About Technological Change. *Energy Economics* 20(5/6): 472–492.

Edmonds, J., H.M. Pitcher, D. Barns, R. Baron, and M.A. Wise. 1995. Modeling Future Greenhouse Gas Emissions: The Second Generation Model Description. In *Modeling Global Change.* United Nations University Press, Tokyo, Japan.

Goulder, L.H., and S.H. Schneider. 1999. Induced Technological Change and the Attractiveness of CO_2 Abatement Policies. *Resource and Energy Economics* 21: 211–53.

Goulder, L.H., and K. Mathai. 1998. *Optimal CO_2 Abatement in the Presence of Induced Technological Change.* Dept. of Economics, Stanford University, Stanford, CA.

Grubb, M., T. Chapuis, and M.H. Duong. 1995. The Economics of Changing Course: Implications of Adaptability and Inertia for Optimal Climate Policy. *Energy Policy* 23(4/5): 417–431.

Grübler, A., and A. Gritsevski. 1997. *A Model of Endogenous Technological Change Through Uncertain Returns on Learning (R&D and Investments).* International Institute for Applied Systems Analysis, Laxenburg, Austria.

Jorgenson, D.W., and P.J. Wilcoxen. 1990. *Reducing U.S. Carbon Dioxide Emissions: The Cost of Different Goals.* Harvard University Press, Cambridge, MA.

Manne, A.S. 1981. *ETA-MACRO: A User's Guide.* EA-1724. Electric Power Research Institute, Palo Alto, CA.

Manne, A.S., and R.G. Richels. 1999. The Kyoto Protocol: A Cost-Effective Strategy for Meeting Environmental Objectives? J.P. Weyant, ed. The Costs of the Kyoto Protocol: A Multi-Model Evaluation, Special Issue of *The Energy Journal.*

Manne, A.S., and C.O. Wene. 1994. MARKAL/MACRO: A Linked Model for Energy-Economy Analysis. *Advances in Systems Analysis: Modelling Energy-Related Emissions on a National and Global Level.* J.F. Hake, M. Kleemann, W. Kuckshinrichs, D. Martinsen, and M. Walbeck, eds. Jülich, pp. 153–191.

Messner, S. 1997. Endogenized Technological Learning in an Energy Systems Model. *Journal of Evolutionary Economics* 7(3): 291–313.

Nordhaus, W.D. 1994. *Managing the Global Commons: The Economics of the Greenhouse Effect.* MIT Press, Cambridge, MA.

Peck, S.C., and T.J. Teisberg. 1992. CETA: A Model for Carbon Emissions Trajectory Assessment. *Energy Journal* 13(1): 55–77.

Rutherford, T. 1992. *The Welfare Effects of Fossil Carbon Reductions: Results From a Recursive Dynamic Trade Model.* Working Paper 112, OECD/GD(92)89, OECD, Paris.

The AES Corporation and Energy and Environmental Analysis, Inc. 1991. *Fossil-2 Energy Policy Model Documented Listing* (6 volumes), Arlington, VA.

12. Hourcade, J.C., K. Halsnaes, M. Maccard, W.E. Montgomery, R. Richels, J. Robinson, P.R. Shukla, and P. Sturm. 1996. A Review of Mitigation Cost Studies. Chapter 9 in *Climate Change 1995: Economic and Social Dimensions of Climate Change.* J.P. Bruce, H. Lee, and E.F. Haites, eds. Cambridge University Press, Cambridge, UK, pp. 297-366. Hourcade, J.C., R. Richels, and J. Robinson. 1996. Estimating the Costs of Mitigating Greenhouse Gases. Chapter 8, ibid., pp. 263-296. More condensed reviews are available in the 1996 *Energy Policy* 24(10/11): Hourcade, J.C., and J. Robinson, Mitigating Factors: Assessing the Costs of Reducing GHG Emissions, pp. 863–873; Richels, R., and P. Sturm, The Costs of CO_2 Emissions Reductions: Some Insights From Global Analyses, pp. 875–887.

13. Technological change that reduces the unit costs of fossil fuels (e.g., improved oil exploration technology), thus making such fuels cheaper to use, increases the use of fossil fuels and makes mitigation more difficult.

14. Barns, et al., 1992.

15. Fossil-2 is a systems dynamics model originally developed at Dartmouth, then revised as Fossil-2 for forecasting and policy analysis at the U.S. Dept. of Energy's Office of Policy, Planning and Evaluation. After modification to the transportation and utility sectors, it was renamed IDEAS, and currently is available through OnLocation, Inc., located in Washington, D.C. (http://www.onlocationinc.com).

16. Global 2100 is documented in Manne, A.S., and R.G. Richels. 1992. *Buying Greenhouse Insurance,* MIT Press, Cambridge, MA.

17. The Canadian Recursive Trade Model (CRTM) is reported in Rutherford, T. 1992.

18. Weyant, J.P. 1997. *Preliminary Results from EMF-14 on Integrated Assessment of Climate Change.* Energy Modeling Forum, Stanford University (April).

19. Hourcade, et al., 1996, p. 322.

20. Burniaux, et al., 1990.

21. MacCracken, C.N., S.L. Legro, J.A. Edmonds, and W.U. Chandler. 1998. *Climate Change Mitigation Costs: The Roles of Research and Economic Reform.* Pacific Northwest National Laboratory, Washington, D.C. The technology cases are described in pages 2–5, the costs at the global level in pages 5–12.

22. Grubb et al., 1995.

23. Grubb et al., 1995.

24. Dowlatabadi, H., and M. A. Oravetz. Understanding Trends in Energy Intensity: A Simple Model of Technological Change, *Energy Policy*, in press.

25. Goulder and Mathai, 1998.

26. Goulder and Schneider, 1999.

27. This model is a computable general equilibrium model with output derived from knowledge capital, regular capital, labor, two forms of energy (carbon emitting and alternative energy), and two intermediate goods (energy intensive and not energy intensive). Knowledge capital can be either specific to the industry (and thus the industry can appropriate R&D benefits) or "spillover" knowledge that benefits other industries as well. The only policy instruments are taxes on output and subsidies on R&D. The model is run to generate base case simulations both with and without induced technological change, and observe the effects of a carbon tax by comparing tax simulations to each of these base cases. Simulations also are undertaken to explore the impact of spillover knowledge and R&D subsidies in various industries. Generally, there are different effects on different industries, both in terms of output and in terms of investment in R&D.

28. Messner, S. 1997.

29. Grübler and Gritsevski, 1997.

goods for services in the United States could provide an explanation for part of the recent energy-use trends (assuming goods are more energy-intensive than services). Preliminary estimates of this shift can be found in the import share of total consumption of goods and services in the economy.

All told, in a world with low-cost information, dis-intermediated energy markets, and a fungible capital stock, the economic costs of climate change policy may be lower than we might otherwise think.

Endnotes

1. This paper focuses on the U.S. economy and climate change policy.

2. Marginal cost is the extra cost of producing an extra unit of output.

3. See Triplett (1998) for more information on the Solow Productivity Paradox.

4. Here, productivity is the extra dollar of GDP output per hour of labor employed.

5. It is important to note that despite the declines in energy intensity measures, total energy use is rising.

6. The Senate testimony can be found at http://www.house.gov/reform/neg/hearings/index.htm. Heated discussions sparked by Mills' arguments can be found at http://www.rmi.org/images/other/E-MMABLInternet.pdf.

7. This is in contrast to "top-down" models that use historical data and aggregate indices to predict how the economy, as a whole, will react to changes in price.

8. To his credit, Romm describes his hypothesizing as a "scenario" as opposed to a forecast, despite the obvious attachment to it. A scenario hypothesizes future outcomes of social, environmental, and economic events to create a number of scenarios defining a future world. A forecast, on the other hand, uses mathematical formulas and/or historical data and extrapolates this information to the future to make predictions.

9. The authors wish to thank Alan Sanstad for summarizing these results. See Sanstad and Laitner (2000) for the entire paper.

10. A futures contract (also known as a forward market) is any transaction that involves a contract to buy or sell commodities or securities at a fixed date at a price agreed in the contract.

11. There was, however, a rapid decrease in funding in the 1980s for energy efficiency measures. Only in the last five years have companies begun to place more resources into energy management operations.

12. 1997 is the most recent data available for estimates of fixed gross U.S. investment. New data published by BEA is expected in early 2001.

References

Bureau of Economic Analysis. 1999. *Fixed Reproducible Tangible Wealth of the United States, 1925–1997.* Washington, D.C. (April).

Bureau of Economic Analysis. 2000a. Fixed Assets and Consumer Durable Goods, 1925–1999. *Survey of Current Business.* Washington, D.C. (September).

Bureau of Economic Analysis. 2000b. *Current Cost Net Stock of Private Fixed Assets, by Industry, 1947–99.* See http://www.bea.doc.gov/bea/.

Bureau of Economic Analysis. 2000c. Current Business. *Survey of Current Business.* Washington, D.C. (September).

Bureau of Labor Statistics. 2000. *Major Sector Productivity and Costs Index.* See http://146.142.4.24/cgi-bin/surveymost?bls.

Byrnjolfsen, E., and L.M. Hitt. 2000. *Computing Productivity: Firm-Level Evidence.* See http://ecommerce.mit.edu/erik/cp.pdf.

Energy Information Administration. 2000. *Annual Energy Outlook.* See http://www.eia.doe.gov/oiaf/aeo/demand.html.

Coase, R.H. 1937. The Nature of the Firm. *Economica* 4: 386–405 reprinted, pp. 331–351. In American Economic Association. 1952. *Readings in Price Theory.* Irwin, Chicago.

Gordon, R.J. 2000. *Does the "New Economy" Measure Up to the Great Inventions of the Past?* See http://faculty-web.at.northwestern.edu/economics/gordon/351_text.pdf.

Greenspan, A. 1999. *Monetary Policy Testimony Report to the Congress, Committee on Banking and Financial Services.* See http://www.bog.frb.fed.us/BOARDDOCS/HH.

Howarth R.B., and A. Sanstad. 1995. Discount Rates and Energy Efficiency. *Contemporary Economic Policy* 13(3): 101–109 (July).

Jorgenson, D.W., and K.J. Stiroh. 1999. *Information Technology and Growth.* Harvard University Working Paper, Cambridge, MA.

Jorgenson, D.W., and K.J. Stiroh. 2000. *Raising the Speed Limit: U.S. Economic Growth in the Information Age.* Mimeo. Harvard University, Cambridge, MA.

Koomey, J. 1999. Personal communication (December).

Koomey, J. 2000. *Rebuttal to Testimony on Kyoto and the Internet: The Energy Implications of the Digital Economy.* See http://enduse.lbl.gov/Info/annotatedmillstestimony.pdf.

Mills, M. 2000. *Testimony before the Subcommittee on National Economic Growth, Natural Resources, and Regulatory Affairs.* See http://www.house.gov/reform/neg/hearings/index.htm.

Oliner, S.D., and D.E. Sichel. 2000. *The Resurgence of Growth in the Late 1990s: Is Information Technology the Story?* See http://www.federalreserve.gov/pubs/feds/2000/200020/200020pap.pdf.

Romm, J., A. Rosenfeld, and S. Herrmann. 1999. *The Internet Economy and Global Warming.* See http://www.cool-companies.com/ecom/index.cfm.

Sanstad, A., and S. Laitner. 2000. *Information Technology and Energy Use if the New Economy Follows the Old Principles: Some Historical Trends.* Paper presented at the Eastern Economic Association, Arlington, VA.

Solow, R. 1987. Review in NY Times Book Review, *New York Times* (July 12).

Stiroh, K. 1998. Computers, Productivity, and Input Substitution. *Economic Inquiry* 36(2): 175–191.

Triplett, J.E. 1998. *The Solow Productivity Paradox: What Do Computers Do to Productivity?* Working Paper. Brookings Institution, Washington, D.C.

Weyant, J.P. 2000. *An Introduction to the Economics of Climate Change Policy.* The Pew Center on Global Climate Change, Arlington, VA.

International Emissions Trading

Jae Edmonds, Michael Scott, Joseph M. Roop, Chris MacCracken

International emissions trading holds the potential of reducing the costs of controlling world emissions of greenhouse gases (GHGs). Because there are large differences in the costs each nation would bear in achieving emissions reductions on its own, a regime allowing nations with higher costs to pay those with lower costs to achieve reductions can lower the costs of emissions control.

However, the potential gains from emissions trading may be very unevenly distributed across the world's participants. While all of the parties to an agreement stand to gain collectively under trade in emissions rights, as compared with independent compliance (i.e., each country meeting its obligations alone), non-participants in a emissions trading agreement may either benefit or not, depending on their own particular circumstances.

A number of global economic models have been used to estimate the effects of emissions trading. Empirical results from these models can be summarized as follows:

- Costs of controlling carbon emissions would be significantly lower if trade in carbon emissions allowances were permitted than if each nation had to meet its emissions reductions alone. The broader the emissions trading possibilities, the lower the costs of control.

- All parties with GHG emissions mitigation obligations — permit buyers as well as sellers — would benefit from emissions trading.

- Parties without obligations may be better or worse off under a trading regime relative to a regime that does not allow trading. However, given the existence of a regime that allows emissions trading, parties without obligations will be better off trading (i.e., selling emissions reductions) than not trading.

- The costs of fossil fuels could be affected by emissions control and emissions trading. For example, the price of fossil fuels is expected to be lower under a regime that prohibits emissions trading, and energy-exporting countries would be worse off relative to a no-control case. Trading would mitigate this effect.

- Gains from emissions trading are sensitive to the difference between the base case and target emissions and to the difference in marginal (incremental) abatement costs among countries.

- The actual cost savings from trade in emissions are likely to be less than the theoretical savings shown in most analyses performed with integrated

This paper is a condensed and updated version of a report published in December 1999 by the Pew Center: International Emissions Trading & Global Climate Change. The Pew Center and the authors appreciate the valuable input of several reviewers of this paper, including Eric Haites and Elizabeth Malone.

assessment models[1] because these models do not include the various measurement, verification, trading, and enforcement costs that would characterize any real trading system.

Details of the trading rules will influence the effectiveness of trading as well as the total gains from emissions trading and the distribution of such gains. Key issues include definitions of the emissions rights to be traded, the rules for crediting carbon sinks, and regulations governing participation in the trading framework. In addition to the above, other issues such as the behavior of countries that have significant market power in supplying emissions credits and the transaction costs associated with trading and enforcement could significantly increase the costs of mitigation, compared to the most favorable case, and could reduce the amount and benefits of trading.

Gains from Trade

Most greenhouse gases (GHGs) mix rapidly in the atmosphere, persist for decades or more, and are expected to affect climate. Because GHGs lead to global effects, it does not matter from where GHG reductions come. Thus the effect of trading on climate is neutral as long as global GHG emissions are the same with or without trading.

Countries and regions differ in their degree of dependence on production activities that emit GHGs, the efficiency with which they produce goods and services per ton of GHGs emitted, and the ease with which they can change their current dependency and efficiency. Therefore, it is only natural that they would experience different marginal (incremental) abatement costs when they attempt to limit their emissions.

The principle of gains from trade states that whenever two or more entities are obligated to produce a fixed amount of a good or service and their marginal costs of production differ, both can be made better off through trade. The gains can be realized if the entity with the higher marginal costs reduces production and pays the entity with the lower marginal costs to increase production. The principle depends on the difference in marginal costs, not the absolute cost. It is equally valid for two low-cost producers or two high-cost producers, so long as costs differ.[2]

GHG emissions control would be less costly overall if those countries and organizations that have relatively high costs of emissions reductions were allowed to pay those with lower costs to undertake more of the reductions. These cost savings (i.e., the gains from trade) would be realized if markets could be established that allowed trading of "permits" or rights to emit GHGs. Of course, the principle is silent on the question of how the savings are actually shared (i.e., who pays whom how much). Nevertheless, the gains from emissions trading are still potentially available to be shared regardless of how responsibility for mitigation is assigned.

Let us begin with an example in which there are two countries emitting carbon and that an agreement to limit emissions exists. One country has higher domestic costs of carbon

Climate Change: Science, Strategies, & Solutions

can benefit from "leakage," in which carbon-emitting activity that is constrained in the industrialized countries migrates to other countries that are not constrained.[9] In effect, the Annex I countries face an economic penalty for using fossil fuels, while non-Annex I countries face no such penalty. Relative to the base case, Annex I fossil fuel-intensive economic activity becomes less profitable and declines, reducing Annex I demand for fossil fuels, putting downward pressure on world fossil fuel prices — especially prices for oil and coal — and shifting some fossil fuel-intensive economic activity to non-Annex I countries where it is relatively more profitable. As a result of lower world fossil fuel prices and reduced economic output in the OECD countries, energy exporting countries (which face reduced prices for their major exports) also show lower economic output relative to the base case. Those non-Annex I nations that are the OECD's principal trading partners (who face declining markets) could also show lower output relative to the base case.

Relative to the base case, the principal economic beneficiaries in the no-trade case would be those countries that use relatively large quantities of fossil fuels and that do not rely extensively on OECD markets. The actual economic impact on a given country would depend on how much carbon mitigation affects international fossil fuel prices. If world oil prices were lowered substantially by carbon mitigation in Annex I countries, several of the non-Annex I countries would stand to benefit from leakage

Table 3

Sensitivity of Economic Activity to the Effects of Carbon Abatement on World Oil Prices: Carbon Control to 1990 Levels with No Carbon Trading

Region	Percent Change in GDP with Base Case Oil Prices		Percent Change in GDP with Oil Prices 10% Below Base		Percent Change in GDP with Oil Prices 20% Below Base	
	2010	2020	2010	2020	2010	2020
United States	−0.24%	−0.29%	−0.24%	−0.28%	−0.24%	−0.28%
Japan	−1.12	−1.41	−1.10	−1.36	−1.07	−1.32
Western Europe	−0.80	−1.41	−0.76	−1.33	−0.72	−1.26
Canada	−1.88	−2.89	−1.86	−2.89	−1.84	−2.89
Australia	−0.73	−0.77	−0.73	−0.76	−0.73	−0.76
Former Soviet Union	0.00	−0.33	0.00	−0.33	0.00	−0.33
Eastern Europe	0.00	−0.99	0.00	−1.00	−0.01	−1.00
China	0.00	0.00	0.02	0.02	0.03	0.03
India	0.00	0.00	0.08	0.08	0.17	0.16
South Korea	0.00	0.00	0.12	0.22	0.23	0.43
Mexico	0.00	0.00	0.04	0.09	0.07	0.12
Rest of World	0.00	0.00	0.04	0.07	0.08	0.14

Note: All figures shown in the table are percentage changes in gross domestic product (GDP) relative to a no-control case. In the base case, oil prices are the same as when no carbon abatement is attempted: prices are 10 percent and 20 percent lower in the other two cases. Non-Annex I countries have a 0 percent change in GDP with base case oil prices due to energy prices in SGM being determined exogenously.

Source: SGM.

of energy-intensive economic activity and from lower oil prices (Table 3). Thus, Korea and India, for example, show higher gross domestic product (GDP). The effect on China and Mexico would be more modest, as China restricts oil imports and Mexico is an oil-exporting country. Because world oil prices are likely to be higher in a regime with carbon reduction and trading than in a regime with carbon reduction and no trading, leakage would be less with trading than without it. Thus, while higher prices would result in less world carbon emissions and could benefit some non-participants, they would not benefit other non-participants.

Annex I Trading

Trade significantly reduces the costs of controlling carbon emissions. Table 4 shows that Annex I countries would reap substantial economic gains from emissions trading. The gains are about $20 billion (1992$) worldwide in the year 2010, a 30 percent reduction in the costs that would have been incurred in the absence of emissions trading.

The effect on carbon permit prices is substantial as well. For example, the market-clearing price of permits (the price that makes quantity demanded equal to quantity supplied and also equalizes marginal abatement costs among regions) is about $106/ton, compared with marginal abatement costs ranging from $0/ton in the Former Soviet Union (FSU) to $304/ton in Japan without emissions trading.

Table 4

Carbon Reduction and Costs Under Annex I Trading
by Region in Year 2010 (1992$)

Region	Domestic Carbon Reduction (million tons C)	Permit Purchases or Sales (million tons C) (negative=sales)	Domestic Total Abatement Cost ($ billion)	Total Direct Mitigation Obligation Cost ($ billion)	Total Direct Mitigation Obligation Cost No Trade ($ billion)	Gains from Trade ($ billion)
United States	386	75	$18.2	$26.2	$27.6	$1.5
Japan	46	49	2.2	7.4	11.8	4.4
Western Europe	129	38	6.1	10.2	11.4	1.2
Canada	28	23	1.3	3.7	5.2	1.5
Australia	20	5	0.9	1.4	1.5	0.1
Former Soviet Union	162	−162	7.2	−10.0	0.0	10.0
Eastern Europe	28	−28	1.5	−1.5	0.0	1.5
Total	800	—	$37.5	$37.5	$57.5	$20.0

Note: Marginal cost = $106/ton C in all regions.

The model solves for a return to 1990 emissions with competitive permit supply. All values in the table except marginal abatement cost are relative to a business-as-usual no-control case. Total direct mitigation cost equals domestic abatement cost plus cost of permits purchased, minus revenues from permits sold. The analysis assumes no restrictions on permit supply or demand; see text for a discussion of such restrictions. Both the FSU and Eastern Europe are treated as though 2010 instead of 1990 is the base year for purposes of emissions reduction requirements, so there are no "base mitigation credits" available. In addition, only the distribution of emissions reduction and total cost, and not the total amount of emissions reduction, is influenced by the availability of trading. Note: Columns may not add to total due to rounding.

Source: SGM.

Total costs in this case include not only the amount spent on domestic emissions control, but also the amount spent or earned (depending on whether a country is a buyer or seller) from permits purchased as a substitute for domestic emissions control. The costs of abatement plus permit purchases are called mitigation obligation costs.

The cost of returning emissions to 1990 levels is met entirely within the Annex I countries in this case since there is no obligation on the part of the rest of the world to reduce GHG emissions.

World Trading

The gains from emissions trading are potentially much greater if the group of nations undertaking reductions could be expanded to include the non-Annex I countries as well as the Annex I countries. Although under the Kyoto Protocol non-Annex I countries currently have no obligation to control GHG emissions, this hypothetical case treats non-Annex I countries as if they agreed to create permits equal to their annual base case emissions and all owed these permits to be traded internationally.[10] Table 5 shows low marginal carbon control costs ($24/ton, in contrast to $106/ton with Annex I trading, or $0/ton in the FSU to $304/ton in Japan with no trading).

Table 5

Carbon Reduction and Costs Under World Trading

by Region in the Year 2010 (1992$)

Region	Domestic Carbon Reduction (million tons C)	Permit Purchases or Sales (million tons C) (negative=sales)	Domestic Total Abatement Cost ($ billion)	Total Direct Mitigation Obligation Cost ($ billion)	Total Direct Mitigation Obligation Cost No Trade ($ billion)	Gains from Trade ($ billion)
United States	110	352	$1.4	$10.0	$27.6	$17.6
Japan	13	82	0.2	2.2	11.8	9.7
Western Europe	38	129	0.4	3.6	11.4	7.8
Canada	8	43	0.1	1.1	5.2	4.1
Australia	6	19	0.1	0.5	1.5	1.0
Former Soviet Union	53	−53	0.6	−0.7	0.0	0.7
Eastern Europe	7	−7	0.1	−0.1	0.0	0.1
China	289	−291	3.1	−4.0	0.0	4.0
India	191	−191	1.7	−3.0	0.0	3.0
South Korea	4	−4	0.1	−0.0	0.0	0.1
Mexico	22	−22	0.2	−0.3	0.0	0.3
Rest of World	57	−58	0.7	−0.7	0.0	0.7
Total Reductions	**800**	—	**$8.6**	**$8.6**	**$57.5**	**$49.0**

Note: Marginal abatement cost = $24/ton C in all regions.

Table represents a return to 1990 emissions and competitive carbon permit supply. All figures in the table except marginal abatement costs are relative to a business-as-usual no-control case. Details may not add to totals due to rounding.

Source: SGM.

The decline in overall cost is the result of engaging the world economy in the search for emissions abatement opportunities. Control costs in developing countries are cheaper and control opprtunities are greater, largely due to relatively low costs of replacing their current outdated and inefficient capital stock (compared to the developed world), their inability to finance new low-GHG-emitting technologies, and their current lack of laws requiring cleaner technologies. In this case non-Annex I nations could search for low-cost abatement opportunities, create an excess in emissions permits relative to their emissions, and sell the excess permits at the world price, which would more than pay for the abatement.

The overall gains from emissions trading are $49 billion with world trading, an improvement of $29 billion relative to Annex I trading. The broader trading regime takes advantage of the more abundant abatement opportunities and greater disparity of marginal abatement costs among the Annex I and non-Annex I countries to achieve greater cost savings. Permit buyers like the United States and other OECD countries achieve their emissions obligations at much lower costs than in the no-trade or Annex I trading cases. The FSU and Eastern Europe would benefit from world trading, but their gains would be smaller than under Annex I trading, where they would be the only net suppliers of permits. In effect, they lose out to less expensive competition. Non-Annex I regions (which include China and the countries listed below it on Table 5) on balance benefit relative to the no-trade case from undertaking emissions reductions and then selling their permits to the Annex I countries.

These calculations estimate the potential value of extending participation to the entire world. Real world gains from emissions trading will likely be smaller. Costs of monitoring and compliance will certainly increase as emissions trading expands from the narrow domain of Annex I nations with extensive monitoring and verification capabilities to encompass the entire world. If, as in the case of the Kyoto Protocol, non-Annex I nations have no formal emissions limitation obligation, mechanisms such as the Clean Development Mechanism (CDM)[11] will have to be used to approximate the case modeled here. With a second-best policy instrument such as the CDM, either the supply of credits and/or the actual environmental benefit (i.e., actual net national emissions abatement, as opposed to emissions abatement that may be achieved for particular projects) would be smaller.

Comparison Among Models of Carbon Trading

The findings described above are not unique to the SGM. Several modeling groups have undertaken empirical analyses to estimate the impact of carbon trading on emissions and costs of GHG abatement, and all have projected substantial economic benefits. Many of these analyses were performed in an attempt to understand the implications of the Kyoto agreement.

oil and gas by the Annex I countries would decline (3–25 percent, depending on the region) under a no-trade scenario, but other regions that are not constrained by the Kyoto Protocol would take up most of this consumption. Output of energy-intensive goods also would decline in the Kyoto-constrained regions (–$159 billion in 1992 dollars), but would expand significantly in the rest of the world (+$116 billion in 1992 dollars).

With Annex I trading in the EPPA model, the domestic price of coal in non-participant nations would fall relative to the base case in those regions where it is used, but the decrease would be only about half as large as under the no-trade scenario. International prices of oil and natural gas would remain virtually unaffected under Annex I trading relative to the base case. Thus, energy-exporting countries would see a much smaller decline in their revenues than when emissions trading is not permitted. With trading, the decline in the production of energy-intensive goods in the Annex I countries would be much smaller than when trade in permits is prohibited (only –$4 billion vs. –$159 billion in 1992 dollars). The rest of the world (mainly China and India) then would show a decline in the production of energy-intensive goods (–$12 billion in 1992 dollars) rather than the major increase shown in the no-trade case.

An analysis of carbon leakage by Charles River Associates (CRA)[20] shows that carbon mitigation by Annex I countries in the absence of emissions trading could have significant effects on the non-Annex I world. Non-Annex I energy-intensive, energy-importing countries that trade mostly with non-Annex I countries would benefit financially from carbon mitigation. Countries that benefit from permit trading tend to have the opposite characteristics. CRA's analysis of non-Annex I countries in the year 2030 shows that a handful of non-Annex I countries would benefit from non-participation in the no-trade case, ranging from Jamaica with 1.5 percent gain in GDP, to Ghana at about 0.1 percent gain. Losses range from small (Poland, –0.1 percent) to significant (United Arab Emirates, –3.3 percent). Because trading mitigates the reduction of energy prices due to carbon control, trade would reduce both gains and losses, but would not necessarily change losers to winners.

Models that include a more sophisticated treatment of international financial flows, such as G-Cubed, show additional effects. In such models, carbon mitigation in the no-trade case would have a negative impact on rates of return on capital in the Annex I countries relative to the non-Annex I countries. This would cause capital outflows to the non-Annex I world, which, in turn, would lead to exchange rate appreciation of these countries' currencies relative to the dollar, yen, and other Annex I country currencies.

This appreciation in exchange rates would strongly limit the non-Annex I countries' advantage in exports of carbon-intensive goods. However, it would have two other beneficial effects: non-Annex I dollar-denominated international debt would be less expensive; and imports of goods and services by non-Annex I countries from Annex I countries also would be less expensive. Both effects would increase domestic wealth in the non-OECD countries. Trade in permits would moderate, but would not eliminate, these effects.[21]

Table 9

Sensitivities of Total Cost of Control for Non-CO$_2$ Trace Gases Under the Kyoto Protocol with No Permit Trading (1992$ billion)

	SGM (Edmonds et al. 1998)			EPPA (Reilly et al. 1999)[a]		
Region	Infinite Marginal Cost	Proportional Cost	$0 Cost	Infinite Marginal Cost Multi-gas Target, CO$_2$ Control	Multi-gas Target and Control	CO$_2$ Target and Control Only
United States	$79	$53	$25	$58	$35	$48
Canada plus Australia	22	15	4	23	10	20
Japan	37	31	17	54	39	43
Western Europe	14	11	1	56	30	38
Eastern Europe	0	0	0	15	12	11
Total	$152	$111	$47	$206	$127	$160

Note: [a]Includes contributions of terrestrial carbon sinks.

Costs shown are total direct costs for control of Annex I GHG emissions to about 5.2 percent below the 1990 level. The column titles for SGM refer to the marginal abatement costs for non-CO$_2$ trace gases (e.g., the marginal cost of control is infinite for these gases). EPPA column titles are self-explanatory. There is no case in the EPPA analysis for which marginal cost of controlling non-CO$_2$ trace gases is zero. Details may not sum to totals due to rounding.

Other Key Sensitivities

Estimation of the gains from trade discussed in this section is sensitive to a number of key assumptions. These include the impacts of non-CO$_2$ trace gases and carbon sinks on compliance costs.

- **Non-CO$_2$ Trace Gases:** Emissions of all non-CO$_2$ trace gases — which include CH$_4$, N$_2$O, PFCs, HFCs, and SF$_6$ — are projected to grow substantially unless they are controlled. Multi-gas control (of CO$_2$ and non-CO$_2$ trace gases) has been explicitly examined in both the SGM and EPPA models.

The EPPA analysis looked at the infinite cost case, but also explicitly examined control of non-CO$_2$ trace gases using a marginal cost relationship for each gas in each region. Table 9 reports some of these cost sensitivities. The cases in Table 9 are:

- Costs of controlling non-CO$_2$ trace gases are proportional to the costs of controlling CO$_2$.

- Non-CO$_2$ trace gases can be controlled at zero marginal cost, so control of all GHGs costs only as much as controlling CO$_2$ alone. This places a lower bound on the cost of multiple-gas control.

- Control of non-CO$_2$ trace gases has an infinite marginal cost, so all commitments must be met from CO$_2$ alone. The effective CO$_2$ control target becomes more stringent and places an upper bound on the cost of multiple-gas control.

Because the marginal costs of control for non-CO$_2$ trace gases are likely to be less than those for CO$_2$, meeting the requirements of the Kyoto Protocol for non-CO$_2$ trace gases may be less expensive without trade, and trading less attractive for these gases than for CO_2. No explicit analysis was found of the cost effects of trading permits for emissions of these gases.[22]

- **Sinks:** Atmospheric GHG concentrations change not only because of emissions due to fossil fuels, but also because of changes in terrestrial sources and sinks from changes in land use or agriculture and forestry practices. The Kyoto Protocol provides credit for new "direct human-induced land-use change and forestry activities, limited to afforestation, reforestation, and deforestation since 1990" — that is, terrestrial carbon sinks established after 1990 (Article 3.3). Sequestration in soils and other reservoirs are not yet considered.

Strict interpretation of Article 3.3 leaves little room for counting sinks toward emissions mitigation in Annex I nations, with the exception of Australia, which has net land-use emissions in 1990. A strict interpretation of Article 3.3 removes an important potential source of net greenhouse gas emissions from the accounts. A full accounting of all net emissions from land use changes could have a significant impact on both marginal and total costs in those cases where a country has significant terrestrial capacity available. In the case of Canada, for example, full allowance for terrestrial carbon sinks could provide a credit equivalent to 80 million tons of carbon emissions, enough to more than satisfy Canada's Kyoto obligations.[23] In the case of the Former Soviet Union and Eastern Europe, sinks offer up to 213 million tons of additional potential baseline credits that could be sold.

In an analysis using the SGM, Edmonds et al. conclude that, overall, full allowance for terrestrial carbon sinks could reduce the Annex I joint trading permit price from $73 to $23 for meeting the goal of emissions 5 percent below 1990 levels.[24] Entering their emissions estimates into the MIT Integrated Global Systems Model (IGSM), which takes account of climate and ecosystem effects as well as natural sources and sinks, Reilly et al. conclude that achieving the same reduction in warming in the year 2100 by control of only fossil CO_2 costs 60 percent more than if other GHGs and terrestrial sinks are considered.[25] The impacts of credits and sinks presumably would be larger still if credit could be taken for non-Annex I carbon sinks.

Some Institutional Issues in Carbon Trading

Numerous issues concerning the structure and details of an international carbon trading regime have yet to be worked out. These institutional issues are of concern because of their potential impact on the effectiveness and cost of trading and on the volume of permits traded.

The actual cost savings from trade in emissions will likely be less than the theoretical savings shown in most analyses performed with integrated assessment models because these models do not include the various measurement, verification, trading, and enforcement costs that would be characteristic of any real trading system. On the other hand, the gains from trade could be greater than the models predict since the models may not anticipate all the control options trade would encourage.

Some of the key issues are:

- whether countries' domestic control regimes will be compatible with trading;
- the effect of restrictions on permit availability or demand;
- impacts of international transfer payments;
- measurement and reporting of emissions, sinks, and costs; and
- accountability and enforceability.

Compatibility of Domestic Control Regimes with Trading

Just because governments have rights that can be traded among themselves does not mean that actual control within each of these countries would take place via tradable allowances allocated to individual carbon emitters such as power plants or companies that supply fossil fuels. Each country would be free to choose its own domestic policy mix for controlling GHG emissions.[26] Emissions trading is only one of three approaches to domestic controls. The others are: (1) taxes on GHGs, carbon, energy, fossil fuels or fossil-energy-related activities (either supply or consumption of fuels); and (2) command-and-control regulations that directly limit emissions or pre-scribe certain technologies or activities (e.g., regulating automobile fuel efficiency). A recent study shows that the full gains to trade cannot be realized unless all of the parties engaged in international emissions trading also employ a domestic marketable tradable permit system.[27]

Restrictions on Permit Availability and Demand

Restrictions on permit availability or demand due to regulation or monopolistic market behavior could drive up the price of permits and reduce the gains achieved from trade. All of the analysis thus far has assumed that the market for emissions permits functions smoothly and without restrictions. A study by Ellerman et al. (1998) notes that if usable world permit supply is low relative to its potential, then the world permit price required to meet Kyoto obligations rises from $31/ton of carbon (1992 dollars) with unconstrained world trading, to $55/ton with 50 percent availability and to $230/ton with only 5 percent availability.[28] Edmonds et al. (1998) examines this issue in the context of Annex I trading only and shows that the permit price rises from $73/ton with unconstrained Annex I trading to $113/ton if no permits were available from the FSU and Eastern Europe.[29] (For example, these countries might "bank" their permits for their own future use.) Some governments could utilize their position to gain monopoly power in the permit marketplace to limit the supply of permits and increase their price.

It is also possible that the kinds of sources that actually participate in emissions reduction and trading could be limited for reasons of administrative convenience, political expediency, or limited technological options, thereby reducing the potential supply of permits. The allocation of permits is extremely important in determining the cost of monitoring and compliance within the system. It is one thing to allocate permits "upstream" — that is, where carbon enters the economy at the point of extraction or import/export — and another to try to allocate permits at the point of combustion. The former has far fewer parties involved in a program of universal coverage than the latter. Systems that try to balance the emissions abatement budget on the backs of a subset of downstream economic activities can be very expensive. To illustrate the cost of narrowly focusing the emissions reduction burden, Edmonds et al. showed that electric utilities' marginal cost of emissions mitigation in the United States would rise to more than 250 percent of the no-trade case if the utility mitigation burden was arbitrarily raised to 70 percent of the total.[30] Although trading would help reduce the impact of such exemptions and technology limitations, the largest possible pool of potential permit suppliers would clearly be advantageous for reducing costs.

Total world mitigation costs would also rise if significant restrictions were imposed on the extent to which imported permits may be used to satisfy mitigation obligations.[31] As the allowed permit import percentage falls from unlimited to a limit of 25 percent, Ellerman et al. (1998) calculated that the price of permits in the world trading case (where the permit buyers satisfied their commitments 71 percent with imports) would fall from $31 to $4, while total world cost of abatement would rise from $14 billion to $70 billion.[32]

International Transfer Payments

While the pattern of trade in emissions depends on the initial allocation of permits, it is likely that there could be substantial transfers of wealth between some countries and regions associated with the trade in permits. The initial allocation and the rules of emissions trading will decide not only the number of permits traded but also whether a given country or region is a net seller or net purchaser of permits.[33] For example, under Kyoto rules, neither the countries with economies in transition (the FSU and Eastern Europe, which are predicted by many models to emit less carbon in the year 2010 than in 1990)[34] nor the non-Annex I countries (which have no obligation under Kyoto) would have to reduce carbon emissions in the year 2010. They might not have permits to sell, and they might not be willing to sell in any case.

If trade in carbon were confined only to the Annex I countries (i.e., excluding economies in transition and the non-Annex I countries), most modeling groups show that the United States would be a net seller of permits to Japan and Europe. However, if the economies in transition and (especially) the non-Annex I countries were allowed to trade carbon permits, the United States would be a net purchaser. One of the consequences, for example, of the United States' desire to purchase large numbers of permits from Eastern Europe and the FSU would be substantial capital flows into those countries from the United States. On the one hand, these flows would provide hard currency reserves necessary to rebuild these economies. On the other hand, the flows probably would strengthen the local currency against the dollar. This would help solidify the local standard of living and make importing easier, but would also make their exports less competitive. These potential large-scale financial impacts of changes in trade are not treated well in many of the current models.

Monitoring, Reporting, and Certification Costs

Transaction costs for monitoring, reporting, and certification could also limit the gains from trade from emissions trading. A structure for emissions monitoring, reporting, and certification must be specified as part of any carbon control system, with or without trading. Each Party included in Annex I (those with responsibility to reduce emissions) must establish a national system for estimation of sources and removals of GHGs. Following are some of the major institutional issues.[35]

Emissions can be monitored either directly using monitoring devices or indirectly using predictive methods (e.g., an emissions factor multiplied times fuel used). There is a trade-off between accuracy and cost. For example, continuous monitoring of smoke stacks provides more accurate measurements but is more costly than occasional air sampling or emissions estimates.

Self-reporting and certification by countries may take place at the national level, while actual emissions reductions and deliberate sequestration of carbon in sinks will occur at the project or company level. Thus, both the quality of the monitoring and certification program within a given country's borders and the accuracy and veracity of its reporting must be considered. Two approaches to the uncertainty created by less-than-perfect monitoring systems are to limit emissions control and trading only to those gases and sources that can be readily and reliably monitored, or to adjust measured emissions using techniques such as presumptive emissions factors. The "presumptive permits" could then be traded. If emissions control and trading are limited to only those gases that can be measured accurately, the potential gains from trade will also be limited. If a presumptive permits system is used, the actual effectiveness of the system may be compromised.

The effects of transaction cost premiums on the volume of trade in permits are illustrated in Figure 1. This figure begins with the volume of emissions permits that would be bought and sold in the year 2010 under an Annex I trading scheme, as in Table 4 (i.e., a total of 190 million tons traded; the U.S. would purchase 75 million tons worth of carbon permits), when

Figure 1

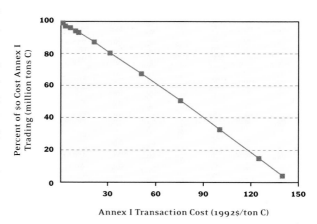

Transaction Cost versus Permit Trading

Percent of $0 Cost Annex I Trading (million tons C) — Annex I Transaction Cost (1992$/ton C)

transaction costs are zero. The figure then shows how, as transaction costs per ton of carbon increase, the volume of permits decreases as a percentage of the volume under the $0 transaction cost case. Any significant level of transaction costs would significantly limit the benefits of a trading program.

Liability, Accountability, and Enforceability

Accountability and enforceability would also be a problem should information concerning permit validity prove inaccurate. Accountability and enforceability are problems that must be solved in all emissions control systems. In the context of trading emissions permits, there is a specific question concerning whether the buyer or the seller is partly responsible (and legally liable) for the integrity or validity of the permit.[36] While trading provides some incentive to obtain accurate information concerning permit validity, the burden of diligence tends to fall most heavily on the party liable for permit validity.

An analysis by the United Nations Conference on Trade and Development notes that strict seller liability is preferable because it enhances the standardization and therefore the tradability of permits.[37] If compliance mechanisms are strong and it is easy to rectify any excess emissions — e.g., by frequent settling of accounts (i.e., many short commitment periods), subtracting emissions allocations in the following period, and adding a penalty — strict seller liability might be all that would be needed. However, with one long commitment period as under the current Kyoto Protocol, there is no way to penalize buyers who buy from suspect sources. Partial or total buyer liability would discourage purchasers from buying permits from suspect sources, but this feature also means that buyers would also discount deeply or not trade emissions because of the market risks involved with buying a permit that was later found to be unallowable. Governments and private parties can provide a useful role as independent verifiers, but such services are neither foolproof nor costless.

Conclusions

Trading emissions permits among nations offers substantial gains from trade because the marginal cost of mitigation differs substantially among the Annex I countries. Costs could be reduced about one-third by trading among the Annex I countries in this case, with more gains from trade to be had, the more severe the emissions target. The differences in costs are even larger between Annex I and the non-Annex I countries. Thus, worldwide emissions trading could reduce the costs of mitigation even more.

Additional effects worth noting include the impact on the economies of non-participants in trading schemes. Some models show that if independent mitigation is undertaken, fossil fuel prices will likely be reduced relative to the no control case. This will adversely affect the economies of energy exporting countries. What happens to other countries' economies is ambiguous. The net effect will depend on: (a) the change in world energy prices and energy intensity of each country's production; (b) the effect of the increases in Annex I countries' demand for imported energy-intensive products; and (c) the overall reduction in demand brought about by reductions in Annex I countries' GDPs.

Most models also show substantial financial wealth transfers from countries that are buyers of permits to those that are sellers, and some models show substantial changes in the terms of trade and in costs of externally held debt that at once add to seller wealth, reduce their costs of imports, and discourage their exports. Relative to independent mitigation, emissions trading moderates the reduction in fossil fuel prices, reduces incentives for carbon leakage, provides net financial flows to the non-Annex I countries, and — because GDPs will be higher — reduces the negative impact on trade.

The paper also notes that there are several critical design issues regarding emissions trading that could substantially reduce the potential gains from trade. Some issues can be resolved with good program design while others appear to be endemic and more difficult to resolve.

In summary, the ability to trade emissions permits offers significant gains from trade. However, various institutional questions, unresolved technical questions of measurement, and the very real costs of transactions and enforcement could substantially limit the extent of trading and the benefits of trade. Thus, policy-makers will have to pay careful attention to actual program design to realize these potential gains.

Endnotes & References

1. "Integrated assessment models" take into account some of the critical features of GHG emissions, the climate system, effects on natural and human systems, and the economy.

2. Because the differences in marginal cost can persist or even emerge as new technology is developed and adopted, trading should be viewed as a permanent option, not as a bridging strategy to be used only until new technologies are available.

3. The authors note the limiting nature of these assumptions. All of the issues enumerated entail costs, and these costs reduce the gains from trade. Their inclusion does not reverse the findings of the simpler case but does affect the magnitude of the gains from trade.

4. Society is accustomed to the situation in which an emissions limitation obligation comes with a set of associated emissions allowances. This need not be the case. Gains to trade occur regardless of the initial allocation of obligations and allowances. Of course, the distribution of emissions and allowances has profound economic implications for the participants.

5. The trade also works if only the high-cost controller initially has permits and the low-cost controller has no obligation to control and therefore, has no permits to sell. The trade could take the form of a purchase of carbon credits, wherein the low-cost controller undertakes an emissions reduction on behalf of the high-cost controller, and the high-cost controller gets a credit for this reduction that functions like a permit. There are important institutional differences between permits and credits.

6. The Second Generation Model (SGM) is described in Edmonds, J., H.M. Pitcher, D. Barns, R. Baron, and M.A. Wise. 1995. Modeling Future Greenhouse Gas Emissions: The Second Generation Model Description. In *Modelling Global Change*. United Nations University Press, Tokyo, Japan, (October). More recent updates may be found in Edmonds, J.A., S.H. Kim, C.N. MacCracken, R.D. Sands, and M.A. Wise. 1997. *Return to 1990: the Cost of Mitigating United States Carbon Emissions in the Post-2000 Period*. PNNL-11819, Pacific Northwest National Laboratory, Richland, WA; Edmonds, J.A., C.N. MacCracken, R.D. Sands, and S.H. Kim. 1998. *Unfinished Business: The Economics of the Kyoto Protocol*. PNNL-12021, Pacific Northwest National Laboratory, Richland, WA. The model's strengths include considerable detail on fuel choice and technology options and decision-making in the energy sector (especially electricity). Because it assumes full employment and does not include a financial or monetary sector, it does not deal with business-cycle issues such as unemployment, inflation, and consequences of international financial capital flows.

7. The Annex I countries are Australia, Austria, Belgium, Bulgaria, Canada, Croatia, Czech Republic, Denmark, Estonia, the European Community, Finland, France, Germany, Greece, Hungary, Iceland, Ireland, Italy, Japan, Latvia, Liechtenstein, Lithuania, Luxembourg, Monaco, the Netherlands, New Zealand, Norway, Poland, Portugal, Romania, Russian Federation, Slovakia, Slovenia, Spain, Sweden, Switzerland, Ukraine,

Climate Change: Science, Strategies, & Solutions

United Kingdom of Great Britain and Northern Ireland, and United States of America. Turkey and Belarus are Annex I nations that have not ratified the Convention and did not commit to quantifiable emissions targets. Some of the analyses in this report were actually done on the so-called Annex B countries. Annex B is the same list of nations as Annex I, excluding Turkey and Belarus.

8. In the real world, many problems confront the extension of participation from those with explicit emissions limitations to those without such limitations. This paper estimates the potential benefit of a successful extension to non-obligated parties' emissions. The Kyoto Protocol includes a Clean Development Mechanism (CDM) that allows non-Annex I parties to create "certified emissions reductions" which may be used by Annex I parties to "contribute to compliance with part of their quantified emissions limitation and reduction commitments." This paper uses the term "credits" to refer to "certified emissions reductions." Some would argue that this case corresponds to very broad utilization of the CDM and perhaps that the inherent limitations of credit trading in general, and of the CDM specifically, would not allow such broad utilization.

9. Leakage is measured as the difference between Annex I emissions reductions and global emissions reductions.

10. This is not necessarily the same thing as a CDM credit, which is created by the Kyoto Protocol. In a perfect world, the CDM and tradable emissions permit might be made to be the same. In the real world, there will be differences that may be large or small depending on a wide array of factors. Depending on how they are created, CDM credits may be either greater or smaller in magnitude than the emissions credits modeled here, and may represent either less or more emissions abatement.

11. Clean Development Mechanism (CDM) projects are defined in Article 12 of the Kyoto Protocol as projects undertaken in developing countries by Annex 1 countries that are intended to meet two objectives: (1) to address the sustainable development needs of the host country, and (2) to generate emissions credits that help Annex 1 countries meet their commitments under the protocol. (See also endnote 8.)

12. The references for the box are as follows:

Bernstein, P.M., W.D. Montgomery, and T.F. Rutherford. 1997. *World Economic Impacts of U.S. Commitments to Medium Term Carbon Emissions Limits*. Charles River Associates, Boston, MA (January).

Burniaux, J.M., J.P. Martin, G. Nicolette, and J.O. Martins. 1990. *The Costs of Policies to Reduce Global Emissions of CO_2: Initial Simulation Results with GREEN*. OECD Dept. of Economic Statistics, Resource Allocation Division, Working Paper No. 103. OECD/GD (91), Paris.

Edmonds et al., 1995. See endnote 6.

Manne, A.S., and R.G. Richels. 1997. On Stablizing CO_2 Concentrations: Cost-Effective Emissions Reduction Strategies. *Energy Journal* 18(3): 31–58.

McKibbin, W.J., R. Shackleton, and P.J. Wilcoxen. 1998. *The Potential Effects of International Carbon Emissions Permit Trading Under the Kyoto Protocol*. The Brookings Institution, Washington, D.C. (November 2).

Tulpulé, V., S. Brown, J. Lim, C. Polidano, H. Pant, and B.S. Fisher. 1998. *An Economic Assessment of the Kyoto Protocol Using the Global Trade and Environment Model*. ABARE, Canberra, Australia.

Yang, Z., R.S. Eckaus, A.D. Ellerman, and H.D. Jacoby. 1996. *The MIT Emissions Prediction and Policy Analysis (EPPA) Model*. Program Report 6, MIT Joint Program on the Science and Policy of Global Change, Massachusetts Institute of Technology, Cambridge, MA (May).

13. Computable general equilibrium models use a computerized solution algorithm to numerically solve a set of simultaneous equations that represent all of the sectors of the economy of each region. In all markets for energy and non-energy goods and services, quantities and prices adjust so that markets "clear" (supply equals demand) in every sector.

14. Results from the hypothetical protocol discussed in the previous section support this result even in the absence of an increase in allowable emissions under trade.

15. G–Cubed initially shows Japan as the highest-cost region; but then clean backstop fuels rapidly penetrate the Japanese market, reducing the required carbon tax. OECD GREEN's marginal abatement cost for Japan in the table appears unusually low. According to the OECD modelers, in the year 2005, the carbon price spikes very high in Japan (about $285/ton C). However, by 2010 in Japan the tax level decreases substantially, "as clean backstops are able to compete successfully due to the high price of energy in Japan. The backstops are not able to penetrate to any significant extent in the other OECD regions ...."

16. In the case of OECD GREEN, results were shown only for real income rather than GDP (which was not reported). In percentage terms, the principal difference is that income includes international payments such as payments for emissions permits, while GDP does not. The OECD GREEN analysis shows that the Former Soviet Union loses 1 percent of base case real income in the No Trade case because, as a major energy exporter, the region suffers from a decline in the price of internationally traded fuels. In contrast, it obtains a 3 percent gain in income in the Annex I case due to sales of permits and the higher export price of energy.

17. That is, in the No Trade case, the EPPA model shows that together, the FSU (whose emissions are not constrained because they are still below 1990 levels) and the non-Annex I countries emit 62 million more tons of carbon than they would if the OECD countries did not undertake emissions reduction. This offsets about 3

percent of the OECD carbon mitigation effort in 2010. See Ellerman, A.D., and A. Decaux. 1998. *Analysis of Post-Kyoto CO₂ Emissions Trading Using Marginal Abatement Curves.* Program Report 40, MIT Joint Program on the Science and Policy of Global Change. Massachusetts Institute of Technology, Cambridge, MA.

18. Ellerman, A.D., H.D. Jacoby, and A. Decaux. 1998. *The Effects on Developing Countries of the Kyoto Protocol and CO₂ Emissions Trading.* Program Report 41, MIT Joint Program on the Science and Policy of Global Change. Massachusetts Institute of Technology, Cambridge, MA.

19. Results on oil and gas prices vary among models and also vary according to how the experiments were done. The articles reviewed for this report did not generally show the effects on oil and natural gas prices; therefore, a more comprehensive assessment was not possible.

20. Bernstein et al., 1997.

21. McKibbin et al., 1998.

22. Reilly, J., R.G. Prinn, J. Harnisch, J. Fitzmaurice, H.D. Jacoby, D. Kicklighter, P.H. Stone, A.P. Sokolov, and C. Wang. 1999. *Multi-Gas Assessment of the Kyoto Protocol.* Program Report 45, MIT Joint Program on the Science and Policy of Global Change. Massachusetts Institute of Technology, Cambridge, MA. Reilly et al. report that they performed an analysis of Annex I trading with multiple gases and concluded that the number of base mitigation permits in the FSU and Eastern Europe are reduced and that the value of trading is reduced, relative to controlling CO₂ alone. In this case, the U.S. marginal cost with trading of multiple gases was almost the same as with CO₂ alone.

23. Edmonds et al., 1998. The analysis was based on estimated land use emissions credit supplied by the U.S. Council of Economic Advisors. The point here is not whether this figure is the "correct" one; rather, the point is that how rules are written regarding countries' ability to trade and take credit for terrestrial sinks could strongly affect compliance costs.

24. Edmonds et al., 1998. Note that the $73 in Table 6 is per ton of carbon equivalent for a 5.2 percent reduction below 1990 levels of all six gases by all Annex I parties. In Table 4 the marginal cost is $106 per ton for stabilization at 1990 levels, but for CO₂ alone. The base case is also slightly different between the two tables.

25. Reilly et al., 1999.

26. Joshua, F.T. 1998. *Greenhouse Gas Emissions Trading After Kyoto: Insights from UNCTAD's Research & Development Project.* See http://www.ecouncil.ac.cr/rio/focus/report/english/Unctad.html.

27. Hahn, R.W., and R.N. Stavins. 1999. *What Has Kyoto Wrought? The Real Architecture of International Tradable Permit Markets.* EPRI Global Climate Change Research Seminar, Columbus, OH (September 27-29).

28. Ellerman et al., 1998.

29. Edmonds et al., 1998.

30. Edmonds et al., 1998.

31. For example, the European Union has proposed that there be a trading rule that establishes, by mathematical formula, a limit on the amount of credits any Annex I nation could sell. The Kyoto Protocol specifies in Article 6 that "acquisition of emissions reduction units shall be supplemental to domestic actions," but does not spell out what "supplemental" actually means.

32. Ellerman et al., 1998.

33. Edmonds et al., 1998; Ellerman et al., 1998.

34. In several of the models, the initial allocation of emissions permits under the Kyoto Protocol means that countries encompassing the FSU and Eastern Europe will have emissions that are below their emissions limitation in 2010 without any explicit abatement efforts. This is due both to their current and expected poor economic performance and to economic restructuring away from energy-intensive industry. As a consequence, these regions are not expected to regain their 1990 emissions levels by 2010. The resulting emissions allowances, which could be sold on the open market, are a potentially important export for the FSU and Eastern Europe.

35. Grubb, M., A. Michaelowa, B. Swift, T. Tietenberg, and Zhong Xiang Zhang. 1998. *Greenhouse Gas Emissions Trading: Defining the Principles, Modalities, Rules, and Guidelines for Verification, Reporting, and Accountability.* United Nations Conference on Trade and Development (UNCTAD), Geneva, Switzerland.

36. For example, it is inherently more difficult to trade emissions credits (which are open-ended and difficult to verify) than to trade permits that represent GHGs measured against a known emissions commitment.

37. Grubb et al., 1998.

A Business Manager's Approach to Climate Change

Forest Reinhardt, Kimberly O'Neill Packard

Extremism and emotion have dominated debates about global climate change. It's time to cool down and take a level-headed business view of the problem: like any other change, climate change presents opportunities as well as risks to well-managed firms.

World business leaders convened in January 2000 in Davos, Switzerland, for their annual executive summit, with the threat of contagious financial panics like those of 1997 still looming, an ongoing financial crisis in Japan, and memories of the World Trade Organization's debacle in Seattle still fresh in their minds. Yet, when they polled themselves on the most pressing issue currently facing the world business community, they selected global climate change.

Although many business leaders recognize that climate change is an important issue, the relationships between climate change and the basic preoccupations of business managers — profit, risk, and asset values — remain incompletely understood. And managers analyzing these relationships have difficulty finding reasonable answers because the climate change debate in the public arena is often polarized, bitter, and unenlightening. The extremes are heard, but the reasonable middle ground is difficult to uncover.

On one extreme, business leaders hear that global climate change is a serious problem: that rising emissions of carbon dioxide in the atmosphere will lead to rising sea levels, increased spread of diseases, and more frequent storms and droughts. But the good news, in this view, is that it will not be expensive to reduce those emissions because potential efficiency gains are widespread, knocking the private cost of carbon reductions down below zero. Better than a free lunch, it is a lunch you get paid to eat!

From the other end of the spectrum, business managers hear that this idea of no-cost carbon reduction is wishful thinking. Of course it will cost money to run the economy on less carbon. If it didn't, more companies would have made carbon reductions already. But, this side continues, the high price of carbon reductions does not matter, because climate change is not going to be a significant global problem. Who cares if the solutions are expensive if there is no need to buy them?

The Pew Center and the authors wish to thank Robert Stavins, Tom Tietenberg, and David Vogel for their review of this report.

These polarized views have two things in common. First, each is basically optimistic: either the solutions are easy and cheap, or the climate problem does not exist at all. Second, each leads to very simple-minded prescriptions for business managers: either business managers are constrained only by their own myopia from devouring a succession of free lunches, or climate change can safely be ignored.

Maybe politicians have to put an optimistic spin on things — gloom is an unelectable platform — but business leaders need to see the problem clearly and steadily. And those on each extreme of the public climate change debate are likely to be right about their bad news and wrong about their good news. Climate change is a serious social problem. It requires attention. But even the most sensible combinations of efficiency improvements, changes in patterns of energy use, and adaptation to shifting climate will end up costing society a lot of money — as much, perhaps, as we have spent on air and water pollution over the past several decades. Further, it is disingenuous to pretend that we have a choice between reducing carbon emissions and living with the consequences of climate change. We will very likely end up doing both.

This means that the simple management prescriptions that arise from either extreme view are also misguided. While climate change cannot be ignored, the notion of free lunches everywhere violates even casual observation. But exactly because the change will be disruptive and difficult at the national and international levels, it presents enormous opportunities — as well as substantial risks — for firms in a wide variety of industries.

From a business strategy perspective, climate change presents managers with three distinct, but overlapping, problems.

The first problem is how to manage the risk of regulatory responses to climate change: the prospect of government interventions like taxes, incentives, and emissions limitations aimed at reducing the speed at which the climate alters. These government initiatives will both push up the expected value of carbon-intensive fossil fuel prices and further increase the already considerable uncertainty about those prices. This issue is especially acute for energy producers; for companies that are intensive consumers of energy, like airlines and trucking firms; and for companies whose products rely on energy, like makers of automobiles and appliances.

The second problem is how to cope with or capitalize on the impact of climate change itself: with rising average temperatures, the possibility of increasingly wide fluctuations around that rising average, and the possibility of more frequent and severe storms and droughts. Naturally this issue is of particular importance to insurance companies, managers of real estate, agricultural producers, tourism firms, and other businesses whose asset values depend on the climate.

The third problem for executives is how they should try to influence the international debate about climate change. While business leaders speak out on a number of public policy issues that affect their businesses, surprisingly few are on the record about climate change. We believe there is significant opportunity for a broader set of businesses to play an influential role in the climate change debate.

Fortunately, business managers know something about how to manage regulatory uncertainty, making investments in technologies that leave their companies hedged across different regulatory outcomes. Business leaders also know something about how to manage variable weather, using such diverse tools as insurance contracts, government transfer programs, and investments that improve buildings and infrastructure. Executives also have experience managing public opinion about their companies and the issues that affect them. By disaggregating the issues surrounding climate change, business managers can turn a single, unfamiliar, and threatening business problem into a set of three familiar problems, each of which is challenging but by no means intractable.

What does a sensible business strategy look like? It anticipates and takes advantage of price changes driven by market forces. It retains flexibility to respond to new scientific and regulatory developments. But, since a perfect hedge is impossible and undesirable, it bets on the most likely scientific and regulatory scenarios. It also seeks to influence the behavior of other actors — public and private — in order to support and reinforce the firm's own actions.

From a business point of view, changes in climate and changes in regulatory structure are fundamentally no different from any other external shocks to the system in which firms compete to create and capture economic value. Any such shock — a change in consumer tastes, the discovery of a new technology, an alteration of government tax policy — will increase or decrease the expected value of particular activities and particular assets. And it will increase or decrease the financial risks associated with any particular activity and the associated assets.

To anticipate and profit either from changes in regulation or changes in climate, managers have a variety of approaches at their disposal. Climate change and the regulatory response to it may:

- Create new opportunities for product differentiation.

- Give rise to new opportunities for a firm to reconfigure its markets so as to impose a cost disadvantage on its competitors.

- Create opportunities for cost savings within the firm. This notion is often called the "Porter hypothesis" after Harvard strategy scholar Michael Porter.[1] Controversy persists as to whether such savings opportunities are widespread across firms, but anecdotal evidence suggests that at least some firms can uncover them.

- Offer chances to improve risk management in ways that will benefit the company in the long term. Since scientific, regulatory, and market uncertainties are pervasive in the climate issue, business planning for climate change is like any risk management problem: executives need to decide which risks to shift, which to reduce, and which to hold.

Regulation-Sensitive Firms: Managing Regulatory Risk and Value

Given the magnitude of the potential effects of global climate change and the degree to which international climate change experts agree that climate change is a serious problem, it is highly likely that the international community will eventually take real action to enforce carbon reductions. Considerable uncertainties persist, however, about when this commitment to change will occur and the precise form that it will take. The prospect of regulation increases the expected cost of energy in the future, and it increases the uncertainty about that future cost. For most firms, it is this regulatory risk that poses the greatest threat to business as usual and that presents the most significant opportunities, and that therefore requires the most attention from senior management.

For some firms, these prospective regulatory changes present an opportunity to differentiate products and create more value for customers and shareholders. In general, a company's ability to differentiate products along environmental lines depends on whether the product market's characteristics allow the companies to find customers willing to pay a premium for environmental quality, to communicate credibly about the environmental attributes of its own offerings, and to protect itself against imitators.[2] For example, Ben and Jerry's and Patagonia have been successful selling high-end ice cream and outdoor wear products to affluent customers by appealing to their concerns for the environment. On the other hand, H.J. Heinz's StarKist subsidiary found, when it tried to reposition its product as "dolphin-safe," that consumers' willingness to pay for environmentally differentiated canned tuna was zero.

At first blush, climate change may seem to offer few opportunities for product differentiation. Energy companies selling gasoline at retail, for example, face markets more like StarKist's than like Patagonia's: the main customer purchase criteria are price and convenience, not "greenness." At the same time, however, government regulation aimed at climate change can create differentiation opportunities by changing the relative costs of energy and other goods.

Consumers, confronted with higher and more volatile energy costs, will want to substitute away from energy while still enjoying the same degree of mobility, comfort, and convenience. And businesses, facing the same increases in energy costs and cost uncertainty, will want to find ways to use less energy while delivering the same level of satisfaction to their own customers. Companies that are able to help consumers and businesses make this transition can benefit from regulatory changes aimed at the climate.

For example, makers of industrial process controls, like Honeywell and ABB, may be big winners, and they are investing in their ability to sell solutions in a world of higher-cost energy. The business logic is straightforward: If government regulations push the price of energy upward, investments in energy efficiency in offices and factories will deliver higher returns. ABB and Honeywell make the capital equipment that will deliver those returns. The key is that their products will lower the total cost that their direct customers must incur to conduct their own business. Given that the products that ABB and Honeywell offer are protected by patents, technological know-how, power in the distribution channels, and other barriers to imitation, they stand to profit handsomely as regulations tighten. This strategy is conceptually similar to that followed by Ciba Specialty Chemicals of Basel, Switzerland, which developed dye products that reduce the environmental impacts of textile dyers' operations and lower the dyers' total costs.[3]

The same argument could be applied to the automobile industry. The core customers of General Motors and Ford buy on price, power, prestige, styling, and so on, and not primarily on environmental considerations. But higher fuel prices and increased price uncertainty may enable the big car companies to shift the basis of competition to energy efficiency in order to beat out less technologically sophisticated rivals. Investments in developing the technologies that could propel cars in a world of more expensive carbon — hydrogen fuel cells and hybrid propulsion systems — are not for the faint of wallet. Government interventions that push the new technologies could allow the biggest automobile firms to carve out a large and growing market segment into which smaller, financially-strapped competitors would find it hard to follow. Manufacturers of appliances and other energy-using consumer goods may find similar strategies to be feasible.

Firms that don't have this product differentiation opportunity, including energy producers like BP, must adopt a different basic business strategy. BP is investing in solar power, which may allow for product differentiation in the future, but for the next several decades its main business will likely be fossil fuels and consumers are unlikely to reward it at the gas pump for its environmental stance. Nevertheless, BP has supported international rules to slow climate change and has announced voluntary cutbacks of 10 percent in its own carbon emissions. Its aim, rather than product differentiation, is a combination of the other three policies enumerated above: management of competitors, risk management, and cost management.

First, BP has invested heavily in understanding how a trading system for greenhouse gases could actually work in practice. Since it is demonstrating the feasibility of a trading system, and sharing expertise on how to operate one with regulators, the company increases the chance that the regulatory apparatus will be one that it understands and can manage. Further, its expertise should help BP to articulate its position when governments design the regulatory regime. Competitors focused exclusively on stonewalling regulation altogether may not be so well-positioned. BP's policies thus combine management of competitors with management of regulatory risk.

Attempting to use regulation to manage competitors is not, of course, a risk-free strategy, and attempting to manage regulatory risk through preemptive investment carries with it its own risks. The most famous example of the strategic use of environmental regulation is DuPont's aggressive shift away from chlorofluorocarbons (CFCs) in 1988. By acknowledging the role of CFCs in stratospheric ozone depletion and by demonstrating the technical feasibility of substitute products, DuPont managers facilitated international regulatory action to curtail CFC production and use. But the development of other "not in kind" substitutes for CFCs, along with continued smuggling of CFCs from developing countries, has depressed the returns on investment in the plants that make the substitutes.[5] Similar technological and enforcement risks confront firms attempting to use climate change for strategic ends.

In the climate change case, too, early movers face the risk that future regulations will set baseline emissions levels on the basis of firms' emissions at the time, penalizing firms that have previously cut emissions voluntarily. Managers need to weigh this risk against the benefits, in the form of potential strategic advantage or the reduction of other regulatory risks, that an aggressive program of emissions reductions might entail.

BP senior managers see other benefits in an aggressive stance on climate change. In particular, they think it allows them to cut costs in other parts of their operations. For example, the national governments with which the company negotiates oil leases may allow it more favorable access because of the reputation it is cultivating as a reliable partner. Further, BP executives think that the public announcement of an ambitious goal like the 10 percent cutback is a way of releasing the creativity of employees and increasing their commitment to their jobs and to the company. "Do not underestimate the power of preemptive, aspirational target-setting," says Dr. Chris Gibson-Smith, BP's executive director for policy and technology. "The role of leadership is to invent actions that naturally have the consequence of transforming people's thinking."[6]

BP's assertion that pressure to reduce carbon emissions can serve as a catalyst for broader cost reductions raises the question, introduced above, of the prevalence of such opportunities in other firms.[7] In general, one would expect such opportunities to be more widespread in companies where cost pressures were historically not too severe, either because of market power or because of a privileged regulatory position. For example, Xerox Corporation, after losing its patent-based monopoly on photocopying equipment, found that environmental management could be a significant producer of private cost savings.[8] Many energy producers face similar circumstances. Resource rents and market power insulated the integrated oil companies from the intense cost pressure found in markets closer to the textbook model of competition. For electric utilities prior to deregulation, the firms' status as regulated monopolies undoubtedly blunted cost pressure relative to what will prevail in the newer, at least partially deregulated regime. In both types of company, then, cost savings driven by new environmental pressure may be more widespread than in a random sample of firms.

Firms with operations that sequester carbon from the atmosphere face a particular kind of regulatory exposure. If regulators provide subsidies or other incentives for carbon sequestration as well as taxes or other disincentives for carbon dioxide production, then agricultural producers and timberland owners whose operations are net removers of carbon from the atmosphere would benefit. Since, however, these entities produce carbon dioxide as well as sequester it, they would suffer under a regime that penalized emissions without rewarding sequestration.

In addition to changes in technological standards, command-and-control regulation, and taxes and subsidies, government can change firms' incentives by requiring the provision of information about the environmental impacts of firms' operations. Regulation of this sort could bring about negative publicity for poor performers and raise the private cost of carbon dioxide emissions.[9]

Weather-Sensitive Firms: Managing Increased Uncertainty

Global climate change — with its predicted shifts in weather patterns and regional temperatures, increases in sea levels and flooding, and possible increases in the frequency of extreme weather events — will likely have direct consequences on asset values for many firms. For example, transportation companies, especially those with complex hub-and-spoke systems, need to assess the impact of possible increases in storm frequency on operations and logistics. In addition, insurance companies will have to adapt their predictive models to ensure accurate pricing and prevent large unexpected losses in the face of more frequent storms and changes in weather patterns. Real estate companies will need to revalue properties in ocean front areas with increased flooding hazards or areas newly prone to natural disasters. Agriculture firms could be left holding infrastructure investments in regions that become too warm to sustain high-yield crops and may need to make large new investments in areas where the climate has warmed enough for viable crop production. In the tourism industry, more frequent storms may depress demand for vacations in some tropical areas, while warmer or drier mountain weather can spell disaster for ski resorts.

For all of the firms, improved management of climate risk is of considerable importance. For some, this risk management includes investing in acquiring better information. For example, the insurance industry is investing heavily to research the consequences of global climate change and revise its predictive models. One consortium of insurers has created the Risk Prediction Initiative, a scientific effort to analyze the changes in hurricane weather patterns. Swiss Re, a leading reinsurer, has a group of in-house specialists who track the latest developments in climate change research to identify emerging risks and trends critical to its business. This information is critical to the success of Swiss Re and its competitors, since their ability to deliver profit to shareholders depends on their acumen in assessing and pricing risk. And the stakes are high. As Dr. Thomas Streiff, head of Swiss Re's environmental management unit, points out, "a single hurricane smashing into Miami, for example, could do property damage worth $60 to 80 billion, of which about 50 percent is covered by the insurance industry. Overall economic loss would be even higher, and could exceed $100 billion."

That, in turn, means that any climate-sensitive company that buys property insurance will be affected. Streiff points out that the effects on insurance customers may not be so crude as a cessation of coverage or even a jump in premiums. Instead, the insurers would prefer to work with their clients to reduce the contingent costs of a storm. They might, for instance, make insurance contracts contingent on investment in real risk reduction activities — tighter building standards, for example — on the part of the insured firms.

Business managers can also manage climate risk by investing in operational changes that reduce weather-related risks. Companies that own commercial timberlands need to be particularly concerned about drier conditions, which can increase fire risk and also make the growing trees weaker and more vulnerable to insect attack. Increased investments in fire and pest management thus may become sensible.

Some executives may choose to manage climate change risk by making adjustments to their portfolio of climate-dependent assets. For example, some insurance companies are trying to reduce the amount of coverage they provide in disaster-prone areas, such as Florida and other coastal areas. These strategies may make sense for some firms in the near term. Abandoning investments completely, however, can generate opportunities for other firms to take advantage of reduced supply and charge premium prices. Given the long time horizon over which the full effects of climate change are likely to occur, drastic short-term actions seem unwarranted.

Some business managers may also find product differentiation opportunities arising directly from changes in climate. Agriculture firms, for example, could invest in options for developing crops that are less sensitive to climate shifts, expecting that demand for such products will rise over time. If protected by technological know-how and other barriers to imitation, investments in such products could deliver high returns. And over the longer term, business managers in a wide variety of industries may see demand for their products shift as climate change affects more people. For example, if tropical diseases migrate into more industrialized nations, as they are likely to do as temperatures warm at higher latitudes, pharmaceutical companies may see their markets expand. Products such as malaria pills, currently demanded primarily in tropical countries, will gain wide demand in richer nations.

Business Leaders' Roles in the Public Debate

A final way for business leaders to manage climate change risk, both for regulation-sensitive and climate-sensitive firms, is to try to influence the international debate about whether and how to mitigate global climate change. Firms with climate-dependent assets can play a key role in moving the debate forward and ensuring some type of regulatory action takes place. And as we have seen, they can find allies among the regulation-sensitive firms, some of which will actually benefit from the competitive shocks created by more costly carbon. Together, these groups will press the debate against parties that stand to lose money from a regulatory change.

Firms can influence the international debate on how to address climate change by carefully choosing the role they play in the press, political forums, and the court of public opinion. For example, Swiss Re and BP have both carefully chosen how they will publicly articulate their position on climate change and how they would like to be viewed by the public. Swiss Re, a firm that could face adverse consequences from the weather component of climate change, has chosen to publicize that it is making investments in research and knowledge about the changing climate. BP, as a regulation-sensitive firm, has made its commitment to reducing carbon emissions widely known, in part because this may help it define the terms of the debate over future regulatory interventions.

As noted above, firms in agriculture and the forest products industries can also be direct beneficiaries of carbon regulation. If it makes sense to tax or otherwise discourage carbon emissions, it may well also be sensible (depending on transaction costs and information issues) to subsidize or somehow encourage activities that remove carbon dioxide from the air. Planting forests may reduce atmospheric carbon more cheaply than reducing emissions, and changes in agricultural practices can sequester more carbon in the soil. Problems of measurement and accounting remain serious, but tree growers and producers of agricultural crops have an interest in seeing their contributions to climate stability recognized by any regulatory system that emerges at the international or national level.

Regardless of what message business leaders wish to communicate to the public, there are three key guidelines all executives should follow. First, business managers should assume that whatever actions they take will ultimately be transparent to the scrutiny of regulators and other interested groups. For example, claims about voluntary emissions reductions should be externally verifiable, and should be measured against baselines that are defensible in the public arena. If firms wish to get credit in some eventual regulatory system for reductions made now, it is prudent to make sure the reductions meet these tests.

Second, it makes sense for firms to participate in efforts to make measurement of emissions and emissions reductions consistent across firms and over time. It's too early to tell whether the accounting rules that emerge for carbon turn out to be written in the private sector, like the Generally Accepted Accounting Procedures (GAAP) established by the Financial Accounting Standards Board (FASB), or whether they will be designed by government officials with input from firms, like the rules for sulfur dioxide trading established under the 1990 Amendments to the U.S. Clean Air Act. In either case, though, well-run firms should participate to ensure that the rules are defensible and that they represent a sensible position in trading off feasibility and objectivity.

Finally, executives should be publicizing the positive actions that they are taking regarding global climate change. Surprisingly few firms make public statements about the issue or devote sections of their company websites to it. Far more companies are making investments to increase energy efficiency, study weather pattern changes, and accommodate change than are currently known in the public domain. Since the climate change debate is only going to increase in intensity and level of public interest, taking credit early for positive actions can only help firms manage their reputation and earn enough credibility to earn a seat at the table when future, far-reaching negotiations take place.

Beyond particular firm-specific regulatory advantages, the business community does have common interests in helping governments develop regulatory systems that are cost-effective, flexible, and far-sighted. A specific example: given business's usual distaste for command-and-control regulation and its championing of markets, it is surprising that business executives are not pushing harder for regulations for carbon control that use tradable permits, as opposed to regulations that reduce carbon through command-and-control mechanisms. The command-and-control way could cost between three and eight times as much as market-based solutions, according to former Clinton Administration economist Janet Yellen.[10] The solutions preferred by many economists — a tax on emissions, or a system in which the government auctions permits for emissions and contributes the proceeds to the general revenue — appear currently infeasible for political reasons, at least in the United States. An alternative, much more likely to be politically feasible, is a system of tradable permits that are allocated gratis to current emitters, like the permits for sulfur dioxide established by the U.S. Congress under the 1990 Amendments to the Clean Air Act.

People on all sides of the debate seem anxious for politicians to reach agreement on how to deal with global climate change. Some want a tough treaty that reduces everyone's emissions starting tomorrow. Others want a commitment that governments will stay out completely and forever. Impatience is understandable, but it can make for bad policy. Executives should recall that, following World War II, the U.S. began pushing its partners down a long, uneasy path toward freer trade. Fifty years later, the system is still imperfect, but trade is much freer, and incomes much higher, than they would be if the U.S. had not been patient and far-sighted back in 1949. Building a workable international system to manage climate change is not a one-shot, one-year project either.

The main lesson here is a simple one: executives need to think through the effects of climate change, and climate-driven regulation, on the values of the assets in which they are now investing. Executives who do not incorporate these changes into their strategic business planning will suffer.

Business leaders also need to know what to tell the governments that represent them. They should want regulatory programs that encourage rather than lock in technology, and that provide businesses (who after all are the ones with the information about costs and technologies) the flexibility to decide how to reduce emissions.

As in other problems involving business-government relations, there is a tension for managers between the pursuit of short-term regulatory advantage and the encouragement of a regulatory framework that will be stable and predictable — and therefore friendly to investment — over the long term. Of course executives can pursue short-term political advantage, in the form of special subsidies or (more likely) in the simpler form of delays in the imposition of regulatory constraints. It's possible, though, that they will purchase this short-term advantage at the cost of a politicized regulatory apparatus that rewards further investment in lobbying. A more stable regulatory system, encouraging the investments in technology and know-how that can ultimately benefit both shareholders and society, is in the long-term best interests of well-managed firms.

Endnotes & References

1. Porter, M.E. 1991. America's Green Strategy. *Scientific American* (April), p. 168; and Porter, M.E., and C. van der Linde. 1995. Toward a New Conception of the Environment-Competitiveness Relationship. *Journal of Economic Perspectives* 9(4): 97–118.

2. Reinhardt, F. 1998. Environmental Product Differentiation: Implications for Corporate Strategy. *California Management Review* 40(4): 43–73.

3. Reinhardt, F. 1999. *Ciba Specialty Chemicals*. Harvard Business School case number 9-799-086. Harvard Business School Publishing, Boston, MA.

4. Reinhardt, F. 2000a. *Global Climate Change and BP Amoco*. Harvard Business School case number N9-700-106. Harvard Business School Publishing, Boston, MA.

5. Reinhardt, F. 2000b. *Down to Earth: Applying Business Principles to Environmental Management*. Harvard Business School Press, Boston, MA, pp. 61–65.

6. Reinhardt, 2000a, p.14.

7. Porter, 1991; and Porter and van der Linde, 1995. Palmer, K., W.E. Oates, and P.R. Portney. 1995. Tightening Environmental Standards: The Benefit-Cost of the No-Cost Paradigm? *Journal of Economic Perspectives* 9(4): 119–132.

8. Reinhardt, 2000b, pp. 79–104.

9. The authors are grateful to Tom Tietenberg for emphasizing this point.

10. Kopp, R.J., and J.W. Anderson. 1998. Estimating the Costs of Kyoto: How Plausible Are the Clinton Administration's Figures? *Resources for the Future Weathervane* (March 12). See http://www.weathervane.rff.org/features/feature034.html, accessed August 28, 2000.

INNOVATIVE SOLUTIONS

Contents

State, Local, & Corporate Climate Actions Enhance Quality of Life

New Hampshire Governor Jeanne Shaheen

The preceding chapters show that the threat of global climate change and increasing climate instability is real, that human activities have contributed to this threat, and that the climate is already changing in ways that have profound economic and environmental consequences for all of us. However, the earlier chapters also demonstrate that many economically sound climate response strategies do exist. Ideally, these actions would occur at a global, if not national, level in order to achieve the most significant impacts. But an important question is: Is it possible, and effective, to proactively move ahead with innovative solutions at the state and local level to address climate concerns?

As a governor, I am pleased to assert that the answer is unequivocally "Yes." In fact, states, cities, and counties are not only setting the pace in reducing climate-altering emissions, they are discovering solutions that provide their communities with extraordinary collateral benefits at the same time, including saving money, improving quality of life, and enhancing economic development and competitiveness. I am extremely pleased that my own state of New Hampshire is among the many states, and hundreds of localities, that are today pursuing innovative, cost-effective ideas to bring these many benefits to citizens.

Greenhouse Gas Emissions Reduction Registries

In mid-1998, a small assemblage of representatives from New Hampshire's business, government, and environmental sectors gathered in Concord to brainstorm about actions that could be undertaken at the state level. The group quickly came to a simple conclusion: Encouraging early reductions in greenhouse gas (GHG) emissions was one of the best things New Hampshire could do to make a positive public statement about climate action.

The history inspiring this conclusion is as follows: Under the federal Clean Air Act Amendments of 1990, emissions reductions achieved by proactive companies prior to 1990 fell under the curse of "no good deed goes unpunished." The emissions reduction requirements imposed by the act — basically percentage cuts — effectively rewarded sources that had been dirtier or slower to clean up, because having done

Governor Shaheen and the Pew Center gratefully acknowledge the assistance of Susan Arnold, of the governor's staff, and Kenneth Colburn, of the New Hampshire Department of Environmental Services, in researching this piece.

nothing, they started off with more, easier-to-reduce emissions. The threat of similar treatment under potential future federal regulations for GHG emissions has understandably led companies to be cautious about making GHG reductions until they would "count." The longer this delay, however, the more the climate will be compromised and the harder it will be to reach ultimate emissions reduction targets.

Although the small group recognized that New Hampshire cannot predict, let alone control, future federal regulatory policy, they believed the state would be willing to stand beside sources that had voluntarily made early GHG reductions in good faith. They decided that the best way to implement this practice would be through a registry: sources would quantify and submit GHG emissions reduction actions to a state database for safekeeping against some future federal day of reckoning. The group secured bi-partisan sponsors for legislation to this effect in the 1999 legislative session, garnered the support of New Hampshire's business and environmental communities, and the bill sailed through to passage. I was pleased to sign this first-in-the-nation measure in July 1999.

That's not the end of the story regarding this innovative solution, however. Environmental officials in Wisconsin caught wind of what New Hampshire had done and began a similar legislative process. This was not surprising because Wisconsin had earlier completed the nation's first statewide climate action cost study, an assessment which showed that implementing all steps that cost zero or less ("negative" cost actions are those that actually save money, like energy efficiency measures) would create over 8,000 new jobs in the state, save nearly half a billion dollars, raise Wisconsin's gross state product, and reduce over 75 million tons of carbon dioxide. Wisconsin passed legislation creating the Wisconsin Air Pollution Emissions Reduction Registry (1999 Wisconsin Act 195) in May 2000. In doing so, the state astutely extended the benefits of the original registry concept to pollutants beyond greenhouse gases. Wisconsin's registry allows voluntary reductions of almost all pollutants of concern — including particulate matter, nitrogen oxides, and mercury — to be recorded with the state.

At the same time, individuals and organizations in California concerned about climate change — led, notably, by a progressive business group called the CEO Coalition — began pursuing the idea of a voluntary registry in that state. After a significant, multi-stakeholder effort, much of it focusing on the efficiency gains that climate actions provide, a bill creating the California Climate Action Registry was passed in September 2000. California again arguably improved upon the registry concept, ensuring that it included third-party audits and whole-corporation reporting. The combination of these two provisions ensures that a source's emissions are viewed from an objective perspective and in a comprehensive manner across all its operations.

These innovative actions by New Hampshire, Wisconsin, and California to recognize and register GHG reductions are already seeding an expansion of both the registry concept and its geographical coverage. In New Hampshire, for example, regulations implementing the Greenhouse Gas Emissions Reduction Registry are already being modified to include two "tiers." The first tier responds to constructive pressure from New Hampshire businesses to implement the registry despite its initial similarity to the federal Climate Wise (1605(b)) program. A second tier, which reflects the more comprehensive emissions assessment and third-party review of the California Climate Action Registry, is now being developed in concert with other states and with the assistance of the Northeast States for Coordinated Air Use Management (NESCAUM).

More importantly, several other states are introducing registry legislation and/or initiating public stakeholder processes with that end in mind. Moreover, this multiplicity of state actions is already provoking active discussions — previously relegated to theoreticians and academics — regarding quantification protocols and potential reciprocity arrangements between jurisdictions. Interestingly, reciprocity is only a short step away from emissions reduction credits trading, so the nation's first interstate GHG trading system could conceivably come into existence without any federal involvement or oversight.

Other Innovative Solutions at the State and Local Level

Greenhouse gas reduction registries are only one example of the variety of innovative activities already underway in the states, cities, and counties. A few such initiatives are covered in the course of this chapter, including Oregon's farsighted new power plant siting law, which imposes aggressive CO_2 emissions limits and/or offset requirements on new plants, and which launched the Oregon Climate Trust. The origin and implementation of the groundbreaking New Jersey Sustainability Greenhouse Gas Action Plan — the first state effort to reduce overall GHG emissions by a specific target amount within a specific timeframe — is also described. Demonstrating that city and county governments are capable of equal — or perhaps greater — accomplishment in recognizing, developing, and implementing local climate solutions, the Austin, Texas, success story is also detailed.

Space constraints, however, make it impossible to highlight all of the innovative solutions now underway at the state and local level. Maryland, for example, has just launched a set of energy efficiency and incentive programs that could be the model for a national program. Amidst the challenges encountered by many states in deregulating the electric utility industry, Pennsylvania has found an effective way to encourage the production and sale of renewable, "green" power. Massachusetts and Connecticut also have provisions on their books regarding CO_2 emissions from power plants. Minnesota's utility regulators ascribe a value to CO_2 emissions in their decision-making, rather than labeling them "externalities" and shifting the costs they impose on society to other venues. That state is also in the midst of a leading effort to generate electric power from the methane that emanates from its major landfills, and is selling both the electricity and the GHG emissions reduction credits produced. Texas established a requirement that 2,000 megawatts (about 3 percent) of new, renewable energy-generating capacity be installed or contracted by 2009, and the state appears likely to meet this target — largely through wind energy — in just three years instead of 10. Vermont's governor has called for explicit action to address climate concerns, and Iowa and other agricultural states are exploring how more carbon can be sequestered — and GHG credits generated — through new "best management practices" like low- and no-till planting.

If these state "soldiers" are marching, the "troops" at the local and county level are positively racing. Under the capable guidance of the International Council of Local Environmental Initiatives (ICLEI), over 75 communities throughout the United States, and hundreds more internationally, have joined its Cities for Climate Protection (CCP) campaign. The U.S. communities alone represent over 28 million people and over 10 percent of U.S. GHG emissions.

ICLEI provides committed communities — like Austin, Texas — with education regarding opportunities for emissions reductions, technical assistance in quantifying them, and aid in preparing a plan to implement them. CCP communities carefully analyze their GHG emissions, establish a reduction target, develop and implement a plan to reach that target, monitor progress, and report the results to the public. The measures implemented cover all sectors — energy efficiency and renewable energy, transportation, solid waste and recycling, government operations, etc. — and are tailored by the community to its particular needs and opportunities. They vary from backyard composting to curbside recycling, from ridesharing and bicycle lanes to light-emitting diode (LED) traffic signals, from ultrasonic building humidification to photovoltaic golf course lighting.

Innovative Solutions from the Corporate Community

Most CCP communities adopt measures to improve the efficiency of transportation systems and government operations, but they also look to their local business community to do its share as well. Together with their corporate citizens, the CCP cities and counties in the United States have implemented over 300 measures that are reducing 7.5 million tons of GHGs, cutting 28,000 tons of traditional air pollutants, and saving $70 million in energy and fuel costs.

Such efforts are rapidly gaining ground because businesses increasingly recognize that civic as well as economic opportunities are associated with climate action. DuPont, for example, announced its plans to reduce corporate GHG emissions by 65 percent from 1990 levels at a September 1999 conference convened by the Pew Center on Global Climate Change. BP — or "Beyond Petroleum" as it now prefers to be called — has also committed to GHG reductions in excess of the U.S. and European Union reduction targets called for in the 1997 Kyoto Protocol. United Technologies Corporation intends to cut its energy and water consumption by 25 percent per dollar of sales by 2007, before the Kyoto Protocol even takes effect.

One of the best examples of corporate environmental leadership is offered by Intel Corporation. Intel illustrates how a world-class company in one of the most competitive, rapidly changing industries in the world can progressively and successfully approach its environmental responsibility. As described later in this chapter, Intel recognizes — and uses — many key aspects of the production-protection interface that easily elude less thoughtful enterprises. Designing environmental improvements into the manufacturing process instead of focusing on end-of-pipe treatment solutions, for example, can not only ensure compliance with regulations, but also often reduces production costs, energy consumption, labor, overhead, and other resources. As a result, productivity — and market image — can soar. Intel also understands the importance of considering the environmental impact of the product itself on the environment, not just the impact of its manufacturing and distribution, and of applying the same sound environmental practices at all its operations throughout the world. This understanding has led Intel to a state-of-the-art "Design for the Environment" program, and its hugely successful "Energy Star" microprocessors, which are bringing the company to the point where it arguably has a net positive impact on the environment.

Intel also recognizes that companies do not exist in a vacuum. They must integrate and coordinate with their suppliers, their host communities, and the government agencies they work with. In the latter context, Intel appropriately argues that in order to address climate change and other global environmental issues effectively, governments need to provide greater cooperation and flexibility, focusing more on holding corporations accountable for their environmental results and less on how companies achieve them. Ultimately, Intel understands that solving such large, looming environmental concerns as climate change requires the combined skills of all parts of society, and that we are better able to achieve these goals if we act as allies instead of adversaries. Intel's approach, and how other progressive companies are going about inventorying and reducing their GHG emissions, are detailed in the last section of this chapter.

Cumulative "Co-Benefits"

The fact that proactive steps to address climate change and other environmental problems provide numerous collateral benefits ("co-benefits") is a crucial element that — while lying at the very heart of climate policy — has often been overlooked in the heat of climate debate. Improved air quality is one major co-benefit. CO_2 is not a pollutant in the traditional sense; it is not emitted as a result of dirty fuel, poor combustion, or inadequate emissions controls. CO_2 is emitted as a predictable by-product of the carbon present in fossil fuels and the oxidation that occurs when they are burned. As a result, the most effective way to reduce CO_2 emissions is to diminish the amount of fossil fuel burned. We can do this by burning fuel more efficiently or by utilizing efficient appliances that do not require as much fuel in the first place.

When we do so, however, we also eliminate emissions of all the other traditional pollutants that accompany carbon in fossil fuels — the pollutants that do result from dirty fuels, poor combustion, or inadequate controls. This is the reason that the same ICLEI measures that are reducing 7.5 million tons of GHGs are also reducing nearly

24,000 tons of sulfur dioxide emissions (which cause acid rain and fine, airborne soot), almost 3,000 tons of nitrogen oxides (which cause ozone smog), over 1,500 tons of poisonous carbon monoxide, over 300 tons of volatile organic compounds, and over 100 tons of other harmful pollutants.

The air quality co-benefits that accompany climate actions are well documented. A year ago, for example, the State and Territorial Air Pollution Program Administrators (STAPPA) and the Association of Local Air Pollution Control Officials (ALAPCO) modeled the air quality improvements that would result from certain climate actions. A set of reasonable CO_2-reducing measures was defined for four distinct areas of the country: New Hampshire; Atlanta, Georgia; Louisville, Kentucky; and Ventura County, California. Each set of measures was different — tailored to the particular electric generation, transportation, and industrial profile of each area. In each case, however, the same steps that reduced CO_2 emissions by 7–15 percent also reduced sulfur dioxide emissions by 2–41 percent, nitrogen oxides by 4–17 percent, particulate matter by 1–12 percent, volatile organics by 3–4 percent, and carbon monoxide by approximately 4 percent.

Co-benefits aren't limited to air quality, however. They are apt to have substantial ancillary benefits in terms of lower public health costs, less need for costly alternative air pollution controls, and less regulatory burden associated with nonattainment of air quality standards. Moreover, fuel that isn't burned doesn't have to be bought, and electricity that isn't used doesn't have to be purchased. As a result, the CCP communities expect to save $70 million in reduced energy and fuel use as noted above. Such savings reduce operating costs for governments, businesses, and citizens — reducing tax burdens, increasing profit opportunity, and putting more money back in the wallet. Moreover, this money is more apt to circulate in the local economy — purchasing goods and services from area merchants — rather than going to energy suppliers outside the community or state. Local purchases help create local jobs and build the local economy. Simply put, communities with stronger local economies, less pollution, lower tax burdens, less vehicle congestion, better public health, and more recreational opportunities are more desirable places to live.

Climate Action as a Quality-of-Life Strategy

Quality of life could be characterized as a combination of a healthy citizenry, a healthy environment, and a healthy economy. The direct and indirect benefits enumerated above suggest that climate action might reasonably be considered a public health strategy, an environmental protection strategy, and an economic development strategy. Collectively, then, climate action could be thought of as a "quality-of-life strategy."

In terms of public health, tens of thousands of premature deaths across the country — dozens in New Hampshire alone — have been linked to fine particulate matter caused by sulfur dioxide emissions from coal-fired power plants that were "grandfathered" under the federal Clean Air Act. The mercury emitted from these and other sources has led to fish consumption warnings in a majority of the states. The ground-level ozone (or "smog") created by nitrogen oxide emissions from power plants and motor vehicles is a potent oxidant — one of the reasons that many of us take anti-oxi-dant dietary supplements like vitamin E. Transportation emissions are also the principal cause of most of the toxic compounds that exceed health risk screening lev-els in New Hampshire and across the country. Finally, many pollution problems are associated with chemical reactions in the atmosphere. The warming created by increasing atmospheric carbon dioxide concentrations will exacerbate these problems by increasing the rate at which the underlying chemical reactions occur.

Regarding environmental quality, these same sulfur dioxide and nitrogen oxide emissions have also acidified our lakes, leached essential nutrients from our forest soils, and mobilized compounds that are toxic to fish. Atmospheric deposition also impacts water quality and contributes to the eutrophication of our coastal estuaries. Ground-level ozone interferes with the ability of all plants — including agricultural crops and forest stands — to make and store food, reducing their yield. Climate change created by human activities will induce species to migrate to more friendly habitats, putting such cultural icons as sugar maples — the backbone of New England's maple syrup industry — and rainbow trout at risk.

The economic opportunities connected to climate action are perhaps most enticing. Some economic gains are obvious, like saving money on energy and fuel costs, or the prospect of improved forest and agricultural productivity with less air pollution and new cultivation techniques. More important, however, are many less obvious effects. In New Hampshire, for example, tourism and recreational activities comprise our second largest industry after manufacturing. New Hampshire's White Mountain National Forest attracts millions of visitors each year, on a par with Yellowstone and Yosemite National Parks. This is not surprising because about one-quarter of the U.S. population lives within one day's drive of the White Mountains. Spending by the approximately 25 million tourists who visit New Hampshire each year creates over $3.4 billion dollars in our state's economy.

Boating, hiking, fishing, hunting, snowmobiling, skiing — they all hinge on the environment in one or more crucial respect. Our fall foliage season, the ability of our ski areas to make and retain snow, the vistas from Mt. Washington (the highest sum-mit in the Northeast), even our native lobster population, all depend on arresting the warming effects of climate change and diminishing the collateral air pollution that accompanies carbon dioxide emissions.

As important as these economic effects are, they are arguably the tip of the iceberg regarding New Hampshire's economic future. Increasingly in the new, knowledge-based economy, companies are locating where the skilled, mobile employees they need want to live. As a result, New Hampshire's quality of life — the public health, envi-ronment, and economy so well-served by concerted climate action — is largely responsible for the state's having achieved the second-highest per capita proportion of high technology workers in the nation. Other elements are also vital, like good schools, and an adequate telecommunications infrastructure. But New Hampshire's natural beauty, landscape diversity, and year-round recreational opportunities have played a key role in state's economic success.

The potential for climate action to enhance quality of life is a good paradigm to supplant some of the crumbling conceptual cornerstones of the old economy. For instance, conventional wisdom has held that energy use — the principal cause of carbon dioxide emissions — is such a critical pillar of the economy that economic growth must occur in lockstep with energy growth. It then follows that environmental controls on energy are, at best, necessary evils whose costs — by making energy more expensive — detract from economic growth. The new reality, however, suggests that environmental quality is better correlated with economic growth than energy use is. In fact, at this point, energy use appears to be negatively correlated with economic growth. Consider the following:

- The U.S. economy grew approximately 4 percent last year, but energy use increased only about 1 percent. Last year was not an exception; the American Council for an Energy Efficient Economy (ACEEE) has calculated that national energy intensity (energy per dollar of gross domestic product) has fallen 42 percent during the last three decades.

- New Hampshire's experience provides a direct indication that the old conventional wisdom is obsolete. In September 2000, ACEEE ranked New Hampshire fourth best overall in the United States at reducing its energy intensity and carbon emissions. Simultaneous with this achievement, however, the state was seventh highest in per capita income, had the fourth lowest unemployment rate, the lowest poverty rate, and high growth rates in jobs, businesses, and per capita income.

- The "Green & Gold" index published by the Institute for Southern Studies illustrates the same point more broadly: "Green states" are more likely to be "Gold states." The strong correlation between states' emphasis on environmental protection and their economic success is evident in Figure 1.

- In today's new electronic and high technology economy, this positive correlation between green and gold is not surprising. Whereas old industries were wedded to specific geographic locations due to the availability of vital natural resources or transportation links, competitive advantage in the new economy hinges on an educated, innovative, and mobile workforce. Increasingly, competitive companies are finding that they have to locate in places where their employees want to live. A key component of the quality of life that today's professionals seek is an enjoyable natural environment. NetworkNH, a business group in New Hampshire, has captured this issue succinctly: "In an economy where physical assets are not as important as they used to be, where intellectual assets

Figure 1

Green and Gold 2000 | **Rankings of States**

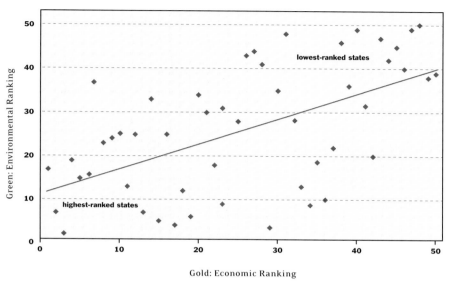

Note: 1= top; 50 = bottom.

Source: Institute for Southern Studies. See http://www.southernstudies.org/992000rankings.html.

dominate, where business can be conducted from anywhere to anywhere, it would seem that place should not matter; in fact, it matters more."

• The need for power quality and reliability is becoming ever more important. Perhaps surprisingly, steps to enhance quality and reliability are often consistent with climate action. The Bank of Omaha needed more reliable power than the grid could provide for its credit card transaction processing operations, so it turned to fuel cells as a precaution. Fuel cells have among the fewest environmental impacts of any electric energy source. Climate-friendly steps like fuel cell installations and combined heat and power (CHP) applications also reduce companies' vulnerability to price shocks and supply disruptions, such as those experienced in California in 2001.

• Recent work done by the Worldwatch Institute demonstrates how significant job creation also follows from energy efficiency measures that benefit the environment. At $300-500 million each, today's power plants represent an extraordinary investment of society's capital resources. Yet, they typically require very few workers to operate, so this commitment of capital provides comparatively few jobs. Investing those capital resources in energy efficiency products and services or renewable energy — like manufacturers of compact florescent light bulbs or wind turbines — would employ (and in several cases, are employing) many more people than constructing an additional power plant to illuminate less efficient incandescent bulbs. Note that many of the largest, state-sponsored renewable energy or technology venture funds (paid for by ratepayers through systems benefits charges) are only 10-25 percent as expensive as a new power plant, but they promise much greater job growth and environmental benefits. They are also more likely to serve us better in the future, binding us to tomorrow's energy technologies and innovations, instead of wedding us to yesterday's obsolete approaches.

• The very foundations of corporate performance ranking and global competitiveness are changing. Companies are increasingly being judged against a "triple bottom line" which assesses their financial, social, and environmental performance. Dow Jones, for example, has created a "Sustainability Group Index" which indicates that companies generate added value by integrating economic, environmental, and social growth potentials into their business strategies. Such strategies are becoming more important too; professionally managed assets in socially and environmentally responsible funds have increased 82 percent since 1997, to $2.16 trillion (13 percent of the total).

The ramifications of this sea change for policy-makers is clear. Whereas it may have once been possible to conceive of a healthy economy and a healthy environment as mutually exclusive goals, it appears today that one can't exist without the other. Proactive, aggressive steps to address climate change and air pollution through the more intelligent, efficient production and use of energy may indeed be the best public health protection, environmental preservation, and economic development — i.e., quality-of-life — strategy possible.

Conclusion

From corporations contemplating their raw materials purchases or production processes to communities considering their transportation systems or electricity grids, forward-thinking entities are discovering that reducing emissions can translate into less waste, that less waste translates into less cost, that less cost translates into higher margins or lower taxes, and that higher margins, lower taxes, and better environmental quality translate into competitive advantage. The field of play of economic competitiveness has shifted dramatically.

Concerted climate action is consistent with the challenge posed by this shift. "Green" and "Gold" are mutually reinforcing attributes. Policy-makers' choices essentially boil down to either ignoring and injuring the environment-economy linkage — generating a "vicious cycle" characterized by environmental and economic decline — or embracing and enhancing the environment-economy linkage — creating a "virtuous cycle" that elevates and sustains both. Advantage will accrue to the best "virtuous cycle" engineers, whether they embody a corporation within an industry, a community within a state, a state within a country, or a nation within the global community of nations.

Some may wonder concerning climate action, "Is it already too late?" There is an odd ambiguity in this question, because it contains two possibilities, "too late for the climate?" or "too late for action?" The latter accurately suggests that the opportunity to secure the competitive advantages bestowed upon early climate actors may already be evaporating. The United States has already lost its market share leadership in wind power manufacturing to Germany, for instance. Most technologically advanced nations are actively investigating and testing innovative solutions to climate concerns, rather than balking at the starting gate. It is not difficult to deduce who is more likely to own the patents of the future.

Regarding the question — "too late for the climate?" — science remains unclear regarding what specific challenges the climate future that we have created will hold for us. One thing is clear, however. The sooner we pursue climate action in earnest, the less profound climate impacts will be, and the more likely we will be able to manage them.

Our response as a nation to the intensifying threat of climate change will either help protect and preserve life as we know it, or it will "merely" provide us with greater competitive advantage long into the future. Either way, we can't afford to dawdle. The case studies in this chapter are already charting the course for us to follow.

+

+

Innovative State Programs:
New Jersey & Oregon Take the Lead

Matthew Brown

In the 1980s the international community started down a long and sometimes frustrating path toward an agreement on what to do about human contributions to the changing climate. The debate has raged since that time, and there still is little promise that the United States will ratify an international treaty to reduce emissions of the gases that appear to contribute to climate change. However, a few states have begun to take steps of their own to tackle the issue. Although early and sometimes seemingly tentative, these steps show some of what states can do to address climate change.

New Jersey and Oregon are two states whose initiatives are worth a close look. Each has announced policies to mitigate climate change — New Jersey's through a set of programs run by the executive branch, and Oregon's through legislation. These programs are new and it is still hard to tell how much they will achieve. But they represent innovation at the state level and may provide models that other states can emulate.

New Jersey

New Jersey has set a goal of reducing greenhouse gas emissions 3.5 percent below 1990 levels by 2005, and has devised a number of strategies to meet it. These strategies include programs to reduce emissions from power plants, increase the efficiency of energy use, reduce emissions from industrial sources, and address emissions from cars, trucks, and buses.

On April 17, 2000, former Governor Christine Todd Whitman and Department of Environmental Protection (DEP) Commissioner Bob Shinn announced the release of the New Jersey Sustainability Greenhouse Gas Action Plan. The goal of the plan is to shave 3.5 percent off the state's 1990 emissions levels by 2005. The plan builds on existing programs for reducing nitrogen oxides (NO_x), sulfur dioxide (SO_2), volatile organic compounds (VOCs), and mercury, and enhances them with further greenhouse gas (GHG) reduction goals. The plan uses GHG reductions as an umbrella indicator for reductions in multiple pollutants. It also represents the first attempt by a state to reduce GHG emissions by a specific target within a defined timeframe. The program is new, however, and early results are hard to find. It is also fully voluntary. Nothing in the plan requires either industry or the state to reduce emissions, nor does it directly

The author and the Pew Center wish to thank Michael Burnett, Christina Rewey, and Michael Winka for their assistance in researching and reviewing the article.

regulate carbon dioxide (CO$_2$). However, Commissioner Shinn has fully committed the NJ DEP to this goal and it is a focal area of the DEP's strategic plan for the years 2000-04.

Background

Commissioner Shinn's announcement of GHG reduction goals continued the state's delicate balance between environmental stewardship and economic growth. This was a balance that previous state policy-makers had dealt with for many years. The state is home to coastal development and popular beaches that give residents and visitors a place to enjoy summer recreation. It is also home to hundreds of energy-intensive and high-technology industries that take credit for jobs and a growing economy over much of the past century. Some of these industries have also been responsible for air and water emissions that give some parts of the state a reputation for pollution problems. New Jersey's policy-makers balance these conflicting interests, and the global warming initiative is no exception to this balancing act.

The state began its effort to reduce emissions with an inventory of GHG emissions. The inventory, completed in March of 1997, showed that as of 1990 New Jersey ranked 36th out of all the states and territories in GHG releases, representing about 2 percent of the nation's total emissions. The inventory also indicated that 82 percent of New Jersey's emissions consisted of CO$_2$ emissions from fossil fuel combustion and, of these emissions, 38 percent came from cars, trucks, and buses in the transportation sector, 24 percent from residential buildings, 22 percent from commercial buildings, and 16 percent from industry. Most of the remaining 18 percent of the state's emissions came from methane that emanated from New Jersey's landfills and from fossil fuel extraction and distribution. (See Figures 1 and 2.)

One of the issues that galvanized the public's opinion that New Jersey had a stake in climate change was the potential for rising sea levels. The same New Jersey report that detailed the sources of its GHG emissions also noted that the sea levels had risen 12 inches in the previous century, and were likely to rise by another 17 inches or more by the year 2100. Building a sea wall large and strong enough to protect a single beach area of the state against rising

Figure 1

New Jersey's **Greenhouse Gas Emissions** by Source (1990 estimates; CO$_2$ equivalent)

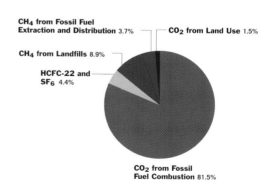

CH$_4$ from Fossil Fuel Extraction and Distribution 3.7%

CO$_2$ from Land Use 1.5%

CH$_4$ from Landfills 8.9%

HCFC-22 and SF$_6$ 4.4%

CO$_2$ from Fossil Fuel Combustion 81.5%

Note: Does not include emissions of CFCs and related compounds, which are under present phase-out requirements.

CH$_4$ = methane
SF$_6$ = sulfur hexafluoride
HCFC = hydrochlorofluorocarbon

Source: New Jersey GHG Emissions Inventory.

seas could cost hundreds of millions of dollars. Many of Commissioner Shinn's early announcements of the climate initiative tied the concerns about climate directly to the dangers of sea-level rise. However, the New Jersey GHG action plan attempts to separate the science of global climate issues from the implementation of cost-effective "no regrets" emissions reduction strategies. Commissioner Shinn has stated that while the debate about the science of global climate change continues, New Jersey must act to implement programs that help the state.

Figure 2

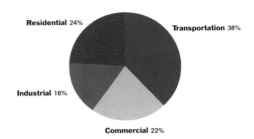

New Jersey's **Fossil Fuel CO$_2$ Emissions** by Sector, 1990

Note: Includes only emissions from operations within NJ borders.

Source: New Jersey GHG Emissions Inventory.

The Greenhouse Gas Plan

New Jersey's GHG emissions have been increasing since 1990, and climate models indicate that levels could reach 151 million tons by 2005. This means the state's goal of cutting emissions 3.5 percent from 1990 levels would translate into a 14 percent reduction from projected levels for 2005, or a reduction of about 20 million tons. With the plan, the state's emissions would be 131 million tons in 2005. All in all, the many programs New Jersey has put forward are expected to avert the release of about 20 million tons of emissions from state sources across all sectors.

The New Jersey Sustainability Greenhouse Gas Action Plan documents that the 3.5 percent reduction goal is reasonably achievable using a "no-regrets" strategy. (See Figure 3.) One of the key findings of the plan is that a reduction in greenhouse gases will result in collateral benefits: Reducing GHG emissions will reduce NO$_x$ and SO$_2$ emissions, and has the potential to reduce VOCs and mercury. These reductions will generate improvements in water quality and may reduce wastewater discharges. New Jersey views GHG emissions reductions as an umbrella indicator.

The plan was developed in a stakeholder process that includes agencies across state government, environmental groups, local government, universities, small and large businesses, and renewable energy and energy efficiency companies. Both the DEP and the Board of Public Utilities (BPU) co-chaired this process, and jointly consolidated the recommendations of the Greenhouse Gas Action Plan stakeholder group.

DEP and BPU officials assembled an array of programs that they felt would yield verifiable emissions reductions. Their analysis convinced them that some of the energy

Figure 3

New Jersey's GHG | **Emissions and Target Levels**

Projected Emissions and Reductions with NJ Sustainability Greenhouse
Gas Action Plan Strategies

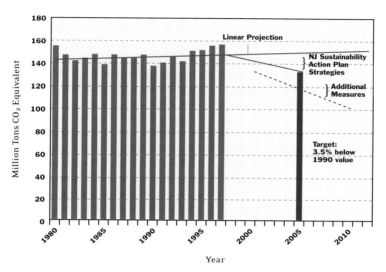

Year

Note: Emissions estimates based on USDOE/EIA data, augmented by N.J. solid waste data.
Projection assumes business as usual.

Source: New Jersey GHG Emissions Inventory.

efficiency and emissions reduction efforts in the state over the past decade had already
begun to bear fruit. As a result, the plan builds on these energy efficiency and emissions reduction programs, pushing them to meet the state's 3.5 percent emissions
reduction goal.

When the state DEP put together its full complement of programs that would
reduce greenhouse gas emissions, targeted areas broke down as follows:

Table 1

NJ GHG Reduction Summary

Measure	Reduction (in tons)
Energy efficiency in industrial, commercial, and residential sectors	6.2 million
Innovative technologies in industrial, commercial, and residential sectors	6.3 million
Energy conservation and innovative technologies in transportation sector	2.2 million
Waste management/recycling from the government sector	4.5 million
Natural resource conservation/open space development from the government and agricultural sectors	0.5 million

Energy Efficiency

The DEP relies heavily on emissions reductions from commercial buildings to reach its goal, with close to two-thirds of the potential emissions reductions from energy efficiency coming from this sector. The largest single measure that the plan has identified in commercial buildings is improvements in the efficiency of lighting systems. The plan suggests that 15 percent of facilities could upgrade their lighting systems to reduce electricity consumption by 40 percent per year per facility. The plan also relies on 15 percent of commercial facilities conducting an annual tune-up of their heating systems, to save 10 percent of CO_2 emissions reductions each year in each facility. These two measures alone account for about two-thirds of the planned reductions in emissions from commercial buildings.

The residential energy efficiency program relies heavily on upgrading heating systems and integrating more energy efficient refrigerators and water heaters into homes in New Jersey. The assumption for the plan appears conservative: that an additional 5 percent of New Jersey residents over business as usual will replace their refrigerators and water heaters each year with high-efficiency units. These measures account for almost all of the improvements to the energy efficiency of the residential sector.

In the industrial energy efficiency sector, the plan relies on ongoing repair and maintenance to steam and air compressor systems, as well as the introduction of newer, high-efficiency motors. These maintenance programs can yield good results; New Jersey asserts that repair of air compressor systems could save 50 percent of the average electricity usage per unit. Still, the plan relies less heavily on improvements in the industrial sector, with an expectation that only about 5 to 10 percent of the total GHG emissions reductions will come from that sector.

Innovative Technologies

New Jersey assumes that between 1999 and 2005 many new and efficient technologies will be entering the marketplace. The emissions benefits from new technologies come from new ways of generating electricity: microturbines, fuel cells, and photovoltaics. They also include the use of geothermal energy for space heating and cooling.

Microturbines and fuel cells are both becoming popular among businesses that want to generate electricity on their own premises. For instance, a McDonald's restaurant in Chicago recently installed a natural gas powered microturbine in a corner of its parking lot. The small turbine, which takes about as much space as one to two parking spots, generates electricity during mid-day for the restaurant, when prices on the utility grid are high. The microturbine also acts as a backup power source should the grid fail to deliver electricity.

Fuel cells usually use hydrogen derived from natural gas, and generate electricity through a chemical reaction that is similar to a typical flashlight battery, but on a much larger scale. Large-scale commercialization of fuel cells is expected by mid-year 2000. A bank in Omaha, Nebraska, installed fuel cells on its premises to generate electricity completely independent of the electric grid. Its installation made sense as a way to secure a highly reliable source of electricity since a power failure could devastate the bank's data processing center. In general, however, fuel cells remain significantly more expensive than grid-delivered power.

Both fuel cells and microturbines tend to be more reliable than power supplied by the electric grid, and hence are attractive to the increasing number of businesses that can little afford an interruption in their power supply.

New Jersey assumes that these new technologies will become more popular and less expensive, and that 10 percent more facilities will adopt these new technologies. Since they release less CO_2 than central station power-generating plants, they will reduce total CO_2 emissions.

Transportation Sector Emissions

Emissions from cars, trucks, and buses account for the largest single source of greenhouse gases. But they are also the most difficult to reduce, since any strategy that a state can pursue necessarily involves changing the habits of millions of drivers. New Jersey's plan relies on an inspection and maintenance program to reduce emissions by 10 percent per year, and increasing mass transit usage for a total of about one and one-half tons of carbon dioxide emissions reductions. New Jersey also counts on new, lower-emitting biofuels made from waste materials beginning to enter the transportation fuel stream, and on some electric and hybrid electric cars beginning to enter the marketplace.

These new transportation technologies play only a small part in New Jersey's total plan, accounting for about three-quarters of a ton fewer CO_2 emissions. They are also among the more speculative elements of New Jersey's plan; it is hard to know how rapidly New Jersey's drivers will adopt these new technologies. The plan recognizes that major reductions in this sector would require new technologies and perhaps some major changes in people's driving habits.

Recycling and Landfill Gases

Finally, New Jersey's plan relies on increased recycling and new facilities to generate electricity from landfill gas. The plan projects that 12 new landfill gas-to-electricity facilities will be developed between 2000 and 2005. These will significantly reduce total GHG emissions. The DEP estimates that if the municipal solid waste recycling rates alone increased from their 45 percent rate in 1995 to a 60 percent rate in 2005, that increase could account for meeting 15 percent of the total GHG goal.

New Programs that Help Meet the Goal

Since announcing the greenhouse gas action plan, New Jersey has crafted some new programs — most notably the environmental permitting programs and the Electric Discount and Competition Program — that could play a significant role in helping the state meet its 3.5 percent goal.

Environmental Permitting Programs

New Jersey's DEP issues permits for companies to emit different levels of pollution into the air. These permits might be issued for a certain period of time — usually five years — and allow the company to release pollutants at a certain rate. These permits apply to the so-called "criteria" pollutants — such as SO_x and NO_x. The DEP does not issue permits for CO_2 emissions since CO_2 is not a regulated pollutant.

For a number of years, New Jersey has been trying to find ways to recognize the companies that consistently meet its emissions standards, go beyond compliance, and require no enforcement actions. As part of this effort, the DEP instituted the Silver and Gold Permitting Track program. This program recognizes "good performers" that meet certain criteria. One of the criteria inserted into the program as a result of New Jersey's GHG action plan was that the companies agree to reduce GHG emissions. The three levels of performance and recognition are as follows:

Silver Track I is the lowest level at which companies receive recognition. To receive this level, companies prepare an environmental management plan and must have a clean record of no violations for the previous five years. They also agree to certain types of community outreach. For instance, a large company might convene a citizens advisory panel to have input on its environmental planning. A small company might agree to post notices or other types of public outreach about proposed expansions or modifications to its facilities. In exchange, the companies get a single point of contact at the DEP, a large plaque, and other types of recognition. This level of the program has no GHG reduction component.

Silver Track II has the same requirements as Silver Track I. In addition, companies must commit to reduce GHG emissions by 3.5 percent below 1990 levels by 2005 — the same as the state's goal. Qualification for this program secures companies the benefits of Silver Track I, plus more flexibility in air, water, and waste permitting. Under Silver Track II, any source with *de minimis* air emissions does not require a permit. This may be particularly helpful to companies with a large number of small sources, each of which would otherwise have required a permit. On the waste side, the program allows flexibility in recycling. The permit flexibility applies to permit renewals, new permits, and permits issued because of modifications to facilities. The greenhouse gas element of this level of the program was a direct result of the state's GHG plan.

The highest level of flexibility is offered to companies that qualify for the Gold Program. Many of the details of this program were still being worked out as of this writing. However, preliminary discussions indicate that in addition to the GHG emissions

many in the state wanted assurance that the new plants would be among the cleanest in the country. This has led to political pressure to develop new laws on the siting of power plants.

An old Oregon law dictated that power plants could only be built after an electric utility went through a detailed planning process called least-cost planning. This process was developed during the late 1970s to 1980s, and originally required utilities to weigh all options that could meet their customers' power needs. The process required utilities to consider the costs and benefits of energy efficiency programs against building different types of power plants. In some states, utilities were also required to quantify the environmental effects of different generating technologies — weighing the environmental effects of efficiency versus gas, coal, hydro, or nuclear facilities.

Least-cost planning focused on utilities. But in Oregon and the rest of the nation, the electric industry began a quick transformation in the early 1990s that meant that few utilities were building power plants. Instead a new breed of power generator surfaced that built a new class of power plant called a merchant plant. Merchant plants are constructed and operated by independent developers that do not face price regulation. Developers earn money by building efficient plants that sell into the competitive power market. If they can produce power more cheaply than their competitors, they thrive. If not, they wither.

Because state regulatory commissions do not regulate these independent companies' prices, many of their activities fall outside the jurisdiction of the public utility commissions. The companies still need to acquire a siting certificate from the proper siting authority, however, and as a result this siting certificate has become one of the means remaining for states to govern new power plants. Siting has therefore begun to acquire new prominence and many states have begun to reexamine their siting statutes.

Such was the case in Oregon. Merchant power plant developers wanted to build facilities to supply not only the Oregon power market, but also the region's market. But the state's siting law, and in particular the requirement that there be a true "need" for new generation established as part of this utility-driven least-cost planning process, effectively made it impossible for the merchant plant developers to build new facilities. They were unable to establish that there was a need for power in Oregon, which prohibited them from building new plants in the state. The political pressure from these power plant developers brought urgency to the debate about power plant siting.

In addition, Oregon had by the mid-1990s conducted a study of how climate change could affect the state, completed a greenhouse gas inventory (a process through which the state determined the sources of its greenhouse gas emissions), and asked electric utilities to consider greenhouse gases when developing least-cost plans. The legislature had also required the state siting board to "consider" greenhouse gases in the siting process for new power plants. These activities had all made the legislature, the executive branch, and citizens of the state more aware of greenhouse gases, and had to some extent galvanized opinion that the time was ripe to do something about greenhouse gas emissions. The siting law appeared to be an appropriate way to address both greenhouse gases and the "need" issue simultaneously.

The Early Debate About Power Plant Siting

Oregon's biennial legislature first faced the issue of siting power plants in its 1995 session. During this session, power plant developers and utilities attempted to have the "need" requirement removed from consideration of whether a power plant could be sited in Oregon. Their attempt to do so failed, and a legislative compromise set up a one-time exemption for a 500-megawatt (MW) power plant to be sited without specific consideration of need.

The seven-member Siting Council decided to award this 500-MW exemption on the basis of a competition among power plant developers. The exemption was to be given to the least-polluting power plant, based on air emissions, water discharges, and land-use effects. Ultimately, CO_2 emissions became the most important factor in the decision process, because the three competitors for the exemption proposed very similar plants.

A plant developed by the City of Klamath Falls won the competition because of its efficient design, its use of co-generation (the use of heat that would otherwise have been wasted), and CO_2 offsets. A CO_2 offset is an action that reduces or absorbs CO_2 emissions in some other location. The Klamath Falls developers planted trees in Oregon, installed photovoltaic systems in Asia, expanded the city's geothermal district

Figure 5

Oregon's **Electric Power Industry CO₂ Emissions** Estimates, 1989-1998

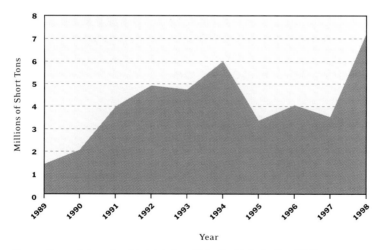

Source: Energy Information Administration, State Electricity Profiles 1998.
See http://www.eia.doe.gov/cneaf/electricity/st_profiles/oregon/.

Climate Change: Science, Strategies, & Solutions

heating system, and set up a system to gather methane from coal mines and sewage treatment plants to generate electricity. These initiatives combined to offset 24 percent of the plant's CO_2 emissions for 30 years.

As a result of this competition, the siting task force that had been appointed by the 1995 legislative action suggested that the state remove the "need" requirement from the siting consideration and add a CO_2 emissions requirement. Both industry and environmental groups agreed that the trade-off met their needs, and the task force submitted a proposal to the state legislature for its 1997 session.

Oregon's New Climate Change Legislation

Oregon's 1997 legislature and governor enacted House Bill 3283 (Oregon Revised Statutes 469.370), which traded a requirement that there be a determination of need for a power plant for a requirement that power plants meet CO_2 emissions standards.

The new statute eliminated any requirement that any party establish a need for a new power plant before receiving a permit to build the plant. Power plant developers would build plants at their own risk, and the state would assume that developers would be unwilling to risk huge capital outlays if they did not expect to earn a return on their investment; the state would no longer be involved in the determination of "need."

The carbon emissions requirements were new, and more complex. They required that any new natural gas-fired power plant in Oregon be at least 17 percent more efficient than the most efficient gas-fired power plant operating anywhere else in the country. As plants around the country become more efficient, the standards will change, but the legislature set the initial CO_2 emissions standard at 0.7 pounds of CO_2 per kilowatt-hour (kWh). The Oregon siting council subsequently revised the standard downward to 0.675 pounds per kWh because of new developments in natural gas-fired power plants. The siting law and rules also distinguish between base-load power plants (those that run all or most of the time), non-base-load facilities (those that run only some of the time), and non-generating facilities. The standards were set at the following levels:

- Base-load gas plants: 0.675 lb. of CO_2 per kWh

- Non-base-load plants: 0.7 lb. of CO_2 per kWh

- Non-generating facilities: 0.522 lb. of CO_2 per horsepower hour

If the power plant could not meet that standard — and it was understood at the time that most plants could not do so — then the developers had three options:

- Use co-generation to more efficiently use the energy from the plant.

- Invest in projects, such as tree planting, energy efficiency, or renewable energy, elsewhere to mitigate their emissions.

- Fund the Oregon Climate Trust, which then would fund offset projects elsewhere.

The Climate Trust

The Climate Trust receives and manages money from developers that decide to pay money instead of attempting to pursue and manage offsets on their own. The Klamath Falls power plant received an initial site certificate under the 1995 competition described above. Under that certificate, the developer decided to invest directly in tree planting and solar energy abroad. Later, when Klamath Falls expanded the size of its power plant from 318 to 484 MW (and therefore needed a new certificate) the developer paid slightly more than 1.5 million dollars to the Trust.

The amount that the Klamath Falls developer paid was based on the 57 cents per ton of CO_2 that the developer needed to meet the emissions target of 17 percent lower than the most efficient gas-fired power plant. The law allows the 57-cent figure to be raised by up to 50 percent every two years. In addition to the 57 cents per ton, the developers pay an additional 5 percent of the total amount to the Trust. This 5 percent helps to fund the process of selecting projects. The Trust then is required to use at least 80 percent of its funds toward direct CO_2 emissions reduction, or for carbon sequestration. Twenty percent of the funds may be spent on monitoring and evaluation of projects and enforcement of the contracts set up to provide the offsets.

The Trust has a two-person staff and a board made up of three people appointed by the Energy Facility Siting Council, a citizens board appointed by the governor, as well as three Oregon citizens appointed by an environmental group and one appointed by the power industry. In addition, each project developer who provides funding gets to appoint a non-voting member.

The Oregon Climate Trust decides how to spend the money it has received through an open competition. In the case of its first competition, the Trust received over 60 applications from 15 states and 15 different countries. The proposals ranged from energy efficiency programs, to renewable energy, to tree planting programs for carbon sequestration and some transportation programs.

The Board allocates the funds based on a number of factors, primarily whether the Trust funding will make a significant difference in whether the project goes forward or not, and the project's realistic potential for CO_2 reductions.

Conclusion

Since the Oregon siting law is so new, and only one facility has been fully sited under the new procedures, it is difficult to assess how well it is working. However, it appears to have potential to link two difficult issues — power plant siting and concern about emissions — through compromise and an open process. Some in Oregon suggest that it has potential to work well because of several factors:

- It only affects new, or greenfield, projects. It does not try to address older, existing projects.

- It is politically palatable because the requirements are not too draconian.

- The initial cost of the emissions reductions is fairly low, and has not placed such a burden on project developers that it has discouraged them from building in the state.

+

+

+

Solutions at the Local Level: Austin, Texas, as a Case Study

Maria Sanders, Abby Young

As the major population and economic centers of the world, urban areas are significant consumers of energy and thus significant emitters of greenhouse gases. As urban areas grow, so do demands for energy to support public infrastructure and the daily activities of urban life and commerce. Most urban areas are also spreading outward, resulting in sprawl-type land use patterns that encourage more driving, thus increasing fuel consumption and producing even more emissions.

Urban waste is another large source of greenhouse gas (GHG) emissions, which result primarily from two processes: The decomposition of organic waste in urban landfills releases methane; and, as large consumers of goods, urban areas are indirectly responsible for a large proportion of emissions related to the mining of raw materials and their manufacture into products.

Local governments have a substantial impact on energy use and waste generation within the communities they govern. As policy-making and governing institutions, cities and counties can use their influence, laws, and purchasing power in ways that increase energy efficiency, decrease dependency on the private automobile, and reduce waste, thereby reducing GHG emissions. By doing so, cities and counties can also reap the added benefits of improving air quality, reducing pollution, saving money, and enhancing the quality of life for their citizens.

In their attempt to create more sustainable communities, several U.S. cities and counties are taking critical first steps to address the issues of energy generation, sprawl, and waste management. These issues, combined with concerns about the impacts of climate change, provide an impetus for action. This chapter highlights one city — Austin, Texas — that is taking meaningful action to reduce GHG emissions and, at the same time, is reaping the many benefits associated with the wiser use of resources.

Local Climate Solutions – Austin, Texas

Located in southeast Central Texas, Austin is the capital of Texas and the state's fifth largest city. It occupies a land area of 225 square miles and has a population of 613,500. Austin is known for its high quality of life. It has an average annual temperature of 67.5°F, is surrounded by rolling hills, and is bisected by the scenic

The Pew Center and the authors wish to thank George Adams, Barsha Cohee, Roger Duncan, Katherine Sibold, and Nancy Skinner for reviewing the paper and providing helpful comments.

Colorado River. It boasts a vibrant and aesthetically pleasing downtown, high job growth, and low unemployment.

Once considered a large town, Austin is now at the center of a fast-growing metropolitan area. It is the focal point of the South's rapidly growing high-tech industry. Dell Computers is the largest employer in Austin. Other large employers include Applied Materials, Motorola, and IBM. Austin occupies most of Travis County, but its area of influence also includes Bastrop, Caldwell, Hays, and Williamson counties. In all, the Austin metropolitan area has a population of 1.25 million, which is projected to grow to 1.6 million by the year 2010. In terms of the built environment, the Austin area has been gaining on average 7,250 homes per year since 1992.

Given these growth pressures, Austin struggles with a fair number of environmental problems. A 1998 Sierra Club report on suburban sprawl listed Austin as the second most threatened mid-sized city in the United States. That same year, Texas A&M Institute for Transportation Studies released a study indicating that traffic had increased 9 percent each year since 1994. Air quality in the Austin metropolitan area, as well as the entire Central Texas region, has been declining to the point that it is now listed as a "non-attainment area" for ozone by the U.S. Environmental Protection Agency.

Austin's Climate Protection Planning Process

Austin's citizens and politicians have demonstrated a decades-long commitment to protecting the environment and the high quality of life in their community. Austin was one of the first cities in the United States to make a political commitment to address global warming at the local level. In 1995, the City Council passed a resolution enabling Austin to become a participant in the Cities for Climate Protection Campaign, which is coordinated by the International Council for Local Environmental Initiatives, a membership association dedicated to helping local governments find solutions to local, regional, and global environmental problems.

As part of this effort, the Austin City Council directed the city manager and staff to draft a Local Action Plan (LAP) to reduce GHG emissions by 20 percent below 1990 levels by the year 2010. Given that Austin occupies most of Travis County, the plan was based on information from Travis County and on strategies that could be implemented throughout the county.

To help ensure public support in the development of the LAP, Austin held a town hall meeting on climate change in 1997. Speakers included Timothy Wirth, then U.S. Undersecretary of State for Global Affairs; former mayor Garry Mauro, former Texas Land Commissioner; Gus Garcia, former Mayor of Austin; and Terry Thorn, Senior Vice President for Enron Corporation. In addition, several scientists and public policy professors from the University of Texas, Rice University, and Texas A&M made detailed presentations on the impacts of global warming on Texas' public health, environment,

Table 1

City of Austin/Travis County GHG Emissions Source Inventory		
Source	Tons CO$_2$ 1990	Tons CO$_2$ 2010
Electricity Generation	5,137,082	9,305,404
Private Ground Transportation	2,902,240	4,720,463
Public Transportation	27,750	50,967
Aircraft	282,681	601,059
Boats	13,525	21,570
Natural Gas	606,210	966,818
Landfills	635,292	991,462
Lawn and Garden Equipment	28,921	46,125
Total	9,633,701	16,703,868

Note: All GHG emissions are calculated in terms of CO$_2$ equivalents, in order to uniformly quantify both methane and CO$_2$ emissions resulting from energy and fuel use and waste decomposition. For carbon dioxide, tons of CO$_2$ equivalent and tons of CO$_2$ are the same thing. For methane, a stronger greenhouse gas, one ton of emissions is equal to 21 tons of CO$_2$ equivalent.

and economy. The town hall meeting was well attended by Austin area residents, business representatives, academics, and policy-makers. It helped establish support for the need to reduce GHG emissions within Austin.

In order to decide how to reduce GHG emissions, Austin policy-makers first had to know what types of activities were the source of the problem. Austin staff prepared a baseline inventory of greenhouse gases, which was based on data from residential, commercial, and industrial electricity use; use of fuel for transportation; and waste generation. The baseline provided information for city staff and elected officials on the sources and quantities of GHG emissions being produced in the community. This allowed the city to craft a local strategy to curb and hopefully reverse the growth of these emissions in the future.

The detailed baseline inventory revealed that in 1990 GHG emissions (as measured in CO$_2$ equivalents) totaled 9.6 million tons. The bulk of 1990 emissions (53 percent) came from electricity consumption by the industrial, commercial, and residential sectors. Emissions from private ground transportation accounted for 30 percent of GHGs, while methane gas from landfills accounted for another 7 percent. Direct uses of natural gas, such as in industrial furnaces and boilers and in residential gas stoves; operation of aircraft, boats and lawn movers; and public transportation comprised the final 10 percent.

Austin planners then forecasted the growth in emissions and determined that, without any attempt to reduce emissions, GHGs would increase by 73 percent to 16.7 million tons by the year 2010. The projected increase was based primarily on population growth, which, at the time of the analysis, was expected to increase 60 percent by 2010. As with the baseline inventory, most GHG emissions will come from a corresponding increase in electricity demand and vehicle miles traveled.

Based on the inventory and forecast, achieving the City Council's goal of a 20 percent reduction would require cutting emissions by 9 million tons — or approximately

Climate Change: Science, Strategies, & Solutions

54 percent below the projected 16.7 million tons in 2010.

Staff identified five types of GHG-reduction strategies that could achieve the 20 percent goal: energy efficiency, renewable energy, recycling, tree planting, and reductions in vehicle miles traveled or increased fuel efficiencies. It is important to note that these strategies result in other benefits for the local community. For instance, energy efficiency provides dollar savings that can be reinvested in the community, while decreasing reliance on private automobiles reduces traffic congestion and parking problems. In addition, all measures that reduce emissions from the burning of fossil fuels will decrease regulated, or "criteria," air pollutants such as sulfur dioxide and nitrogen oxides as well as carbon dioxide.

Each strategy was evaluated for its technical potential to make reductions consistent with the 20 percent goal. After the approximate technical potential of these various strategies was estimated (see Table 3), renewables were assumed to supply the remaining reductions — about 40 percent of the total goal.

Figure 1

City of Austin/Travis County 1990

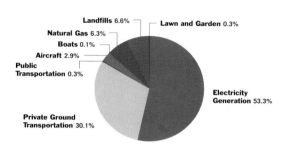

Source: Austin Energy Services, City of Austin.

Figure 2

City of Austin/Travis County 2010

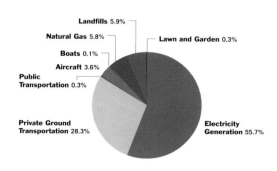

Source: Austin Energy Services, City of Austin.

City staff recognized that while reducing GHG emissions by 9 million tons was technically possible, it would be difficult to achieve given population growth and funding limitations. Therefore, the LAP also contained a second scenario outlining strategies that would reduce GHG by 10 percent, or 4.5 million tons, by 2010.

Table 2 provides the list of strategies proposed in the LAP when it was developed in 1997. It also estimates the GHG-reduction potential of each strategy under the 20 percent and the 10 percent goals. Since that time, new technologies and opportunities for reducing GHGs have become available. Several of these new measures, such as cost-effective recovery of landfill methane and public acceptance of Smart Growth land-use planning, will be discussed in the next section.

Austin's Climate Protection Success Stories

Austin has chosen to exercise a great deal of control over the types of operations, policies, and programs that result in GHG reductions. First, the city owns and operates its electric utility, Austin Energy, which has been able to use utility revenues to finance community-wide energy efficiency programs and to increase the share of renewable energy sources as part of its electricity mix. Second, the city has instituted curbside recycling programs and installed infrastructure that prevents landfill methane from escaping into the atmosphere. Finally, Austin has recently adopted new land use controls and permit fees intended to promote compact and transit-accessible development.

Energy Efficiency Programs

Currently, Austin's primary source of electricity is coal, which accounts for 54 percent of the total resource mix. Nuclear, natural gas, and a small percentage of wind, solar, and biogas (methane from landfills and wastewater treatment plants) comprise the remaining supply.

In 1982, the City Council called for the creation of energy conservation programs to delay the construction of a new power plant. It followed suit by amending its Building Code to include an Energy Code that established minimum standards for the efficiency of windows, insulation, air infiltration, lighting, and heating and cooling systems in all buildings. To date, Austin's energy efficiency programs have avoided the generation of an average 47,713 tons of CO_2 per year. In addition to CO_2 reductions, these programs have had other important air quality benefits: they have reduced annual emissions of sulfur oxides (185.8 tons), nitrogen oxides (176.2 tons), volatile organic compounds (1.6 tons), carbon monoxide (9.0 tons), and airborne particulate matter (4.9 tons).

Examples of these energy efficiency efforts, some of which are among the most innovative in the country, follow.

Municipal Buildings. The Municipal Energy Conservation Program was created in 1983 to implement cost-effective energy conservation improvement projects in city-owned or leased facilities. As customers of the electric utility, Austin's municipal departments spend an average of $17 million (1996 dollars) in annual utility operating costs. Through such improvements as lighting system changes, air-conditioning system replacements, and utility water treatment process improvements, the program has helped the city avoid an estimated $1.33 million per year in energy costs.

Table 2

City of Austin GHG Reduction Scenarios

Actions to Meet 20% Goal (9 million tons)	Annual Tons of GHG Reduced	Actions to Meet 10% Goal (4.5 million tons)	Annual Tons of GHG Reduced
Energy Efficiency	1.9 mil.		1 mil.
• Increase energy efficiency programs to save 47 MW per year		• Maintain energy efficiency programs at their current levels to save 32 MW each year	
• Dramatically expand public education, Energy Code requirements, and additional mechanisms to promote energy efficiency		• Dramatically expand public education, Energy Code requirements, and additional mechanisms to promote energy efficiency	
• Invest in cost-effective efficiency improvements to the electric power system		• Invest in cost-effective efficiency improvements to the electric power system	
• Encourage implementation of all cost-effective cogeneration projects		• Encourage implementation of all cost-effective cogeneration projects	
Renewable Energy Resources	3.7 mil.		1.9 mil.
• Meet 97% of electricity requirements above current level with renewable resources (40% of 2010 capacity)		• Meet half of electricity requirements above current level with renewable resources (20% of 2010 capacity)	
Transportation	2.9 mil.		1.3 mil.
• All Travis County workers take some action to reduce vehicle miles traveled		• 45% of Travis County workers take some action to reduce vehicle miles traveled	
• Expand public transit to 18% of total transportation level		• Expand public transit to 5% of total transportation level	
• Assume federal government increases fuel efficiency requirements to average 45 mpg		• Assume federal government increases fuel efficiency requirements to average 37.5 mpg	
Recycling	0.5 mil.		0.3 mil.
• Divert 50% of Travis County waste from landfill		• Divert 30% of Travis County waste from landfill	
Tree Planting	0.05 mil.		0.03 mil.
• Increase entire city to "well-canopied" by planting 1.2 million trees		• Aggressively expand current programs to plant more than 200,000 trees	
Totals	9.05 mil.		4.53 mil.

Private Sector Buildings. The high growth in home building in Austin presented a big opportunity for the city to reap significant energy savings by getting builders to include energy efficiency as a major component of newly constructed houses.

- *The Austin Energy Star Program.* The program, which ran from 1986 to 1993, rated the energy performance of newly constructed homes. Homes that voluntarily exceeded the minimum Energy Code requirements by a certain percentage were awarded between one and three stars, depending on the percentage exceeded. Builders of those homes then received free marketing assistance through the Austin Energy Star Program. This included special newspaper advertisements, brochures, public displays, an "Eco-Home Buyer's Guide," and other public outreach efforts.

In addition to the marketing assistance, two crucial elements of the program included technical assistance for builders on how to construct energy-efficient homes and flexible guidelines on how they could exceed the Energy Code requirements. While not an incentive-based program itself, Energy Star linked builders to such energy-efficiency incentives as the appliance efficiency rebates.

More than 6,000 homes were rated under this program between 1986 and 1993. Energy Star ratings demonstrably boosted the salability of a home, effectively pulling the larger housing market toward energy efficiency. In terms of program success, the average energy savings per home nearly doubled between 1986 and 1992. The Austin Energy Star program was the only U.S. recipient of a United Nations Local Government Initiatives Honor, presented at the Earth Summit in Rio de Janeiro in 1992.

- **Austin Green Building Program.** In 1993, the city rolled out a much more comprehensive approach to building development by expanding the Energy Star Program to include water conservation, environmentally sensitive building materials, and construction waste reduction. The city also expanded its scope to provide similar services to commercial building developers and to require a green building standard for all city-funded projects.

The expanded residential program, now called the Austin Green Building Program, continues to rate residential buildings, now on a scale of one to four based on the number of "green" features found in the home. It provides a flexible checklist for builders to consider ways to improve water and energy efficiency, reduce the environmental impacts of building materials, and reduce construction waste. Again, it is meant to help interested homebuilders voluntarily exceed the local building and land development codes by providing technical and marketing assistance. The program has recently expanded to include multi-family residential structures.

For commercial buildings, given their increased complexity and variability, the Green Building Program provides a checklist outlining measures to be considered during all phases of the development process — from site selection and project programming through construction management and building operation. Projects that go through the process are nominated for the Austin BEST (Businesses for an Environmentally Sustainable Tomorrow) Awards. Under the program, design teams can qualify for a cash bonus. The cash bonus is tied to energy efficiency rebates, with total rebate and bonus payments subject to a $150,000 cap per project.

The Austin Green Building Program has captured the imagination of many municipalities and regional Home Builders Associations throughout the United States. Since its inception in 1993, similar local programs have been or are being developed in almost every region of the country. It has also inspired a national rating system developed by the U.S. Green Building Council in conjunction with leading firms, trade associations, and companies involved in residential, commercial, and institutional building development.

Renewable Energy

Austin's Local Action Plan forecasted that even with its current energy conservation and green building programs, approximately 1,260 megawatts (MW) of new capacity would be needed to meet the demand for an additional 5,747 million kilowatt hours (kWh) of electricity in 2010, or about 40 percent of the entire 2010 capacity. It determined that meeting all this additional demand with renewable energy sources such as wind, sun, or methane gas from landfills (as assumed under the 20 percent GHG reduction target) was technically feasible if the cost of renewables continued to decline, the cost of fossil fuel power continued to rise, and subsidies for renewables were more widely available. However, the plan also indicated that current costs make this infeasible and that supplying 20 percent of the 2010 capacity with renewable technologies was more attainable.

In a separate action, the Austin City Council adopted a resolution requiring that 5 percent of Austin's electricity come from renewable sources by the year 2004. To meet this goal, Austin Energy has made significant investments in renewable energy resources and has devised innovative green purchasing programs to increase the share of renewables in Austin's overall energy mix.

Austin Energy is allocating an annual expenditure of $1 million or 2 percent of actual net income, whichever is greater, to purchase up to 100 MW per year of power generated from renewable energy sources. In addition, the utility is actively marketing a Green Choice Program in which citizens and businesses can opt to pay a premium (the difference between the cost of renewables and the cost of the traditional mix) for electricity from renewable sources in order to push Austin Energy's annual purchase of green power even higher. In this case, Austin Energy matches revenues from Green Choice subscribers on a dollar-for-dollar basis. In addition, the program gives consumers the option of paying 5 percent extra on their electric bills now in return for a freeze on that portion of their bills for the next 10 years. As of summer 2000, this strategy is paying off for Green Choice customers. The cost of electricity generated from fossil fuel rose, closing the gap between the Green Choice premium and regular electricity rates to one-tenth of one cent per kWh (down from four-tenths of one cent when the program started).

Green Choice subscriptions have helped raise $7.8 million. Austin Energy is hoping that the Green Choice Program can help dedicate up to 5 percent of actual net income each year to the purchase of renewable energy-generated power. At 100 MW per year, renewables could provide about 4 percent of Austin Energy's generation mix by the end of 2001.

In 1998, one half of 1 percent of Austin's energy mix came from renewable sources. By the end of 2000, this mix will increase to 2.5 percent, with a capacity of about 250 million kWh, or enough energy to supply 27,000 households. Total CO_2 reduction from renewables will be 187,500 tons per year. This reduction has the added benefit of reducing sulfur oxides by 730.2 tons, nitrogen oxides by 692.4 tons, carbon monoxide by 35.3 tons, particulate matter by 19.4 tons, and volatile organic compounds by 6.3 tons.

Austin's portfolio of renewable energy sources includes:

- 31 MW from wind-generated electricity;
- 24 MW from electricity from biogas (e.g., landfill methane); and
- 750 kW from solar panel installations.

Austin Energy has been operating a 3-MW methane-to-electricity facility at its Sunset Farms Landfill, which produces two million cubic feet of methane gas every day. In order to take advantage of this resource and to keep the harmful gas from reaching the atmosphere, the city entered into an arrangement with a private energy company to establish a methane-to-electricity facility in 1995. The facility has been so successful that Austin Energy is entering into long-term, turnkey contracts with private energy companies to build and operate similar facilities at six other landfills throughout southern Texas.

Austin Energy is developing its solar panel installations in conjunction with a nationwide consortium of utilities dedicated to increasing demand for solar-generated power and reducing its production costs. As a part of this consortium, Austin launched its Solar Explorer Program, offering residents and businesses in Austin Energy's service area the opportunity to underwrite the cost of 50-watt "blocks" of solar energy for as little as $3.50 a month. During its first year of operation, the program enrolled over 450 sponsors for a new 32-kW photovoltaic system that was installed at a park-and-ride lot and later moved to the taxi stand at the Austin-Bergstrom International Airport. The system, mounted on ten-foot pillars, does double duty by providing shade for parked cars as well as generating solar energy. Additional solar installations are located at the Austin Public Library (producing 10 kW) and on buildings at the airport (producing 111 kW). Total production capacity of these three facilities is 153 kWh.

Solid Waste Management – the Four Rs

Landfills produce methane gas as food and plant waste, wood, paper, and other organic materials buried in them decompose. Methane is a very powerful greenhouse gas, contributing 21 times more per unit to global warming than carbon dioxide. Local governments have jurisdiction over the municipal solid waste stream. Thus, they can have a significant impact on landfill emissions in two ways: first, by diverting waste from landfills through recycling efforts, and second, by recovering the methane that is emitted from the landfill. Austin has chosen to do both.

Financed by Austin Energy's Solar Explorer program, this 32-kW photovoltaic array provides shade for up to 40 waiting taxis at Austin-Bergstrom International Airport..

Photo courtesy of Juan Carrasco at Austin Energy.

Reduce, Reuse, Recycle

Approximately 1.5 million tons of waste goes to landfill in the Austin area each year. The city's Solid Waste Services Department has greatly expanded its recycling efforts in recent years, focusing on increasing the amount of residential and commercial waste being diverted from landfills. The recycling rate in 2000 is 27 percent, avoiding the generation of 40,000 tons of GHGs per year.

Austin's Curbside Recycling Program serves single-family homes, duplexes, and multi-family building residences. Materials collected at the curb include newspapers, magazines, catalogs, tin, steel and aluminum cans, glass and plastic containers, and corrugated cardboard.

The city picks up residential yard waste and tree trimmings. These materials are chipped, and then mixed with composted wastewater sludge to produce "Dillo Dirt," a garden soil enhancer that is sold to garden stores, nurseries, and offered to the public. The city also provides a backyard-composting program to residents. Currently, nearly 5 percent of all single-family households are participating in this program.

In addition, recycling goals have been incorporated into the Green Building Program by giving points for new homes that recycle construction waste or include recycling and composting storage areas.

Recovery

As discussed under Renewable Energy, Austin Energy is developing methane-to-electricity facilities at six landfills in Texas. By reusing landfill methane to produce electricity, these projects are not only keeping methane out of the atmosphere, they are offsetting the need to burn fossil fuels for energy, thus further cutting GHG emissions.

Transportation and Land Use

Local government decisions that determine the type of land uses and infrastructure that are or are not provided within a community contribute to the level of a community's dependence on automobiles. In U.S. cities, energy use in the transportation sector is typically four times higher than in Western European cities, largely due to land-use decisions that create sprawl and a lack of effective public transportation options.

In a high-growth city such as Austin, managing growth, let alone reducing the number of miles people drive, is particularly challenging. The more widespread sprawl becomes, the more people become dependent on their cars to get around. Recognizing this fact, the Austin City Council has made a commitment to the Austin Smart Growth Initiative. The key goals of the Initiative are to:

- Invest in the central city and older suburbs to restore community vitality; and

- Promote patterns of development that have a balanced, more compact mix of land uses that can support a transportation system that accommodates pedestrians and transit.

In particular, Austin's Smart Growth Initiative seeks to focus growth toward a Desired Development Zone (DDZ) and discourage growth in environmentally sensitive areas, especially the watersheds from which Austin gets its drinking water — the Drinking Water Protection Zone (DWPZ).

Unveiled in 1998, the Initiative accomplishes its goals through a combination of zoning and incentives. There are several levels, or tiers, of Smart Growth Incentives. These incentives are available only within the DDZ and include reductions in fees the city charges for zoning, subdivision, and site plan applications, and for water and wastewater capital recovery fees. Within the DDZ, these fees are reduced on a sliding scale based on where the project is located. Within the DWPZ, development application fees are not reduced and capital recovery fees are slightly increased.

In addition, certain developments within the DDZ may receive additional incentives to help offset the high cost of developing in urban areas. These incentives may include a waiver of development fees and public investment in improving such infrastructure as water and sewer lines, streets or streetscape improvements, or similar facilities. These incentives require City Council review and approval. To assist the City Council in analyzing how well these development proposals meet the Initiative's goals, a "Smart Growth Matrix" was designed to assign a priority or weight and measure the location of the development, its proximity to mass transit, its compact development characteristics, and compliance with nearby neighborhood plans.

Finally, the Smart Growth Initiative also provides a variety of incentives for large employers to establish their businesses in and relocate to the DDZ. In many cases, these types of employers generate significant levels of growth, both within their specific project and in the surrounding area, as new residences, commercial and retail services, and other related businesses follow. By directing these employers to the DDZ, the city can have a significant impact on long-term growth and commuting patterns. Incentives include fee waivers, new water and sewer lines, transportation improvements, and expedited processing of development applications.

Conclusion

Austin's energy efficiency and Green Building programs, its renewable power purchase, waste reduction efforts, and Smart Growth Initiative are examples of the types of policies and programs that the city is implementing as part of the Cities for Climate Protection Campaign. The measures Austin chose were selected as much for their ability to address existing community concerns — like air quality and suburban sprawl — as for their potential to reduce GHG emissions.

While many of these measures (especially those related to energy savings) can be quantified to demonstrate a direct GHG reduction, indirect reductions in emissions are more difficult to document. For instance, while decreased demand for products with high-energy inputs will be a result of the recycling and green building programs, it is difficult to calculate attributable upstream reductions in GHGs. In addition, some strategies selected for the LAP (such as vehicle trip reduction) have not been consistently tracked to allow sufficient data for accurate quantification of emissions reductions. That being said, the measures Austin has been able to track (including various measures not described here), have reduced CO_2 emissions an average 292,900 tons per year since 1990.

Preventing climate change is not the only or, in some cases, the primary reason local governments such as Austin are acting to reduce GHG emissions. Local governments understand that reductions in GHGs are indicators of other important benefits, such as improved air quality, lower energy bills and thus more disposable income, and more livable communities.

Intel Corporation: Environmental Leadership in Industry

Timothy G. Higgs

It is a well-known fact that human activities in our modern technological society create wide-ranging environmental impacts. This is true of our methods of transportation, generation of electric power, and the manufacturing processes that create the products we consume. In fact, it becomes increasingly true each year as the world economy grows and more of its citizens struggle to achieve higher living standards. In the last few decades, world population and economic output have both grown dramatically. This growth places increasing strain on our precious natural resources.

At the same time, public awareness of environmental issues has increased greatly and, as a result, public scrutiny of the environmental performance of the business world has increased with it. It is no longer acceptable for companies to focus simply on profits, without considering environmental costs. Shareholders, neighbors, employees, and others now rightly expect that the companies they invest in, live near, or work for run their businesses in a responsible manner.

This growing awareness has led to many changes in methods of doing business. The modern environmentally responsible company now strives to be a leader in driving environmental improvements, rather than waiting for those improvements to be forced on them by government agencies.

It would be hard to deny the positive results that have come from this increased focus on environmental issues. They include dramatic improvements in U.S. air quality. In just the last 10 years alone, lead levels have been reduced by more than 75 percent, while carbon monoxide and sulfur dioxide levels have been cut by more than 37 percent, according to U.S. Environmental Protection Agency (EPA) data. Since 1970, overall levels of the six "criteria" pollutants in the Clean Air Act are down by more than 32 percent.

This progress is particularly notable when the growth of population and pollution-generating activity are taken into consideration. Since 1970, for example, the number of vehicle miles traveled (a gauge of the growth in the use of the automobile — traditionally a major source of many of the criteria pollutants) has more than doubled.

The author and the Pew Center would like to thank Tim Mohin and Greg Slater for their review of early drafts of the report, and Dave Stangis for his efforts in compiling information from documents previously published by Intel.

Making a Commitment to Environmental Responsibility

Significant environmental issues still remain, and resolving these issues will require the combined talents of all segments of society. The business world, with its vast resources and large number of well-educated and talented employees, is in a unique position to take a leadership role on these issues. Many companies recognize this and have developed very progressive environmental programs. At Intel, our company environmental, health, and safety policy states, "Intel Corporation and its subsidiaries are committed to achieving high standards of environmental quality and product safety, and to providing a safe and healthful workplace for our employees, contractors and communities." The policy goes on to state Intel's commitment to conserving natural resources and reducing the environmental burden of its operations. This policy has been widely distributed. It serves to notify the public of Intel's commitment and to set a standard that presents a challenge for the company to meet.

Of course, a policy by itself does not produce environmental improvements. Those only come when the beliefs stated in the policy are practiced in everyday operations. Though simple in concept, this is the real challenge of environmentally responsible manufacturing. Developing improved processes that lessen environmental impact requires significant commitment and dedication of resources. In the past, managing environmental impacts typically meant installing various sorts of pollution treatment equipment at the back end of a process, thereby reducing pollution levels before discharging to the environment. In most cases, these approaches were mandated by government regulation.

While end-of-pipe treatment represented a good first step in reducing the environmental impacts of industrial production — and will always have a role in the process — it clearly has limits. The treatment systems themselves consume energy and resources, and many generate byproduct pollutants in the course of treating the pollutants of concern. Even the most efficient systems cannot remove or destroy 100 percent of the pollutants being sent to them, so a better solution is to design environmental improvements into the manufacturing process itself, thereby eliminating or reducing the need for after-the-fact treatment. The best-run, state-of-the-art-pollution control system may be capable of removing 99 percent or more of the pollutants sent to it, with the remaining 1 percent escaping to the environment. If a pollutant is never generated in the first place, however, even the 1 percent that would have passed through the treatment process is prevented from entering the environment.

The process of developing such innovative solutions is difficult, requiring extensive experimentation and significant resources. Government regulations must support this new approach by providing industry with the flexibility it needs, rather than focusing strictly on prescriptive end-of-pipe mandates and lengthy reviews of manufacturing changes. Rather than acting as adversaries, government and industry need to explore more ways to be partners in finding environmental solutions, with each side drawing from the extensive expertise of the other. The potential of this approach has been demonstrated in recent years.

The challenge for industry is to develop means of meeting the ever-expanding needs of a growing world while reducing the environmental impact of the production processes that make this happen. This concept is often referred to as sustainable development. Because of its role as a large consumer of energy and other resources, and a producer of the goods that society requires, business is on the front line of making sustainable development a reality. Doing so requires taking a hard look at two complementary aspects of sustainability — the impact that occurs from manufacturing and distributing the product, and the impact of the product itself upon the environment.

At Intel, a "design for environment" (DFE) program was developed years ago with a goal of reducing the overall impacts from the manufacturing processes. In the rapidly changing semiconductor industry, where new products and manufacturing processes are developed and introduced about every 18 months, there is excellent potential for making ongoing environmental improvements this way. The DFE program has been very successful in making environmental improvements to the manufacturing process. Emissions of volatile organic compounds (VOCs) were reduced by 50 percent per unit of production over a few years, and company-wide VOC emissions have actually been decreasing for the past several years in spite of greatly increased manufacturing, as illustrated in Figure 1. In addition, certain chemicals of concern, such as chlorofluoro-carbons and certain glycol ethers, have been eliminated from the manufacturing process, and careful review of chemicals before they are introduced into the manufacturing process has facilitated an increase in the ability to recycle chemical waste.

Figure 1

VOC Emissions from Intel Facilities Worldwide

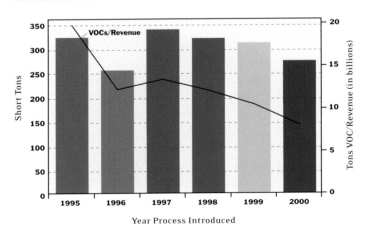

Global Warming – A New Challenge

More recent, and possibly even more challenging, projects have focused on reducing greenhouse gas (GHG) emissions from the manufacturing process. This issue is especially difficult, because semiconductor companies rely very heavily on a class of compounds known as perfluorocompounds, or PFCs, which are believed to have significant global warming potential. PFCs have long been used in the industry and have a number of beneficial properties. For starters, they are essentially non-toxic, so they are very safe to handle and use. PFCs and other fluorine-containing compounds are useful both for etching circuitry into silicon wafers and for cleaning the quartz reaction chambers in which many of the process steps occur. PFCs are ideal for these uses because they are highly stable and can be broken down at a very controlled rate using a high-energy plasma. The breakdown produces the fluorine needed to etch silicon, and the tight control over the rate is essential in the highly precise world of semiconductor manufacturing. For these reasons, addressing global warming concerns that result from the use of these compounds is a significant challenge.

Many scientists consider PFCs to be important global warming compounds, even though they account for only about 4 percent of total greenhouse gas emissions in the United States. Of those, only about 3 percent are from the semiconductor industry, so in total, PFC emissions from semiconductor manufacturing account for well under 1 percent of total U.S. greenhouse gases.[1] However, these materials persist in the environment for a very long time (see Table 1), and Intel and other companies realize that good environmental stewardship requires taking action regardless of the ultimate outcome of the ongoing global warming debate.

Government and public focus on PFCs and global warming in general increased after the 1992 Rio Earth Summit and the 1993 release of former President Clinton's Climate Change Action Plan. At about this time, the semiconductor industry began a series of meetings with the EPA to discuss possibilities for reducing these emissions.

Table 1

Commonly Used PFCs in the Semiconductor Industry

Chemical	Atmospheric Lifetime*	Global Warming Potential
C_2F_6 (hexafluoroethane)	10,000 yrs.	9,200
NF_3 (nitrogen trifluoride)	50 – 740 yrs.	8,000
CF_4 (carbon tetrafluoride)	50,000 yrs.	6,500
SF_6 (sulfur hexafluoride)	3,200 yrs.	23,900
CHF_3 (trifluoromethane)	264 yrs.	11,700

*Atmospheric lifetime and GWP values for all chemicals other than NF_3 are taken from "Inventory of U.S. Greenhouse Gas Emissions and Sinks," published annually by the U.S. EPA. Since NF_3 has not yet been listed in this report, its values are derived from estimates by the chemical manufacturer. Global warming potential is relative to CO_2 — i.e., a GWP of 1,000 represents a material that is 1,000 times more potent a global warmer than CO_2, on a per-molecule basis.

The ultimate result was a memorandum of understanding signed by the EPA and the industry, which established goals to identify global warming solutions and called for periodically reporting progress to the EPA. As this process was occurring, Intel and other semiconductor companies began aggressive programs to redesign their manufacturing processes to reduce GHG emissions.

Since PFCs are critical to the manufacture of semiconductors, reducing their use and resultant emissions presented significant technical challenges. In addition, since PFCs had not been used in large quantities compared to other industrial chemicals and had not been the focus of past regulation, readily available treatment systems did not exist. Although Intel's emissions reduction program has required many different approaches, and a great deal of trial and error that continues to this day, it already has been very successful. For example, the newest manufacturing process at Intel generates only about 10 percent of the amount of PFCs per unit of production generated by the process that was introduced in 1996 (see Figure 2). In 1999, the effort was expanded into a worldwide agreement among other semiconductor companies to reduce emissions to 10 percent below 1995 levels by the year 2010. Since worldwide semiconductor manufacturing is expected to be many times larger in 2010 than it was in 1995, this represents an enormous challenge. Nonetheless, it is one the industry believes is worth pursuing.

Expanding the Scope of Environmental Responsibility

In many ways, the program described above provides a good example of new approaches to dealing with environmental problems. The accomplishments so far have

Figure 2

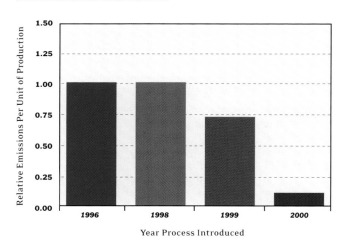

Relative PFC Emissions by Manufacturing Process

been voluntary — not driven by government regulation. The solutions have been varied; many are focused on reducing pollution generation rather than on end-of-pipe treatment. The projects discussed so far, however, are typical of traditional environmental approaches in the way they are focused strictly on the environmental impact of the manufacturing process.

Environmental management must now look more broadly and consider the impact of the manufactured products after they are sold for their intended purpose. In the early 1990s, the EPA estimated that 5 percent of all commercial electricity was being used to run computers — a figure they expected to double by 2000. Since a large percentage of the world's computers contain Intel chips, this was an issue that interested us greatly. Therefore, when the EPA implemented the Energy Star Computers program in 1992 to promote the manufacture of more energy-efficient computers, Intel was the first microprocessor maker to join. The product of this effort was a computer that enters into a "sleep" or standby mode when not in use, and cuts energy consumption during those times by 80 percent or more. According to more current EPA estimates, Energy Star computers are now estimated to save up to 10 million tons per year of CO_2 emissions.[2]

Minimizing the environmental impact of the manufacturing process and the product itself represents a great achievement, but it would be even better if the product could end up having an overall positive impact on the environment. Computer chips offer just such a possibility. Semiconductors and electronics play a significant role in increasing the energy efficiency of industrial processes, automobiles, residential heating and cooling systems, and many other amenities that we use on a daily basis. As the use of personal computers and other electronic communications devices has increased, telecommuting has become increasingly popular. This has significant potential to reduce emissions caused by vehicles. In fact, the U.S. Department of Transportation has estimated that telecommuting has the potential to reduce travel by up to 35 million vehicle miles per year, which would equate to about 20 million tons of global warming emissions. The role of electronic technologies in reducing global warming emissions was recently spotlighted in a World Resources Institute (WRI) report, "Taking a Byte Out of Carbon."

Being a Leader in the Community

Even after the environmental impact of the finished product has been considered, the responsibilities of the modern environmentally responsible company are not over. The business world is an important part of a community, and large employers especially have a very visible presence. As leaders, they have an opportunity to play a substantial role in local environmental issues and activities.

These activities can take many forms. Many Intel facilities have participated in local programs focusing on ride sharing and trip reduction, which have the dual benefits of reducing both local smog and greenhouse gas emissions. In the Phoenix metropolitan area, Intel was one of the key participants in a program aimed at avoiding violations of the national air quality standard for ozone and at preventing the area from being redesignated as a "severe" ozone non-attainment area. The Intel New Mexico facility worked with the city of Rio Rancho, NM, to promote a local water conservation program, the facility has implemented very aggressive water recycling and conservation programs.

Semiconductor manufacturing employees inside the "cleanroom" working on a manufacturing tool.

In some of the international locations where Intel operates, the environmental regulations and technologies are less developed than they currently are in the United States. In those areas, Intel applies the same environmental standards to its operations as those applied in the United States or other locations with stricter standards. We have also actively worked with local governments in those areas to promote environmental improvements and to share key lessons learned from programs in the United States. At all Intel locations, employees are frequently involved in local programs to promote environmental awareness or to work on specific environmental projects. All of these examples illustrate ways in which the modern company can be a leader on environmental issues, and not limit its focus to meeting the bare minimum regulatory requirements. As more focus is placed on complex problems like global warming, which promise to defy traditional narrow regulatory fixes, more of this kind of leadership will be needed.

Solving Problems Through Partnerships

Needless to say, the results that can be achieved by any company or individual working alone are limited. To truly work toward sustainable development, the operations need to be fully integrated and coordinated with other businesses, the community, and government agencies.

Manufacturing complex products like semiconductors requires the use of many raw materials and chemicals, as well as a large number of manufacturing tools. It would be impossible to manufacture environmentally friendly products, or manufacture them in a way that minimizes environmental impacts, without putting significant effort into selecting the right materials and tools in the first place. For that reason, partnerships with equipment and chemical suppliers have been important in driving environmental improvements in the industry. At Intel, new chemicals are subject to extensive environmental and safety review before use is allowed. Manufacturing equipment is required to meet emissions and discharge limits. The U.S. semiconductor industry as a whole is involved in partnership programs with its suppliers of both chemicals and manufacturing tools that establish safety and environmental requirements and identify areas for future improvements.

Suppliers and semiconductor manufacturers have worked very closely to develop manufacturing processes and tools with lower PFC emissions. The solutions have taken many forms, including using new chemicals, optimizing manufacturing tools, and developing new technologies for treating emissions. Obviously, such a complex set of solutions requires the expertise of all these parties, as well as research agencies and collective industry bodies such as the Semiconductor Industry Association (SIA) and Sematech, a consortium of U.S. semiconductor companies.

Innovative Solutions – Working with Government

The idea of partnership should be extended to government agencies as well. In the global warming example, the U.S. EPA and the semiconductor industry began discussing possible voluntary actions to reduce emissions many years ago. This ultimately evolved into the initial voluntary agreement to study means of reducing emissions (the first memorandum of understanding, or MOU, discussed above), and that in turn led to the worldwide reduction agreement among semiconductor manufacturers. The U.S. industry is now working on formalizing that agreement in a second MOU with the EPA. The agreement should lead to significantly reduced global warming emissions, and has already resulted in new technologies and techniques for managing emissions that didn't exist as recently as a few years ago. Examples include recycling and treatment systems for PFCs and new techniques for optimizing PFCs used in the process so emissions are minimized.

The semiconductor industry has long believed in this cooperative approach and has therefore been among the leaders in the drive to reinvent environmental regulation in ways that provide more flexibility while maintaining environmental protection. This has largely been inspired by the industry's need for constant innovation, which requires numerous and rapid changes in manufacturing processes. A large "fab" (semiconductor manufacturing plant) contains hundreds of individual manufacturing tools, and can make hundreds of minor changes and adjustments in the manufacturing process over the course of a year. Traditional regulatory approaches that require extensive review of even minor process changes are simply incompatible with modern, fast-changing high technology industries. In fact, these regulations can actually impede progress by subjecting to review and delay even those changes that may reduce emissions. In an industry where delays can mean missed market opportunities, this may prevent worthwhile emissions reduction projects from being done.

Concerns like these prompted Intel to become involved with regulatory reinvention efforts like Project XL and the Pollution Prevention in Permits Pilot (P4) program. These programs start with a core belief that environmental protection and economic competitiveness are not incompatible goals. In fact, the projects have demonstrated that more flexible regulation can even help drive environmental improvements by allowing companies to experiment with new solutions.

Under the P4 program, Intel's Oregon facility received a "major source" air permit with special provisions that increased the ability to make process changes without regulatory delay, in exchange for a commitment to incorporate pollution prevention into its operations and report on the success of the pollution prevention program. The program was so successful in reducing emissions that, upon expiration of the permit, the company was able to repermit the facility as a "minor source" of emissions, which carries with it less regulatory burden than the major source program.

Under Project XL (for "excellence" and "leadership"), Intel entered into an agreement to demonstrate superior environmental performance in exchange for increased operating flexibility and public disclosure of environmental impacts. Environmental goals were set for such things as waste recycling, overall air emissions, and water use. Results against these goals are regularly reported to the public; significant environmental improvements have been seen as a result, and the public has far more information about the facility's environmental performance than it would have under a traditional permit program. The underlying concept in both these programs is the same: companies should be held accountable for their environmental results but allowed flexibility in determining how best to achieve those results. Intel's view of the subject was well stated by Chairman Emeritus Gordon Moore who said: "What I see is a real paradigm shift. The old way of doing business was that government dictated every move a business must take to protect the environment. The new system, envisioned by Project XL, is to work cooperatively and focus on the results: a cleaner environment; a faster, less costly system; and more input from the local community."[3]

The same concepts apply to the recent global warming emission reductions efforts as well. It is significant that much of Intel's experimentation and technology development efforts to reduce PFC emissions were done at the Oregon facility that was operating under the P4 permit. As discussed above, a significant amount of experimentation was needed as was simple trial and error. New equipment and various types and combinations of process chemicals were evaluated before final solutions were developed in a process that continues today. Many of these actions may have triggered a permit review, and potential delays, under a traditional, less flexible regulatory approach and this clearly would have limited the number of solutions evaluated and ultimately, the success achieved. Instead, the more flexible approach allowed the innovation that was needed to reach environmental improvements.

Intel is trying to apply many of these same lessons to key environmental issues that arise at its overseas locations. Many European nations have been very active in developing legislation aimed at minimizing waste disposal problems by requiring manufacturers to take back products from the consumer at the end of the product's life, and by setting strict product content rules. Intel is participating in the development of these programs and believes we have an opportunity to help shape them in a manner that is efficient and flexible for industry while doing what makes sense for the environment.

Partnering with the Community

Members of the local community also have an important role to play in these efforts. In essence, when a company is asking for increased regulatory flexibility, it is also asking for less oversight when it comes to reviewing the impact of manufacturing process changes. It is the opinion of Intel and many other companies that this flexibility should be allowed, provided the company can still demonstrate to the agency and the public that it is meeting its emissions limits and other regulatory requirements. Some are concerned about this approach because they view the regulatory agencies as their means of ensuring that companies around them are following the rules. To them, reducing oversight sounds like an opportunity for those companies to bend or even violate the rules. This places a burden on the company to prove to the public that it is operating in an environmentally sound manner, and that it can be trusted to do so in the future. Any company wishing to achieve increased flexibility, and establish itself as an environmental leader, must recognize this obligation for public disclosure and accountability.

Building this sort of trust with the public and the agencies has been critical to the success of programs like Project XL and P4. Regular public meetings were held during the development of Project XL, and environmental emissions data for the factory are routinely placed on a publicly available website. There is also a community advisory panel that continues to meet to review environmental progress under XL. In this way, the community has played a very active role in the project's implementation.

Though this process can sometimes be difficult, it is clearly preferable to the old style of industry and the community facing off on opposite sides of a public hearing to argue the relative merits and problems associated with the project. The communication that results from these situations is invaluable. In many cases, members of the public may be concerned about an issue that the industry never would have thought would be a big concern. In other cases, members of the public may find that some of their worst fears were unfounded. Through efforts like these, industry has become much more aware of the types of concerns the public has, and many members of the public have come to realize that industrial production and environmental protection are not incompatible goals.

The same willingness to work openly and proactively is also an important part of current efforts to reduce GHG emissions. One of the reasons the EPA approached the semiconductor industry about a voluntary global warming effort was because the industry had in the past established itself as one that took environmental issues seriously, such as when it aggressively pursued CFC elimination. Many in the public have also recognized the industry for such efforts, though of course there are some who remain unconvinced that this (or perhaps any) industry is serious about environmental matters. The global warming issue provides another opportunity for environmental leaders in industry to demonstrate their commitment and be a key part of the solution.

Summary

Solving complex environmental problems like global warming will require new ways of thinking about environmental problems, and a willingness to try new approaches. Moving forward from here will require the combined skills and expertise of all parties with a stake in environmental protection — which is to say, all of us.

Industry must recognize its opportunity to be an important leader on these issues. Customers, employees, shareholders, and neighbors want environmentally responsible manufacturing and environmentally friendly products. It is no longer sufficient to feel that the company's responsibility ends once environmental requirements are met. Instead, industry must lead the way in creating sustainable processes and products, as well as play an active role in important environmental issues in the community.

Government agencies, and citizens in general, must also be willing to look at industry in a new way. Some believe that industry will only behave in a responsible manner when forced to do so by rigid regulations, or when threatened with citizen lawsuits. Unfortunately, one can still find examples today that would support this view. However, even more examples exist of companies showing a commitment to environmental leadership and to developing innovative solutions. This new reality must be recognized by regulatory approaches that provide the freedom to innovate (as the development of future solutions will require) and by a willingness to trust that industry is indeed committed to these issues. We in industry must recognize that developing this trust requires an open approach and willingness to share data about the environmental impacts of our operations and engage in routine communications with the communities around us.

Global warming and other new environmental concerns are far more complicated than most of the issues that have been dealt with in the past. Solving problems like these will require the combined skills of all parts of society, and will require all of us to realize that we must be allies, instead of adversaries, in achieving our mutual goals.

Endnotes

1. Figures taken from the U.S. EPA publication, *Inventory of Greenhouse Gas Emissions and Sinks, 1990–1998.*

2. From a model developed by the U.S. EPA to estimate personal computer (PC) energy consumption and the environmental savings attributable to the Instantly Available PC.

3. Quote from Intel Chairman Emeritus Gordon Moore was previously published in *Environmental Champions*, a supplement to *Chemical Engineering and Environmental Engineering World.*

Taking Inventory of Corporate Greenhouse Gas Emissions

Christopher Loreti, William Wescott, Michael Isenberg

At a Pew Center conference on early action held in September 1999, DuPont announced plans to reduce its greenhouse gas (GHG) emissions 65 percent from 1990 levels by 2010. BP also has announced plans to reduce GHG emissions — by 10 percent from 1990 levels by 2010 — and has implemented an internal emissions trading system across all of its businesses. United Technologies Corporation (UTC) plans to cut energy and water usage by 25 percent per dollar of sales by 2007.

Motivated by factors ranging from a desire to monitor and reduce energy consumption to concern for the environment to anticipation of future requirements to cut emissions that contribute to climate change, a growing number of companies are voluntarily taking action to reduce their emissions. This report provides an overview of how GHG emissions are estimated and reported in emissions inventories. It highlights a variety of approaches taken by companies to identify and track their emissions, and provides insights from their experiences.

By properly accounting for their GHG emissions and removals (sinks), corporations have an opportunity to establish a foundation for setting emissions reduction goals and measuring progress in attaining these goals. Emissions inventories also allow corporations to: evaluate cost-effective greenhouse gas reduction opportunities; clearly communicate with their stakeholders; contribute to the development of accurate national inventories; and provide data that support flexible, market-oriented policies.

The 1997 Kyoto Protocol, under which industrialized countries pledged to collectively reduce their greenhouse gas emissions to roughly 5 percent below 1990 levels during the period 2008–2012, is one impetus for conducting emissions inventories. Although the Protocol has not been ratified by any major industrialized nation, and it is unclear how corporations would be affected by any such plan to reduce emissions, knowledge of current emissions (and means for their reductions) is essential for companies to understand how the policy options currently being debated might affect them and how they should participate in the debate.

Inventorying GHG emissions and emissions reductions is necessary to document the effects of voluntary actions taken to reduce emissions and to enable companies to

This paper is a condensed and updated version of a report published in August 2000 by the Pew Center: An Overview of Greenhouse Gas Emissions Inventory Issues. The authors and the Pew Center wish to thank the companies featured here for their stories and insights, and acknowledge the members of the Center's Business Environmental Leadership Council, as well as Janet Raganathan, for their review and advice on an early draft of the original report.

claim credit for these reductions. Even if marketable credits are not a primary reason for conducting an inventory, the historical documentation of inventories may be useful in ensuring that companies are not penalized in the future for any voluntary reductions they make today. By accurately inventorying emissions, companies will be in a better position to count current reductions toward those they might be required to make under a future regulatory regime.

Other reasons cited by companies for conducting emissions inventories are primarily financial. Conducting inventories in conjunction with energy measurement and conservation programs enables companies to identify opportunities to save energy and, at the same time, cut costs and emissions associated with energy usage.

This report provides an overview of key issues in developing greenhouse gas emissions inventories, with particular emphasis on corporate-level inventories. The purpose is not to develop or propose a protocol for use in conducting inventories, but rather to illustrate the range of approaches being taken by different organizations and corporations in inventorying and reporting their emissions. No particular approach or methodology for conducting inventories is advocated, nor are any particular policy positions taken. Instead, because potential future requirements for reporting emissions are uncertain, and at present reporting is a voluntary activity, pragmatic considerations for dealing with different possible future scenarios are emphasized.

Specifically, the paper addresses these major areas:

- Effective inventories: purpose, scope, and principles;
- Greenhouse gases, baselines, and metrics;
- Drawing the boundaries; and
- Conducting the inventory.

Effective Inventories: Purpose, Scope, and Principles

Companies face a range of options in how they conduct their GHG inventories. While no two companies may address these options in exactly the same way, experience to date indicates that effective inventories share several attributes: a clear purpose, a well defined scope, and adherence to the principles of flexibility, transparency, simplicity, and innovation.

Purpose

The purpose of conducting an inventory differs from company to company. Since the purpose of the inventory will determine how it is approached, a company needs to have a clear understanding of the purpose of its inventory at the outset, taking into consideration: the nature of its emissions, how the inventory will be used, and which emissions to include in the inventory.

Companies with experience in conducting emissions inventories have found the following guidelines useful in determining the best approach to take:

- Start by understanding your emissions. Knowing the relative magnitude of emissions coming from various sources is necessary to understand whether or not they are material contributors to a firm's total emissions. Understanding the nature and the number of the emissions sources will facilitate the use of the inventory development guidance that is becoming available.

- Understand the likely uses of the emissions inventory. Companies conduct GHG emissions inventories for purposes that range from internal goal-setting to external reporting to obtaining financial benefits. These different uses of the inventory information imply different levels of completeness, accuracy, and documentation. Each organization will need to reach its own conclusion as to the cost/benefit balance of developing its inventory, depending upon its set of likely uses.

- Decide carefully which emissions to include by establishing meaningful boundaries. Questions of which emissions to include in a firm's inventory and which are best accounted for elsewhere are among the most difficult aspects of establishing GHG emissions inventories. Since the purpose of conducting an inventory is to track emissions and emissions reductions, companies are encouraged to include emissions they are in a position to significantly control and to clearly communicate how they have drawn their boundaries.

Inventory Scope

GHG emissions inventories may be conducted to report on emissions on a facility, entity-wide (corporate), or project-specific basis — or to report on the emissions of a product over part or all of its life cycle. These types of inventories are not mutually exclusive, and many companies conduct more than one type. By clearly defining the scope of their inventories from the start, companies will be in a position to report their emissions in whichever way they desire.

Company-wide emissions inventories are usually derived from facility inventories. Fossil fuel consumption and other activities that lead to GHG emissions are usually tracked at the facility level, and thus GHG emissions are estimated for individual facilities. The corporate inventory is simply the sum of the emissions from the facilities that make up the corporation.

Project-specific inventories are used by organizations to track and report specific emissions reduction projects, and firms may report on emissions reduction projects without conducting inventories of their entire operations. Indeed, the vast majority of the reports that are submitted to the U.S. Department of Energy's voluntary reporting program developed under Section 1605b of the Energy Policy Act, which allows for both project-specific and corporate-wide reporting, are for specific emissions reduction projects.

Emissions may also be inventoried and reported on a product life-cycle basis, which means the total emissions for a product from its design phase, through its manufacture, use, and disposal (or recycling) are quantified. While particularly important for products that have large GHG releases over their working lives, the estimation of life-cycle emissions can be quite complicated. Life-cycle emissions inventories, discussed in greater detail below, are much less common than the other types.

Principles for Conducting Successful Inventories

A standardized protocol for conducting GHG emissions inventories has yet to be fully developed and widely implemented. Nevertheless, many corporations are already conducting inventories and reporting their results. A review of experience to date and of issues surrounding emissions inventories suggests several general principles for developing successful inventory programs:

- *Maximize flexibility.* Since requirements to report or reduce GHG emissions under a future climate policy regime are uncertain, companies should prepare for a range of possibilities. By maximizing the flexibility in their emissions inventories — for example, by being able to track emissions by organizational unit, location, and type of emission or by expressing emissions in absolute terms or normalized for production — organizations will be prepared for a wide range of possible future scenarios.

- *Ensure transparency.* Transparency in reporting how emissions and emissions reductions are arrived at is critical to achieving credibility with stakeholders. Unless the emissions baseline, estimation methods, emissions boundaries, and means of reducing emissions are adequately documented and explained in the inventory, stakeholders will not know how to interpret the results.

- *Encourage innovation.* Now is the time to try innovative approaches tailored to a company's particular circumstances. Learning what works best through experimentation — and doing so before regulatory requirements are in place — will provide invaluable experience.

Simplicity and cost-effectiveness are also important at a time when GHG reporting is voluntary because the success of corporate programs is dependent on obtaining broad acceptance and support within the organization.

Clearly, tensions among these attributes exist. Having flexible boundaries, for example, means that a facility or firm has the flexibility to decide what to include or not include in its inventory. If all firms within an industry decide boundary questions differently, comparisons among firms will not be as meaningful.

This is not to suggest that comparability should be a primary goal of emissions reporting. Experience with other reporting schemes has shown, however, that once information on emissions is made public, comparisons will be made whether they are valid or not. To address potential inconsistencies in reporting and assure comparability, the Greenhouse Gas Measurement and Reporting Protocol collaboration[1] is in the process of developing a voluntary, common standard for company reporting of GHG emissions.

Greenhouse Gases, Baselines, and Metrics

National and international programs influence companies' selection of gases to include in their inventories and the baseline year and metric system used to measure their progress in voluntarily reducing emissions. As with other aspects of emissions inventorying, companies have to determine which approach to these issues works best for them.

Gases

The Kyoto Protocol covers six greenhouse gases or categories of gases, as shown in Table 1, and these are the gases typically included in facility level emissions inventories. The potential for these gases to cause warming relative to that of carbon dioxide (a property referred to as global warming potential, GWP) ranges widely — from 21 for methane to 23,900 for sulfur hexafluoride — indicating that companies must choose carefully which gases to include in their inventories.

The Kyoto list of greenhouse gases is not exhaustive, however, and some reporting schemes include other gases. For example, the DOE's Section 1605b program includes reporting of emissions and emissions reductions for chlorofluorocarbons (CFCs),

Table 1

| Greenhouse Gases | Included in Various Reporting Schemes

Greenhouse Gas		Reporting Scheme		
Name	**Abbrev.**	**Kyoto**	**IPCC**	**DOE 1605b**
Direct GHGs				
Carbon Dioxide	CO_2	√	√	√
Methane	CH_4	√	√	√
Nitrous Oxide	N_2O	√	√	√
Hydrofluorocarbons	HFCs	√	√	√
Perfluorocarbons	PFCs	√	√	√
Sulfur Hexafluoride	SF_6	√	√	√
Indirect GHGs				
Nitrogen Oxides	NO_x		√	√
Carbon Monoxide	CO		√	√
Non-Methane VOCs	NMVOCs		√	√
Montreal Protocol Compounds				
Chlorofluorocarbons	CFCs			√
Hydrochlorofluorocarbons	HCFCs			√
Halons				√
Carbon Tetrachloride	CCl_4			√
1,1,1-Trichloroethane	1,1,1-TCA			√

hydrochlorofluorocarbons (HCFCs), Halons,[2] carbon tetrachloride (CCl_4), and 1,1,1-trichloroethane (1,1,1-TCA), but none of these compounds is included in either the Kyoto Protocol or in guidelines developed by the Intergovernmental Panel on Climate Change (IPCC). They were intentionally left out of both because reporting commitments and schedules for ending their use are part of the Montreal Protocol.

In the aggregate, the CFCs, HCFCs, and other compounds covered by the Montreal Protocol phase-out are relatively small contributors to greenhouse gas emissions and whether or not they are included in emissions inventories has little effect at the national level. At the facility level, however, they may account for the majority of GHG emissions, depending on the nature of the operation and the other types of sources present.[3]

The IPCC reporting guidelines and the U.S. DOE voluntary reporting guidelines include several conventional air pollutants due to their indirect effects as greenhouse gases. These compounds are included because of their role in the formation of tropospheric ozone, another greenhouse gas. These pollutants,[4] which are not included in the Kyoto Protocol, are oxides of nitrogen (NO_x), carbon monoxide (CO), and non-methane volatile organic compounds (NMVOCs). These compounds are typically not included in corporate greenhouse gas emissions inventories because their global warming potentials are highly uncertain (IPCC, 1996).

Few companies actually report on exactly those compounds covered by the Kyoto Protocol or the IPCC. Table 2 illustrates the wide range of GHGs that are included in the emissions inventories of selected companies. While some firms limit their reporting to CO_2 emissions (and may have no other significant emissions), others, such as ICI and Shell International, report on the whole suite of GHGs. Reporting on less than the suite of Kyoto gases could lead to questions about why some gases have been left out. Conversely, including compounds covered by the Montreal Protocol could lead to criticism that companies are trying to take credit for emissions reductions they would legally be required to make anyway, even though the reductions represent a real decrease in GHG emissions.

Baselines and Metrics

In addition to determining what will be included in their inventory, companies also need to decide how they will track trends in emissions and progress toward any emissions reduction goals they may have established. In making this decision, they are setting the baseline against which future emissions and their voluntary emissions reductions will be evaluated (see Box 1).

There are two important aspects to setting a baseline: timing and the way emissions are represented. Timing refers to the year in which a company begins to track its emissions and the year against which future progress is measured. In many cases these are the same. The year 1990 is often set as the base year by companies that plan to reduce emissions. 1990 is the base year used in several international and national

Box 1

DuPont's GHG Emissions Reduction Goals

DuPont's plan to reduce its GHG emissions 65 percent from 1990 levels by 2010 illustrates two important points about tracking emissions and achieving reductions:

- The selection of the baseline, both in terms of the gases it includes and its timing, affects the magnitude of reductions that can be achieved.
- The path chosen to reduce emissions can greatly affect the potential for further reductions in the future.

For most companies, nitrous oxide (N_2O), hydrofluorocarbons (HFCs), and perfluorocarbons (PFCs) account for negligible fractions of their greenhouse gas emissions. Often, these compounds are left out of corporate emissions baselines. For DuPont, however, these compounds are critical components of its GHG emissions inventory, accounting in 1997 for approximately 75 percent of total corporate emissions of gases covered by the Kyoto Protocol on a carbon dioxide equivalent basis.

DuPont intentionally based its GHG emissions reduction goal on the set of gases described in the Kyoto Protocol. Had it chosen to include other greenhouse gases or to use another date for its baseline based on the UNFCCC and the Climate Wise partnership program, it could have claimed even larger emissions reductions for the same endpoint. If chlorofluorocarbon (CFC) and hydrochlorofluorocarbon (HCFC) emissions (the "additional other" shown in the figure) were included in the baseline, DuPont could have stated its GHG emissions reduction goal as roughly 73 percent from 1990 levels. CFCs and HCFCs were specifically excluded from the Kyoto Protocol, however, because they are already scheduled to be phased out under the Montreal Protocol.

DuPont could also have announced a somewhat larger emissions reduction goal,

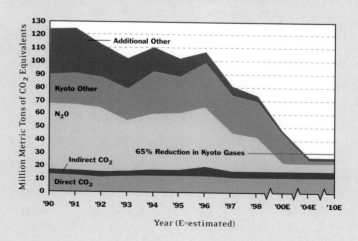

DuPont | Global Greenhouse Gas Emissions

(continued on following page)

Box 1

DuPont's GHG Emissions Reduction Goals (continued)

68 percent, had it selected its peak year as its baseline for emissions of the Kyoto gases (1996) rather than 1990, even with the final target in 2004 being the same. Doing so, however, would have meant that its base year would have been inconsistent with the Kyoto Protocol, and possibly raised questions about what its reductions were versus the more common base year of 1990.

The risk that DuPont faces is that having made a substantial investment to voluntarily reduce its emissions, further reductions may be required under some future regulatory scheme that does not account for its voluntary efforts. If 1990 is selected as the base year for the future regulatory program,[5] then through its voluntary reductions DuPont may have already met those requirements. If, however, a later base year — such as 2000 or later — is selected, then the company will be faced with making additional, and likely more expensive reductions from a baseline that has been substantially reduced. Financially, the company would have been better off not to have voluntarily reduced emissions, but instead to have waited for the reductions to be required. Had DuPont delayed action, however, the environmental benefits of the early reductions would be lost.

Table 2

Greenhouse Gases Included in the

Inventories of Selected Companies

Company	Greenhouse Gases Included in Inventory								
	CO$_2$	CH$_4$	N$_2$O	HFCs	PFCs	SF$_6$	CFCs	HCFCs	Others
AEP	√					*	√**		
Air Products	√	√	√	√	√	√			
Baxter	√						√	*	CO, NO$_x$, under consideration
BP Amoco	√	√							
DuPont	√	√	√	√	√				
ICI	√	√	√	√	√		√	√	CH$_2$Cl$_2$, CH$_3$Cl, CCl$_4$, 1,1,1-TCA, CO, NO$_x$, VOCs
Niagara Mohawk	√	√							
Shell International	√	√	√	√	√	√	√	√	Halons, TCE, VOCs, NO$_x$
Suncor	√	√	√						
Sunoco	√	√	√				√		NO$_x$
UTC	√								

Note: CH$_2$Cl$_2$ = methylene chloride, CH$_3$Cl = chloroform, CCl$_4$ = carbon tetrachloride, TCA = trichloroethane, CO = carbon monoxide, NO$_x$ = nitrogen oxides, VOCs = volatile organic compounds, TCE = trichloroethylene.

AEP = American Electric Power, UTC = United Technologies Corporation; The ICI Group is one of the world's largest coatings, specialty chemicals, and materials companies.

*to be tracked in future

**inventoried but not counted toward reduction targets

programs, including the Kyoto Protocol, the U.S. Climate Change Action Plan, and 1990 is recommended in the DOE's reporting program. Target years for achieving announced emissions reductions vary considerably from firm to firm, though generally they are not later than the 2008–2012 target period used in the Kyoto Protocol.

Of the 11 corporations listed in Table 2, six have set explicit targets for controlling GHG emissions. Five — American Electric Power, BP, Niagara Mohawk, Shell International, and Suncor — have selected 1990 as their base year. One — ICI — has selected 1995, the baseline year recommended by the Coalition to Advance Sustainable Technology[6] (CAST, 1998). A key challenge in setting baselines for past periods is that of data availability. In order to construct an inventory of emissions for the base year, historical data on the quantities and types of fuel combusted are needed.[7] If the inventory is to include emissions from purchased electricity, information on electricity consumption and the source of that electricity (how much CO_2 and other GHGs are emitted per kilowatt-hour produced) is required. Information is also needed on process-related emissions, such as the level of activities and the emissions factors for these activities during the base year. The availability of these data will depend on company's record-keeping policies.

In addition to selecting a base year, companies must choose the metric they will use to track and communicate their emissions and emissions reduction goals. These methods can be divided into absolute and normalized measures (also referred to as rate-based measures). Absolute reductions refer to specific mass or percentage reductions in emissions, or to specific caps on emissions, most commonly expressed in terms of metric tons per year of carbon dioxide equivalents (the sum of emissions of each greenhouse gas multiplied by its global warming potential).

Companies that normalize emissions do so in one of two ways: (1) emissions per dollar of revenues or expenditures, or (2) emissions per unit of product, potentially ranging from kilowatt-hours to stereo components to whatever else the company produces (CAST, 1998). The CAST proposal promotes a rate per dollar of sales approach, while one bill introduced in the U.S. Congress for providing credit for early emissions reductions (H.R. 2520) takes an emissions per unit of product approach. Each approach offers advantages and disadvantages to understanding trends in reducing greenhouse gas emissions, and neither approach is applicable across all industries. (See Table 3).

To date, the use of rate-based metrics is still in an experimentation phase, and most companies report on an absolute basis. For example, Shell International has announced an absolute goal of reducing its GHG emissions by 10 percent from 1990 levels by 2002. In contrast, Interface, Inc. has chosen an emissions rate based on annual sales figures. Another exception is UTC, which has announced targets to reduce energy and water usage by 25 percent per dollar of sales. Baxter International has based its goals for reducing energy consumption on physical units of output, rather than on the value of those outputs.

Table 3

Comparison of [**Normalized Emissions Factors**]

Approach	Advantage	Disadvantage	Applicable Industries
Emissions rate per revenue or expenditures	• Can adjust the rate by allowing for real price* adjustments • Provides a benchmark to compare GHG intensity of different industries with different products • Eliminates the need to adjust the baseline as the business changes	• Creates difficulties for companies in sectors with real product price* declines over time (e.g., electronics, renewables) • Requires price adjustments across industries • Prices may be not be uniform across regions • Requires high quality public information about pricing trends	• Industries with products with well-established price levels • Commodity industries with high levels of public pricing information
Emissions rate per unit of output	• Provides flexibility in changes to market share • Focuses on the efficiency of the sources	• Many companies have a wide variety of products • It is difficult to account for product enhancements	• Mature industries with lower rates of new product introductions • Companies with uniform product offerings

*Real prices are adjusted for inflation.

+

Although a total tonnage approach is clear, simple, and directly applicable to the goals of the Kyoto Protocol and other international agreements, a normalized approach allows companies that are growing rapidly to participate in emissions reduction efforts without being penalized for their success and avoids the issue of granting companies credit for reduced output. On a national level, normalizing emissions (e.g., on a per capita or per GDP basis) also helps to understand GHG emissions absent population trends.

In addition to selecting a base year for measuring progress and choosing whether to express their emissions on an absolute or normalized basis, companies must also decide whether and how they will adjust their baseline over time. When significant divestiture or outsourcing of production occurs, most companies reduce their baselines to avoid indicating emissions reductions when the emissions are merely transferred to another entity and not necessarily reduced on net. A similar situation would exist if a

+

company loses market share to a competitor having comparable emissions. If the company reports its emissions on an absolute basis, its inventory might be misinterpreted to indicate a reduction in total GHG emissions when no net reduction actually occurred, unless it adjusted its baseline.

Drawing the Boundaries

Deciding how to clearly, consistently, cost effectively, and equitably draw the boundaries around emissions sources (i.e., determining which sources to include and

+

how to include them) is one of the most difficult aspects of developing and maintaining a GHG emissions inventory. Important questions relevant to the issue of setting the boundary for a corporate emissions inventory are:

- Ownership — who "owns" or is responsible for the emissions?

- Indirect emissions — which such emissions should be included in an inventory?

- Life-cycle emissions — when is it appropriate to track emissions over part or all of the life of a product?

- Materiality — which emissions are significant enough that they must be included in the inventory and which are so insignificant that they can be ignored?

Ownership and Control of Emissions

The determination of who "owns" GHG emissions is complicated by the range of ownership options for corporations and other organizations. Corporate ownership can have a wide variety of structures from wholly owned operations to joint ventures incorporated by other companies to non-incorporated joint ventures.

To resolve the dilemma of accounting for ownership of emissions, various approaches have been taken, including ones based on majority ownership of the source, equity share, managerial control, and share of output. Majority ownership refers to the firm owning most of the operation or source taking responsibility for all of its emissions. Equity share, as the name implies, is the accounting of emissions based on the fraction of the emitting source each party owns. For example, if one firm owns 60 percent of an enterprise and another owns the remaining 40 percent, the firm owning 60 percent would account for 60 percent of the emissions from the enterprise and the firm owning 40 percent would account for 40 percent of the emissions. A combination of the latter two approaches based on Financial Accounting Standards Board (FASB) rules for managerial control has also been suggested (PWC, 1999). The share of output taken by the owner has been used to scale emissions from jointly owned power plants.

Most companies account for and report emissions from operations they partly own by scaling the operation's emissions by their equity share or by accounting for all of the operation's emissions if they own half or more of the operation or if they are the operator, as shown in Table 4. (The distinction between majority ownership and operational control is important because in some industries, such as the oil industry, operators without majority ownership are common.) The other methods listed do not appear to be in widespread use.

Each of the methods listed above has advantages and disadvantages, as shown in Table 5. While equity share might seem to be the appropriate way of sharing the emissions, it raises several practical problems. A minority owner wishing to include its share of the operation's emissions might not be able to obtain the necessary data from the majority owner or verify the accuracy of the data it receives. If reporting is left to only a majority owner, emissions from facilities with no single majority owner may not be counted at all.

Changing business operations can complicate the issue of emissions ownership even when there are no questions about ownership or control of the facility. If a manufacturing plant begins to manufacture a part that it formerly purchased, it would then own the associated emissions. Conversely, a facility would no longer be responsible for

Table 4

Inclusion of Greenhouse Gas Emissions from Partially Owned Operations in the Inventories of **Selected Companies**

Company	Means of Accounting for Emissions from Partially Owned Operations
AEP	Scale by amount of energy taken from jointly owned assets
Air Products	Scale by equity share
Baxter	Include 100% of emissions if ≥ 50% ownership, otherwise none
BP Amoco	Scale by equity share
DuPont	Include 100% of emissions if ≥ 50% ownership, otherwise none
ICI	Include 100% of emissions if ≥ 50% ownership, otherwise none
Niagara Mohawk	Scale by equity share
Shell International	Include 100% of emissions if under operational control, others may be included if Shell HS&E policy has been implemented and external verification of data is permitted
Suncor	Include 100% of emissions if under operational control, otherwise none
Sunoco	Under investigation
UTC	Include 100% of emissions if ≥ 50% ownership, otherwise none

Table 5

Approaches to Addressing Emissions from Partially Owned Operations

Majority Ownership	Advantages	Disadvantages
Majority Ownership Report 100% of owned (≥50%) equity) joint ventures	• Simple • Clearly defined	• May overstate (if minority owners also report) or understate emissions (if there is no majority owner)
Equity Share Sharing according to the share of equity	• More fully represent GHG emissions • Control might not be clearly defined	• More complex and detailed • Information may not be readily available
FASB Managerial Control Accounting Approach 100% of emissions for ≥ 50% ownership Scale by equity share for 20–50% ownership 0% of emissions for <20% ownership	• Follows established accounting procedures	• Emissions not accounted for at all if more than 5 equal owners • Emissions may be double-counted (e.g., 2 owners, one with 60%, the other 40%) • A consensus on what constitutes a *de minimis* level has not been defined

emissions when it stops producing a part in-house and instead purchases it from outside vendors.

A similar process takes place in the event of acquisitions and divestitures. When a company buys or sells an ownership stake in another entity, GHG emissions need to be added to or removed from inventories for the emissions and emissions reductions to be properly accounted for. In practice, most companies adjust their baselines to remove those emissions that have been divested and to add those assets that have been acquired.

Indirect Emissions

Indirect emissions are emissions from sources not owned or leased by a company but which occur wholly or in part as a result of the company's activities (Hakes, 1999). Emissions resulting from purchased electricity are one example. Including such emissions in GHG inventories — a subject of widespread interest and concern — has advantages and disadvantages that companies should consider in drawing meaningful boundaries.

Including indirect emissions in an inventory has one principal disadvantage — the potential for double counting the emissions. This would occur if both the electricity producer and an electricity consumer reported the emissions as their own. BP's protocol for accounting for emissions avoids this problem by counting emissions from net energy consumption. The company includes emissions resulting from steam or electricity it produces for its own use as well as emissions from the energy it purchases. It does not count as its own emissions from steam or electricity it produces for outside sales.

Purchased electricity and steam are often included in emissions inventories because firms have a large degree of control over the consumption of these energy sources. This is especially true in countries like the United States, where due to deregulation of the electric power industry, firms have increasingly greater choice about their electricity supplier, and thus may select generators with more or less carbon intensive power (CO_2-equivalent emissions per kilowatt-hour). Firms also have control over the energy efficiency of their processes, which affects the amount of electricity they consume. Corporations typically do include indirect emissions from electricity consumption in their inventories, just as electricity is counted in energy audits (see Table 6).

The subtraction of emissions from exported energy is handled inconsistently among corporations. All of the oil companies listed in Table 6 (except Suncor) exclude emissions from exported power, while most of the chemical companies include these emissions. The rationale for excluding the emissions for exported power is that the power is being used by someone else, not by the company in its own operations, and thus should be accounted for by the purchaser, in the same way that the company accounts for emissions resulting from the electricity it purchases. Though internally consistent, if this form of accounting were applied to electric power companies, they would report only on emissions from electricity they consume (such as for pulverizing coal and operating electrostatic precipitators) but not from electricity they produce for

Table 6

| Indirect Greenhouse Gas Emissions | Included in the Inventories of

Selected Companies

| | Include GHG Emissions Associated With: | | | |
Company	Purchased Energy?	Energy Sold to Others?	Business Travel?	Employee Commuting?
AEP	no	yes	no	no
Air Products	yes	yes	no	no
Baxter	yes	na	no	no*
BP Amoco	yes	no	no	no
DuPont	yes	no	no	no
ICI	yes	yes	no	no
Niagara Mohawk	yes	yes	no	no
Shell International	no	no	no	no
Suncor	yes	yes**	no	no
Sunoco	yes	no	no	no
UTC	yes	no	yes	no

Note: na = not applicable.

*While Baxter does not include emissions associated with commuting in its GHG inventory total, it does estimate and report them separately.

**If the energy generated and exported by Suncor is of a lower emissions intensity than the electricity being displaced and that would otherwise have been consumed by the same users, then Suncor would also credit to its account the associated emissions reductions.

sale to others. In actuality, power companies report all of their emissions regardless of where the power is sold, and many other business that sell power do as well.

The accounting of indirect emissions becomes more contentious when credit for emissions reductions is being considered. For example, while a firm may wish to receive credit for emissions reductions resulting from the improved efficiency of one of its operations, a utility may have played a role in these reductions through demand side management programs. Other factors such as government policy or consumer choice also may have played a role, raising questions about who should get credit for the reduction.

The concept of "causing emissions" is inherently more ambiguous than "owning a stack," and extends beyond the consumption of secondary energy sources. Other examples include emissions from employee business travel (or emissions from employees commuting to and from work), emissions from shipments of raw materials and finished goods, and emissions from waste disposal. Relatively little guidance on inclusion of these indirect sources exists. Voluntary reporting guidelines published by the United Kingdom's Department of Environment, Transport and Regions (DETR, 1999) recommend including long-distance business travel, but not commuting or short-distance business travel. In contrast, the U.S. DOE reporting guidelines for the 1605b program include measures for reducing employee commuting as examples of legitimate emissions reduction projects.

Life-Cycle Assessment

Companies that report their GHG emissions typically account for emissions from their own operations. Yet in some industries, the principal emissions and opportunities to reduce emissions come not from the direct or indirect emissions associated with manufacturing but rather from the use of the manufactured products. For example, appliances such as washing machines and refrigerators produce most of their GHG emissions through their use rather than through their manufacture. The same is true for motor vehicles. The GHG emissions reported by General Motors for all GM vehicles in operation in the United States accounts for 23 percent of transportation-related emissions (EIA, 1997), an amount far greater than the emissions from GM's factories.

The evaluation of greenhouse gas emissions throughout the full product or service system life cycle (see Figure 1), a process known as life-cycle assessment (LCA),[8] is an evolving area. The greater emissions from the use of products compared to their manufacture suggests that greater benefits could be achieved through improving the energy efficiency of the products than through reducing emissions during manufacture, although presumably such features could support a premium price or expanded market share.

If, under some future greenhouse gas emissions reduction scheme, a company is required to reduce only its own emissions, then it would have little incentive to produce lower emitting products. For this reason, manufacturers have suggested that they be allowed to take credit for the emissions reductions that result from the improved efficiency of their products (see Box 2).

Figure 1

Product/Service   **Life Cycle**

Materiality

An important aspect of drawing the boundaries around a firm's emissions is deciding which emissions are large enough to be included in the inventory and which can be ignored without any significant effect on the overall results. The answer to these questions varies with the industry, and even the particular firm that is conducting the inventory. For a service company, for example, emissions resulting from employee business travel may be significant, while for an electric utility they are unlikely to be. These emissions may also be material for firms that are primarily manufacturers. For

Whirlpool Corporation and Life-Cycle Emissions

Certain industrial sectors can make more of a contribution to reducing greenhouse gases by modifying the products they produce than by reducing the energy they consume in producing those products. This is because the products themselves are responsible for GHG emissions, particularly those that consume energy. The home appliance industry is an example of one such sector.

In the appliances industry, the greenhouse gas emissions at manufacturing sites are small when compared to the emissions associated with the power generated to run those appliances over their long useful lives. Whirlpool has estimated that its clothes dryers use 20 times more energy over their working lives than the energy used to produce them, and its washing machines 50 times more. Thus, limiting a GHG inventory to on-site emissions might misdirect the efforts of some industries, including appliance manufacturers, away from the areas where they can do the most to reduce greenhouse gas emissions.

By taking a life-cycle perspective, Whirlpool has identified an opportunity to retire inefficient appliances in Brazil and replace them with more efficient models. Refrigerators in Brazil account for 27 percent of residential energy consumption. Current models are 30 percent more efficient than products made four years ago, and an additional 20 percent efficiency improvement will occur by 2002 as a result of a voluntary industry agreement. Through financing of an incentive program in which utilities legally must direct 1 percent of revenues toward energy efficiency, up to 28.7 million refrigerators could be retired over a 10 year period. Whirlpool has calculated that this could reduce CO_2 emissions by more than 3 million tons annually, even accounting for the fact that most of Brazil's electricity comes from hydropower.

example, UTC, a major provider of high-technology products and services to the aerospace and building systems industries,[9] has found emissions associated with business travel (as well as emissions from the testing of its jet engines) to be material.

Most firms conducting inventories today do not have strict rules on what qualifies as material. Only two of those participating in the survey conducted for this report, UTC and ICI, had any rules at all. Rather, materiality has been treated as a matter of professional judgement for those performing the inventory. Where firms have set materiality thresholds, they have typically done so on the basis of a minimum size of emissions (e.g., X metric tons of CO_2 equivalents/year)[10] or a minimum percentage completeness for the inventory (e.g., at least 99 percent of corporate emissions will be accounted for). ICI is an example of one firm taking the latter approach. UTC, whose GHG emissions accounting is an outgrowth of its energy and water tracking and conservation program, employs a variation of the former approach. Water and energy consumption (and thus GHG emissions) from non-manufacturing facilities are ignored if their combined cost is less than $100,000 per year. Even drawing the line at this amount, UTC inventories 229 facilities worldwide, including its corporate headquarters.

The issue of materiality is further complicated by the difficulty of estimating certain types of emissions. Emissions associated with employee commuting generally fall into this category, and typically are not included in corporate GHG emissions inventory. Emissions due to product transportation can also be difficult to assess, and thus are not always included in inventories. BP, for example, includes emissions from commercial shipping in its inventory if it owns the vessels or has a charter for more than a year. If the transporter is also shipping the products of other companies, however, the activity is judged as too difficult to reasonably assess the fraction of the emissions attributable to BP. Thus, the inclusion of these emissions sources is based as much on the ability to estimate them as it is on their materiality to the total corporate emissions.

Emissions sinks — activities or operations that remove GHGs from the atmosphere — are another example of an item that may or may not be material to an inventory. Sinks can range from the absorbance of carbon dioxide by grass and trees around an office building, to absorbance by forest areas set aside for conservation, to sequestration through reforestation, to reinjection of CO_2 into depleted oil wells or other geological formations. In addition to the complexities of accounting for these sinks, their significance to a company's greenhouse gas inventory needs to be assessed. Typically major sinks such as reforestation programs or CO_2 reinjection into oil reservoirs for enhanced oil recovery are included in emissions inventories, while minor sinks such as growing trees around office buildings are not. CAST addresses the question of materiality directly, at least for domestic sinks, by giving firms the option of including domestic sinks in their corporate inventories. However, firms must include all of their sinks if they choose this option (CAST, 1998).

The inventorying of emissions sinks provides special challenges since the methodologies for estimating the amount of carbon sequestered are generally less well-developed than for the major greenhouse gas emission sources. Quantification of carbon sequestration is often left to experts in the field (see Box 3).

Boundary issues present a multitude of options for the way firms may report their GHG emissions inventories, and a multitude of questions they must answer in conducting their inventories. On some of these questions there is general agreement. There is widespread consensus that an organization should include within the boundary of its inventory all material sources owned and operated by the organization, which typically corresponds to sources located within its facilities' fence lines and mobile sources that it owns. A consensus is also growing to account for electricity usage because of its ubiquity and the degree of control possessed by organizations to modify their electricity consumption.

Box 3

Entergy's Approach to Carbon Sequestration

Entergy is a major global energy company in the field of electric power production, distribution operations, and related services. Entergy's approach to reducing the GHG impacts of its operations includes forests and wetlands projects to offset its emissions. The projects are managed by forest management consultants and include analyses of the net amount of CO_2 removed from the atmosphere and stored as carbon in the forests and wetlands. Entergy believes the analyses will help it understand, and possibly improve, removal rates and methods for estimating carbon stored. The analyses also enable the company to track projects and report results under the Department of Energy's 1605b program. Finally, the analyses help address stakeholder concerns regarding the validity of the environmental benefits of the projects.

Entergy's analyses follow the Department of Energy's *Guidelines for Voluntary Reporting of Greenhouse Gas Emissions, Reductions, and Carbon Sequestration*. It is important to note that each project is different and requires project-specific estimates for growth rates, conversion factors, and baselines. Neither Entergy nor DOE supply detailed rules for determining inputs for the reports. Entergy looks for clear, simple, logical analyses and relies on the best professional judgment of its consultants for specific factors. Each analysis estimates the net amount of carbon removed from the atmosphere (stored in the forest or wetland) by the project. Estimates of net carbon removal can be made using tables of factors provided by the DOE, factors from other credible sources, or the consultant can develop its own. Field studies may be conducted, but this is not required.

For one set of projects, Entergy planted pine and hardwood forests on 24 different sites, for a total of 4,096 acres. Each site had different characteristics. On one of the sites, a forest had been clear-cut, while the other sites had been farmlands and pastures. The amount of carbon stored was estimated following DOE guidelines specific for site characteristics and species planted. These guidelines provide moderate accuracy, but assume, for example, ideal planting conditions. While field studies would be more accurate, the guidelines provide a quick, economical means to estimate carbon stored. Entergy's forest projects are estimated to remove 7,500 tons of CO_2 from the atmosphere per year. The wetlands projects are estimated to remove 60,500 tons of CO_2 per year.

Further expanding the accounting boundaries, emissions resulting from employee business travel, product and raw material shipments by third parties, and employee commuting could be included. An additional widening of the accounting boundaries would be an assessment of the complete life-cycle emissions of a product. The extent to which sources beyond the fence line are included in the inventory depends on the degree of control the company has over these sources, its stakeholder expectations, and its ability to implement emissions reduction programs and initiatives. Box 4 describes how one firm, UTC, has addressed these questions.

Box 4

United Technologies' Approach to Boundary Issues

United Technologies Corporation (UTC) provides a broad range of high-technology products and services for the aerospace and building systems industries. UTC has developed a worldwide inventory of its energy use and the associated greenhouse gas emissions as part of its commitment to reduce energy consumption. The primary focus is energy use that UTC has direct control over and is material to the company. The UTC program is a dynamic, evolving effort. It is expected that as the program matures, additional refinements and enhancements will be made to improve the program based on changing circumstances and experience gained through implementation.

In designing the initiative, a number of boundary decisions were made that have defined the scope of information collected, analyzed, and reported. In general, decisions on what to include or exclude were driven by the desire to:

- Focus attention on major opportunities for improving energy efficiency;
- Recognize the diversity, complexity, and magnitude of the organization's operations;
- Ensure that the program parameters were understandable;
- Avoid duplication of other ongoing efforts; and
- Prevent overly burdensome reporting or management procedures.

The application of these principles can be illustrated by examining four boundary issues: ownership, direct/indirect emissions, materiality, and life-cycle emissions.

- **Ownership**. UTC requires facilities that manufacture products to report energy and water consumption data, and requires non-manufacturing sites that have a combined annual energy and water cost of more than $100,000 (U.S.) to report. An exemption is provided for manufacturing sites that have less than $500,000 (U.S.) in annual sales. The reporting requirement applies to all joint ventures where the UTC ownership is 50 percent or higher.

- **Direct/Indirect Energy Emissions.** UTC indirectly causes the generation and release of carbon dioxide through its use of purchased electric power. UTC measures a facility's direct electrical energy consumption and, using this figure, calculates the energy required and carbon dioxide emissions produced by the utility to deliver such electrical power to the facility. The facility is therefore responsible for inefficiencies in the utility's generation and transmission of electric power.

- **Materiality — Energy Sources.** UTC identified the need to collect data on its usage of electricity, propane, natural gas, butane, oil, gasoline, diesel, and coal. In addition, the company included jet fuel as a key energy source since it is used in significant amounts in various UTC activities including corporate aircraft operated by the company, jet engine and helicopter testing, and employee travel on commercial airlines. UTC decided not to track certain fuels, such as kerosene, since initial measurements proved that not enough was consumed to warrant the tracking.

- **Materiality — Travel/Transportation.** UTC products consume energy as they are shipped around the world, and UTC employees consume energy as they travel on business. As a global corporation with significant travel and transportation activities directly under its control, the corporation decided to include employee travel on company business and transportation of its products by company

(continued on following page)

vehicles in its inventory. On the other hand, a decision was made to specifically exclude employee commuting from the program since commuting is not controllable by UTC, and the collection of this type of information raised invasion of privacy issues. Similarly, non-UTC commercial trucking is excluded since there were questions regarding who "owns" any energy reduction associated with this activity.

- **Life-cycle Emissions.** UTC decided not to account for the life-cycle emissions of greenhouse gases associated with its products. Even though these emissions are orders of magnitude greater than the CO_2 emissions from all of its facility operations combined, efforts to reduce these emissions are managed as part of UTC's Design for Environment, Health, and Safety program. Therefore it was decided not to duplicate these efforts and to focus exclusively on operations and travel related activities.

Conducting the Inventory

In preparing to conduct an inventory, firms need to determine: what protocols and tools they will need to collect and report emissions data, how they should measure or estimate emissions, and how often they should collect and report emissions data.

Tools for Conducting the Inventory

Most firms have developed their own protocols for collecting data and reporting emissions. The calculation of emissions is typically based on emissions factors that have either been developed for company-specific operations, or more likely, are available as published guidance (see Table 7).

Some firms, such as Sunoco, have used computerized accounting and reporting tools, such as the U.S. EPA's Climate Wise software for tracking their emissions.

Estimates vs. Measurements

Emissions of greenhouse gases may be measured or estimated. Which approach is taken depends on the availability of emissions-related data, the cost of developing it, and the accuracy needed for the inventory. In practice, most organizations use a combination of measured and estimated parameters to calculate their emissions. Except for carbon dioxide emissions, which are measured by the electric utility industry, the direct measurement of GHG gas emissions is relatively uncommon.

The reason greenhouse gases are not typically measured is that current air pollution regulations generally do not require them to be, and doing so is expensive. If the

Table 7

Resources for Estimating Greenhouse Gas Emissions

Revised 1996 IPCC Guidelines for National Greenhouse Gas Inventories	Estimation methods for the major sources of gases listed in Table 1.	Developed for national level inventories, but may be useful for company- level estimates in the absence of other data. Available at: http://www.ipcc-nggip.iges.or.jp/public/gl/invs1.htm
U.S. DOE, Lawrence Berkeley National Laboratory	Guidelines for monitoring, evaluation, reporting, verification, and certification of energy-efficiency and climate change mitigation projects.	Successors to International Performance Monitoring and Verification for energy efficiency projects. Other related reports also available. Focus is on energy-related emissions and reductions, rather than industrial emissions of GHGs. Available at: http://eetd.lbl.gov/ea/ccm/ccpubs.html
Australian Greenhouse Office: • National Greenhouse Gas Inventory Committee Workbooks; and • Greenhouse Challenge Vegetation Sinks Workbook	Workbooks on national emissions inventory with supplements for state and territory governments. Step-by-step procedures for estimating carbon sequestration. Focuses on forest-based sinks.	National workbooks available at: http://www.greenhouse.gov.au/inventory/methodology/method_content.html State and territory supplements available at: http://www.greenhouse.gov.au/inventory/inventory/stateinv/statemethod.html Summary of vegetation sinks workbook available at: http://www.greenhouse.gov.au/pubs/sinks.html
U.S. EPA Emissions Inventory Improvement Program, Volume 8, Greenhouse Gases	Fourteen chapter volume designed to provide guidance to states on estimating emissions of each of the Kyoto GHGs.	Available at: http://www.epa.gov/ttn/chief/eiip/techreport/volume08/index.html
Global Environmental Management Initiative (GEMI)[11]	Overview of the corporate GHG emissions inventory process; contains links to other resources.	See "Measurement and Metrics" section of: http://www.businessandclimate.org
UK Department of Environment, Transport, and Regions Guidelines for Company Reporting on GHG Emissions	Manual on GHG emissions reporting for voluntary reporting by companies. Covers Kyoto gases.	Provides guidance on boundary questions as well as emissions estimation for fossil fuel combustion. Includes lists of guides for sector-specific emissions. Available at: http://www.environment.detr.gov.uk/envrp/gas

(Continued on following page)

Table 7

Resources	for Estimating Greenhouse Gas Emissions *(continued)*

Resource	Scope	Comments
U.S. DOE 1605b	Guidance for participants in the DOE's 1605b program on the estimation and reporting of GHGs emissions and emissions reduction projects.	Estimation methods focus on emissions from fossil fuel combustion (including transportation), forestry, and agricultural sectors. Available at: http://www.eia.doe.gov/oiaf/1605/guidelns.html
U.S. EPA AP-42	Compilation of conventional and GHG air pollutant emissions factors for stationary sources.	Available at: http://www.epa.gov/ttn/chief/ap42/
U.S. EPA Climate Wise	Software for tracking GHG and conventional pollutant emissions, energy use, and costs at the process unit, facility, and company level.	Distribution of software is currently limited to participants in the Climate Wise Program.
World Business Council for Sustainable Development/World Resource Institute Collaboration	Standardized, international, GHG emissions reporting protocol under development. Web site contains a wide range inventory resources and related materials.	See "Resources" section of: http://www.ghgprotocol.org
Winrock International Institute for Agricultural Development	Methods for inventorying and monitoring carbon in forestry and agroforestry projects.	Publications, bibliography, and case studies available at: http://www.winrock.org/REEP/forest_carbon_monitoring_program.htm
World Bank Greenhouse Gas Assessment Handbook	GHG emissions assessment methodologies for energy, industrial, and land-use projects.	Designed for evaluation of World Bank-sponsored projects. Available under "Tool Kit for Task Managers" at: http://www-esd.worldbank.org/cc/
American Petroleum Institute Petroleum Industry GHG Emissions Estimation Protocol	Methods to estimate emissions of carbon dioxide and methane from petroleum industry sources.	Under development; expected to be completed in 2001.
Gas Research Institute GRI-GHGCalc™	Personal computer program to calculate methane, carbon dioxide, and nitrous oxide emissions from natural gas operations.	Software description and ordering information available at: http://www.gri.org/pub/content/jan/20000117/115155/ghgcalc.html

emissions are from a distinct point, such as CO_2 emissions from a smokestack, the same measurement methods as used for conventional air pollutants may be applied — the concentration of the pollutant in the flue gas is measured, and this concentration is multiplied by the measured flow rate of the flue gas to arrive at a mass emissions rate. The mass emissions rate is then annualized to give the emissions for an entire year. The main shortcoming of this approach is its cost, particularly if the sampling is done frequently or continuously. In the case of CO_2 emissions from combustion sources, direct measurement of emissions may be no more accurate than emissions estimates based on fuel use (see Box 5).

Fortunately, for emissions of most greenhouse gases — and in particular for CO_2 from combustion, the largest source of emissions in industrialized countries — emissions can be estimated indirectly. This is so because when fossil fuels are combusted, the amount of CO_2 released is directly proportional to the amount of carbon in the fuel. As discussed in Box 5, it cannot be assumed that calculated CO_2 emissions based on fuel consumption are necessarily less accurate than direct measurements. For both this reason and the added cost associated with direct measurement, at both the national and corporate levels, calculation of emissions based on fuel composition and consumption can be expected to remain the primary means of inventorying CO_2 emissions from fossil fuel use (UNCTD, 1999).

For GHG emissions that do not emanate from a single point, such as fugitive emissions of methane from pipeline systems or nitrous oxide emissions from agriculture, taking direct measurements is more difficult. For these types of emissions, estimates are typically made based on extrapolations from studies of similar operations.

Frequency

Greenhouse gas emissions inventories, like those of other air pollutants, are generally reported on an annual basis. This frequency has been adopted by convention rather than for any particular reason.

In most cases, inventories are conducted soon after the end of the calendar year. The actual frequency of inventorying should, however, be based on the reason for conducting the inventory. If the inventory is being conducted to calculate emissions reductions that are involved in an emissions trading program, then the frequency of inventory should, at a minimum, correspond with the frequency with which the organization wishes to accumulate tradable credits. For example, if a party wishes to accumulate credits (and if necessary have them certified) on a quarterly basis, then the inventory would have to be conducted at least quarterly to establish the number of available credits.

Where companies have targeted particular parts of their operations for emissions reductions, or merely want to track certain emissions more closely, they may wish to inventory these operations more frequently than others. BP, for example, which has set aggressive targets for greenhouse gas emissions reductions, requires quarterly reporting frequency for its exploration and production operations, the primary area that it has targeted for reductions, while its refining, chemical, and other divisions are required to report annually.

The current, dynamic state of affairs in inventorying emissions, and the experimentation and range of approaches that companies are taking will ultimately lead to more uniform and recognized procedures that have already been tested in the field. Ensuring that current emissions reduction efforts are recognized under future policy regimes is important and only likely to occur if reductions are found to be real, quantifiable, and verifiable.[12]

Box 5

American Electric Power's
Experience Measuring Emissions

American Electric Power (AEP) is a leading supplier of electricity and energy-related services throughout the world. AEP operates a diverse fleet of coal-fired generating units representing approximately 21,500 megawatts of total rated capacity. The emissions from these units, as at virtually all fossil fuel electric utility power plants in the United States are required by the 1990 Clean Air Act Amendments to be monitored with Continuous Emissions Monitoring Systems (CEMS).

After the monitoring program began in 1993, a discrepancy was noted between the CEMS values and emissions calculated by the traditional mass balance method of calculating emissions based on fuel consumption. For the AEP system, it was estimated that the CEMS values were approximately 6 percent high on average, and as much as 30 percent high for individual stacks. At current market prices, the high emissions readings could cost AEP around $6 million annually in over-consumption of sulfur dioxide (SO_2) emissions allowances. The CEMS-derived values were immediately suspect, as the fuel-based information has had an extensive history.

The traditional, fuel-based method for calculating heat input analyzes daily fuel samples to determine the heating value of the fuel (Btu/lb coal). Fuel flow meters, which weigh the coal as it is fed to the boilers, measure the number of tons of coal burned each day. Multiplying the heat value (expressed in Btu/lb) by the amount of coal burned (lbs) provides the heat input to the power plant boiler. SO_2 emissions are calculated by multiplying the heat input rate by the amount of sulfur in the coal (lbs/Btu),

which is also measured daily. CO_2 emissions are calculated similarly, though the carbon content of the coal is analyzed less frequently. Since coal carbon analyses are highly accurate, and little variability has been observed in the carbon content of the coal (lbs/Btu), emissions calculated in this way are considered accurate.

Emissions derived from CEMS are calculated by multiplying the concentration of the pollutant in the stack gas by the volumetric flow rate of the stack gas. Problems with the volumetric flow rates used to calculate the emissions with the CEMS were found to be the cause of the overestimate of the SO_2 emissions. AEP has worked with the vendor that installed its flow monitors to improve the calibration of its flow meters. The result has been the development of an automated method for performing required test audits, which AEP believes has improved the accuracy of the flow measurements.

The current method of calculating volumetric flow does not properly account for non-axial flow within the stack (swirl) and wall effects, contributing to the inaccuracies. In May of 1999, after extended field studies, EPA promulgated optional flow measurement techniques to account for these flow effects. AEP will be implementing these techniques in 2000. With the combination of the automated auditing system and the correction to the methodology used to calculate volumetric flow, AEP believes that the CEMS-measured emissions should more closely match the fuel-derived measurements. Nevertheless, AEP would choose the fuel method over the CEMS for CO_2 emissions measurements because AEP believes it to be more reliable.

Endnotes

1. In May 1999, the World Resources Institute and the World Business Council for Sustainable Development convened an open, international, multi-stakeholder collaboration to design, disseminate, and promote the use of an international corporate protocol for reporting business greenhouse gas emissions. Both the Pew Center and Arthur D. Little have participated in this effort. The core operations module of the protocol is being road-tested with projected release in summer 2001.

2. Halon is a DuPont trade name for a group of bromofluorocarbons, which are used almost exclusively in fire protection systems.

3. Manufacturers of HCFCs and their substitutes, and firms that have used CFCs as blowing agents in the manufacture of foam are examples of the types of companies where questions of whether to count CFC and HCFC emissions are key to accounting for GHG emissions and emissions reductions.

4. The IPCC reporting guidelines also include sulfur dioxide (SO_2) as a precursor to sulfate aerosol, which is believed to have a cooling effect.

5. Because CFCs are not typically included in emissions baselines and their replacements — HFCs and CFCs — were not produced in significant quantities until the mid-1990s, 1995 could be chosen as the base year for HFC and PFC emissions. If 1990 were chosen as the base year, firms that emit the CFC replacements would have an unfairly strict baseline for these compounds. Neither the CFCs, which were emitted in 1990, nor the replacements, which were emitted later, would be included in the baseline.

6. The Coalition to Advance Sustainable Technology (CAST) is a public policy organization of CEOs who share the view that environmental stewardship is compatible with sound and competitive business practices.

7. Information on the efficiency of the combustion process in converting carbon to carbon dioxide is also needed; default values from the literature are available if equipment-specific values have not been determined.

8. Life-cycle environmental accounting or analysis, also abbreviated LCA, are other terms for life-cycle assessment.

9. UTC's best-known products include Pratt & Whitney aircraft engines, Carrier heating and air conditioning systems, Otis elevators and escalators, Sikorsky helicopters, and Hamilton Sundstrand aerospace systems.

10. The metric ton (or "tonne") — 1000 kilograms — is used in this report because it is the most common unit for reporting GHG emissions.

11. The Global Environmental Management Initiative (GEMI) is a nonprofit organization of leading companies dedicated to fostering environmental, health, and safety excellence worldwide through the sharing of tools and information.

12. This paper does not address the verification of emissions inventories and emissions reductions. Verification is the subject of another paper prepared by Arthur D. Little, Inc., for the Pew Center.

References

CAST. 1998. *First Movers Coalition Climate Change Early Action Crediting Proposal.* Coalition to Advance Sustainable Technology, Portland, Oregon.

DETR. 1999. *Environmental Reporting: Guidelines for Company Reporting on Greenhouse Gas Emissions.* Department of the Environment, Transport and Regions, London, UK.

DOE. 1994. *General Guidelines: Voluntary Reporting of Greenhouse Gases under Section 1605b of the Energy Policy Act of 1992.* U.S. Department of Energy, Washington, DC.

EIA. 1997. *Mitigating Greenhouse Gas Emissions: Voluntary Reporting.* U.S. Energy Information Administration, Washington, DC. See http://www.eia.doe.gov/oiaf/1605/vr96rpt/home.html.

Hakes, J. 1999. *Testimony on The Voluntary Reporting of Greenhouse Gases Program* before the House Government Reform Committee, July 15. See http://www.eia.doe.gov/neic/speeches/htest715/testmony.html.

IPCC. 1996. *Climate Change 1995: The Science of Climate Change.* J.T. Houghton et al., eds. Cambridge University Press, Cambridge, UK.

PWC. 1999. *Developing an Internationally Accepted Measurement and Reporting Standard for Corporate Greenhouse Gas Emissions.* Paper on Boundary Issues submitted by Karan Capoor, Price Waterhouse Coopers, for consideration by the WRI/WBCSD collaboration (August).

UNCTD. 1999. *International Rules for Greenhouse Gas Emissions Trading.* United Nations Conference on Trade and Development, New York and Geneva.

CURRENT DEVELOPMENTS

Contents

Climate Negotiations at a Crossroads

Eileen Claussen

The issue of climate change received a great deal of attention in the waning months of 2000 and early 2001. Unfortunately, much of the attention was due to events that raised questions about the capability of the global community to address this serious issue in a collaborative and decisive way. The failure of the climate talks in The Hague and the emerging U.S. position suggest that it will be some time before clear forward momentum develops.

Background: From Kyoto to The Hague

In assessing the current state of affairs, it is important to remember that the Kyoto Protocol was negotiated for a very important reason — because the voluntary emissions reduction objectives outlined in the 1992 United Nations Framework Convention on Climate Change were not effectively limiting atmospheric concentrations of greenhouse gases. Few developed countries were undertaking the necessary actions to reduce their emissions to 1990 levels by 2000, as they had agreed to do under the Convention.

Under the Kyoto Protocol, industrialized countries agreed to binding emissions reductions during the period from 2008 to 2012, with countries' targets averaging about 5 percent below 1990 levels. The Protocol also began to outline how countries could achieve their targets — for example, by trading emissions credits and by using sinks such as forests to remove and store carbon from the atmosphere. However, further elaboration of the rules for these and other important provisions was still needed.

In particular, the parties still had to work out a series of politically charged issues, as well as a large number of more technical ones. Issues of political concern included whether there would be a limit on the emissions reductions that countries could undertake outside their borders through trading and other flexible mechanisms; how to define forest and soil management — and how these activities could be counted toward a country's target; and what kinds of projects would be eligible for the Clean Development Mechanism (CDM), which provides industrialized countries with credits for emissions reductions achieved in the developing world. In addition to these issues, negotiators still had to resolve the question of how to ensure a proper level of assistance for developing countries by supporting capacity building, technology transfer, and efforts to adapt to the effects of climate change.

Also undefined on the eve of the meeting in The Hague were the rules for exactly how the Kyoto mechanisms, such as the CDM and emissions trading, would work.

And, negotiators still had to structure an effective compliance regime to ensure that countries were living up to their commitments.

Reflections on COP6

Organizers of the meeting in The Hague — officially known as the Sixth Session of the Conference of the Parties to the U.N. Framework Convention on Climate Change (COP6) — had hoped to reach agreement on these and other issues. By the second week of the talks, however, it became clear that the issues were all very difficult to resolve, both politically and technically, and that the parties still were far apart in their judgments about exactly how to structure the Kyoto framework.

Press accounts of the talks' demise suggested that the negotiators would have gone home with a complete and final agreement if not for one issue: exactly how much credit countries should receive for carbon sinks. But, in reality, there were many other reasons why the talks did not bear fruit. Chief among these was the fact that COP6 was burdened from the beginning by an overly ambitious agenda. Among the warning signs: 275 pages of text, each with a myriad of issues needing resolution; a set of political issues with technical underpinnings, none of which had been resolved in advance; and clear and strongly held differences of view between the European Union (EU) and the "Umbrella Group" of countries, including the United States.

An underlying factor in the failure of COP6 was a philosophical gulf between the United States and Europe. The EU takes as its starting point the need to effect wide-spread behavioral changes through government policies aimed at limiting and eventually curbing the use of fossil fuels. The United States, on the other hand, comes down on the side of short-term, cost-effective actions, combined with a concerted effort to develop the technologies that will deliver solutions over the longer haul. In The Hague, the negotiating positions that reflected these distinct philosophical approaches proved too far apart to bridge.

A New U.S. Administration

With questions over the future of Kyoto still fresh, the Bush Administration's rejection of the Kyoto Protocol in March 2001 created additional uncertainty and unease. The administration's rejection of the Protocol was based on the view that its implementation would be excessively costly and that it did not include developing countries. However, at the time this publication went to press, it was unclear what options for an international response to climate change would be proposed by the Bush Administration.

The Protocol's ambitious, short-term reduction targets and lack of developing country participation had been key concerns even before President Bush took office.

However, while few countries are scheduled to meet these targets, many countries remain strongly supportive of both the Protocol's targets and framework.

The subject of intense activity on the part of the global community for almost 10 years, the Protocol's signature contribution is its framework. This framework includes binding targets, the use of market-based strategies such as emissions trading, the use of sinks, and the inclusion of non-carbon dioxide gases. But the Protocol also has significant weaknesses. The targets negotiated in Kyoto were agreed to before it was clear what mechanisms would be available to meet those targets. This has encouraged countries to distort the mechanisms in ways that would allow countries to meet their targets, a contributing factor in the demise of the negotiations in The Hague. Additionally, the timeframes for meeting the targets are virtually impossible for some countries, including the United States, to meet because emissions have continued to rise and the time remaining for implementation has grown increasingly short. For example, emissions in the United States are already almost 15 percent above 1990 levels, meaning that the minimum reduction required would be 22 percent. If administrative procedures (and in the United States these are complex and time-consuming) further delay the time for ratification and implementation, the 22 percent reduction could easily become 30 percent.

The future of the Protocol rests in large measure on what alternative the Bush Administration proposes, and how the global community responds to that proposal. But what is clear is that the increasingly strong science demands a global framework that: (1) achieves real emissions reductions, (2) is market-based and cost-effective, (3) can be sustained over a long period of time, and (4) includes, at some future point in time, all nations that emit significant quantities of greenhouse gases.

Looking Ahead: Sustaining the Momentum

No matter how the global community chooses to address climate change (either by continuing with the Kyoto Protocol or by choosing a different vehicle), the need for a global response is pressing. Many governments, at all levels, are taking measures to begin delivering on their Kyoto commitments, and many businesses are working hard to reduce their emissions. These activities, however, do not obviate the need for a global framework, or for strong actions on a national basis. Nor will the current state of uncertainty assist those in the private sector who must make significant investments in the technologies of tomorrow — technologies that will form the core of any successful strategy to address climate change.

The reality is that we must move much further than our current programs suggest. For, unless we are prepared to renounce our faith in science, we must believe what it tells us. And the scientific consensus says that the world is warming, and that human activity is at least partly to blame.

The ingenuity and the capital that will be needed to address this enormous challenge must come, ultimately, from the private sector. But experience has shown that voluntary action alone is not enough. Governments, working together, must provide the goals, framework, and support in the form of a strong and effective international agreement.

In judging any approach to this problem, our benchmarks should be the three objectives embedded in the Framework Convention on Climate Change: environmental effectiveness, cost-effectiveness, and equity. Looking ahead, we will achieve these objectives only if we work hard, rise above politics, and recognize that what works best for the global environment will ultimately work best for us all.

+

+

+

Climate Change
Legislation in the United States

Vicki Arroyo Cochran, Manik Roy

In spite of the increasing scientific consensus that human-induced climate change is occurring, U.S. action to address this problem is still in its early stages. A number of legislative bills that have implications for climate change have been introduced in the U.S. Congress in recent years; however, none of the bills has been enacted. Debate in the Congress and among various stakeholder groups has often focused on concerns about the cost of reaching reduction targets for greenhouse gas (GHG) emissions and the lack of similar targets for developing countries.

Many countries, including the United States, signed and ratified the 1992 United Nations Framework Convention on Climate Change, pledging to reduce greenhouse gas emissions to 1990 levels. A later protocol to the convention, crafted at a 1997 conference in Kyoto, Japan, has not yet entered into force. The Kyoto Protocol establishes emissions reduction targets for 38 developed countries to meet during the period 2008–2012.

Given its limited action to date,[1] the United States — like other countries featured in the Global Strategies chapter — is unlikely to meet its Kyoto target of reducing greenhouse gas emissions to 7 percent below 1990 levels by 2008–2012. Nor is it likely to reach the goal under the Framework Convention of stabilizing emissions at 1990 levels. Prospects for meeting either commitment appeared even more remote following an announcement by the Bush Administration in March that it would not implement the Protocol.

This paper examines the background and context of recent efforts in the U.S. Congress to address climate change.

Background: The U.S. Legislative Process

The U.S. Constitution provides that all legislative power is vested in the U.S. Congress, comprised of a Senate and House of Representatives.[2] The Senate is composed of 100 members — two from each state — elected for terms of six years. The House is composed of 435 members elected every two years from separate election districts, each comprised of approximately half a million people. Each senator and representative has one vote in his or her respective chamber. A Congress lasts for two

The authors and the Pew Center would like to thank Bryan Hannegan and Chris Miller for their review of this paper and their helpful suggestions.

years, commencing in the January following a biennial election. The 107th Congress convened on January 3, 2001.

The chief role of Congress is to make laws. Among other things, Congress passes the 13 annual appropriations acts that fund the federal government. A bill that has been agreed to by both houses of Congress becomes a law only upon: presidential approval, presidential inaction for a period of 10 days following the bill's passage, or the override of a presidential veto by a two-thirds vote of each chamber.

In addition to making law, Congress provides oversight of federal activities and can pass non-binding resolutions. Also, consent must be received from the Senate for treaties and certain presidential nominations. Treaties must receive at least 67 votes in the Senate for ratification.

During most of the Clinton Administration, climate change was one of many issues the Republican-controlled Congress associated with the Democratic Clinton, and therefore treated with suspicion. With climate change, as with all major issues, the members of Congress have a wide variety of views that evolve as understanding increases. This evolution accounts for the amount of time it can take for a major policy issue to be addressed in legislation. The recent evolution of the climate change issue in Congress has taken place in three phases: first, Senate advice to the president on negotiations over the Kyoto Protocol; second, Congress's attempt to prevent "backdoor implementation" of the unratified Protocol; and third, Congress's attempt to begin developing climate change legislation despite lack of agreement over the Protocol.

The Byrd-Hagel Resolution

U.S. climate change legislation since 1997 should be viewed in the context of the Kyoto Protocol and Senate Resolution 98 of the 105th Congress (commonly referred to as the Byrd-Hagel Resolution).

In June 1997, anticipating the December 1997 meeting in Kyoto, Sen. Robert C. Byrd (D-West Virginia) introduced, with Sen. Chuck Hagel (R-Nebraska) and 44 other co-sponsors, a resolution stating "it is the sense of the Senate that ... the United States should not be a signatory to any protocol to, or other agreement regarding, the United Nations Framework Convention on Climate Change of 1992, at negotiations in Kyoto in December 1997, or thereafter, which would —

(A)   mandate new commitments to limit or reduce greenhouse gas emissions for the Annex I Parties [i.e., industrialized countries], unless the protocol or other agreement also mandates new specific scheduled commitments to limit or reduce greenhouse gas emissions for Developing Country Parties within the same compliance period, or

(B)   would result in serious harm to the economy of the United States ....

Resolutions of this sort merely state the "sense of the Senate," and are not enforceable as law. However, under the U.S. Constitution, the president can only enter into a treaty with the "advice and consent" of the Senate. Senators Byrd and Hagel intended their resolution as Senate advice on the form of any protocol that might be negotiated as an extension of the U.N. Framework Convention.

Prior to introduction of the Byrd-Hagel Resolution, the Clinton Administration had not insisted on developing country commitments to limit GHG emissions in the same time frame. In fact, the international community had agreed upon a statement known as the "Berlin Mandate," adopted in April 1995 under the U.N. Framework Convention. The Berlin Mandate called for commitments to limit greenhouse gas emissions only by industrialized countries as listed in Annex I of the Framework Convention. The Berlin Mandate explicitly avoided the topic of commitments by developing countries in the same time frame.

Given its agreement with the Berlin Mandate, the Clinton Administration might have been expected to oppose the Byrd-Hagel Resolution, urge that the resolution be amended by striking any reference to developing country commitments, or urge that the resolution be amended to state that mandates on developing countries would only come into force during a later compliance period. With the last amendment, the Byrd-Hagel Resolution would have called for a Kyoto Protocol similar in structure to the Montreal Protocol on Substances that Deplete the Ozone Layer.[3]

Instead of pursuing any of these three options, the Clinton Administration signaled its willingness to accept the language of the resolution without amendment, leading Senate Democrats to join Republicans in voting for it. The Byrd-Hagel Resolution passed 95–0 in July 1997, without substantial amendment. Despite the Senate's advice, in December 1997, the administration agreed to a Kyoto Protocol that mandated GHG reductions for industrialized countries, but none for developing countries — effectively defying the Byrd-Hagel Resolution.

The Kyoto Protocol has never been submitted to the Senate for ratification. Nevertheless, the incongruity between the Protocol and the Byrd-Hagel Resolution has already complicated negotiations over the Protocol, enactment of climate change legislation, and even the funding of longstanding energy efficiency programs.

The Knollenberg Appropriations Amendments

After the 1997 conference in Kyoto, the Clinton Administration's primary legislative effort in the climate change area was to request large budget increases for climate research and programs promoting research into and use of energy efficient technologies. The administration dubbed the effort to expand these longstanding programs the "Climate Change Technology Initiative." While Congress did not grant the requested increases, the programs continued to receive funding and, in some cases, moderate increases.

In addition, however, Congress wrote provisions into several appropriations acts restricting federal climate change activities. These provisions embodied the Republican Congress's concern that the Democratic president might implement the Kyoto Protocol without Senate ratification — so-called "backdoor implementation." The first of these provisions was enacted in 1998 as part of the U.S. Environmental Protection Agency's (EPA) appropriations act for fiscal year (FY) 1999 (known as the VA-HUD-Independent Agencies Appropriations Act, 1999). The provision, initially written by Rep. Joseph Knollenberg (R-Michigan), effectively prohibited the EPA from proposing or issuing rules, regulations, decrees, or orders implementing the Kyoto Protocol. Language similar to Knollenberg's subsequently appeared in six of the 13 FY 2000 appropriations acts, including those for the EPA, the Department of State, the Department of Energy (DOE), the Department of Agriculture, and the U.S. Agency for International Development (USAID), and in eight of the 13 FY 2001 appropriations acts.

Similar provisions, along with other restrictions on the ability of the administration to address climate change, were introduced as a bill (H.R. 2221) in 1999 by Rep. David McIntosh (R-Indiana). The bill was co-sponsored by Rep. Knollenberg and 31 others, but not acted upon.

The Knollenberg provisions created uncertainty regarding the appropriate role of federal agencies. In February 2000, believing the conduct of certain agencies' personnel to be in violation of the prohibition, Rep. Knollenberg requested an investigation by the inspectors general of the EPA and DOE to determine whether the agencies' actions could be interpreted as illegal implementation of the Protocol. These actions had a chilling effect on activities undertaken by federal agencies. Eventually, the inspectors general found that the agencies' activities were within the proper bounds, but it was clear that the appropriations language needed clarification.

In June 2000, during House deliberations over the EPA's FY 2001 appropriations act, Rep. John Olver (D-Massachusetts) offered an amendment clarifying that "any limitation imposed under this Act on funds made available by this Act for the Environmental Protection Agency shall not apply to activities specified in the previous proviso related to the Kyoto Protocol which are otherwise authorized by law." In other words, under the Olver amendment, ongoing efforts by the EPA and other agencies to work toward the terms of the Rio Convention or to implement existing domestic statutes, such as the Clean Air Act, would be deemed appropriate. The amendment passed by a vote of 314–108. A similar Olver amendment was offered during debate on the appropriations bill for the State Department and USAID. Again, the amendment prevailed by 217–181. These votes clarified the U.S. climate change position. The United States could implement climate change programs under existing law, but could not otherwise implement the unratified Kyoto Protocol.

Climate Change Measures in Congress

Though not prepared to support the Kyoto Protocol in its current form, Congress has shown an increasing willingness to draft and debate climate change legislation. In the 105th Congress (1997–1998), aside from appropriations measures, only seven bills were introduced to address the climate change issue, while 25 such bills were introduced in the 106th Congress (1999–2000).[4] (For a list of these bills, plus some significant bills on energy efficiency, see Box 1.) The rest of this chapter describes these measures.

Credit for Early Action

Legislation to provide marketable credits for early GHG reductions was first introduced in October 1998, then reintroduced in the 106th Congress by the late Sen. John H. Chafee (R-Rhode Island) and Senators Connie Mack (R-Florida) and Joe Lieberman (D-Connecticut). Representatives Rick Lazio (R-New York) and Calvin Dooley (D-California) introduced similar legislation in the House in 1999. Their bills would issue credits to businesses and other entities for GHG emissions reductions made before the effective date of a future domestic GHG reduction program — thus creating an incentive for early action. The credits, which would be used to comply with such a program, should one be enacted, could also have value to companies in a domestic or global greenhouse gas market. Progress on the early action bills stopped with the death of Sen. Chafee in late 1999 and with Rep. Lazio's decision to run for the Senate.[5]

Research and Voluntary Action

Other members of Congress promoted an alternative approach that would expand investment in long-term energy technology research and development. Bills such as S. 882 by Sen. Frank Murkowski (R-Alaska), S. 1776 by Sen. Larry Craig (R-Idaho), and H.R. 3384 by Rep. Joe Barton (R-Texas) — all introduced in 1999 — would increase investment in scientific research, focus on commercialization of new energy technologies and international technology transfer, and expand and consolidate the existing voluntary reporting system managed by the Department of Energy.

Sen. John McCain (R-Arizona), who held hearings on climate change science as chairman of the Senate Commerce, Science, and Transportation Committee, sponsored S. 3237, the International Climate Change Science Commission Act, which would provide for a scientific commission to assess changes in global climate patterns and to conduct scientific studies. It is unclear how this commission would relate to existing efforts by groups such as the Intergovernmental Panel on Climate Change to determine the status of climate change science and impacts.

Box 1

Legislative Proposals in the 106th Congress (1999–2000) Regarding Global Climate Change

Research and Voluntary Action
- S. 882, Energy and Climate Policy Act of 1999, by Sen. Frank H. Murkowski (R-Alaska)
- S. 1776, Climate Change Energy Policy Response Act, by Sen. Larry E. Craig (R-Idaho)
- S. 3237, International Climate Change Science Commission Act, by Sen. John McCain (R-Arizona)
- H.R. 3384, Energy and Climate Policy Act of 1999, by Rep. Joe Barton (R-Texas)
- H.R. 3385, To strengthen provisions in the Federal Non-nuclear Energy Research and Development Act of 1974 with respect to potential Climate Change, by Rep. Joe Barton (R-Texas)

Credit for Early Action
- S. 547, Credit for Voluntary Reductions Act, by Senators John H. Chafee (R-Rhode Island), Connie Mack (R-Florida), and Joseph I. Lieberman (D-Connecticut)
- H.R. 2520, Credit for Voluntary Actions Act, by Representatives Rick Lazio (R-New York) and Calvin Dooley (D-California)

Power Plant Emissions
- S. 1369, Clean Energy Act of 1999, by Sen. Jim M. Jeffords (R-Vermont)
- S. 1949, Clean Power Plant and Modernization Act of 1999, by Sen. Patrick J. Leahy (D-Vermont)
- H.R. 2569, Fair Energy Competition Act of 1999, by Rep. Frank Pallone, Jr. (D-New Jersey)
- H.R. 2645, Electricity Consumer, Worker, and Environmental Protection Act of 1999, by Rep. Dennis J. Kucinich (D-Ohio)
- H.R. 2900, Clean Smokestacks Act of 1999, by Rep. Henry A. Waxman (D-California)
- H.R. 2980, Clean Power Plant Act of 1999, by Rep. Thomas H. Allen (D-Maine)
- H.R. 4859, Great Smoky Mountains Clean Air Act of 2000, by Rep. Charles H. Taylor (R-North Carolina)

Carbon Sequestration
- S. 1066, The Carbon Cycle and Agricultural Best Practices Research Act, by Sen. Pat Roberts (R-Kansas)
- S. 1457, Forest Resources for the Environment and Economy Act, by Sen. Ron Wyden (D-Oregon)
- S. 2540, Domestic Carbon Storage Incentive Act of 2000, by Sen. Sam Brownback (R-Kansas)
- S. 2818, Food Security and Land Stewardship Act of 2000, by Sen. Tim Johnson (D-South Dakota)
- S. 2982, International Carbon Sequestration Incentive Act, by Sen. Sam Brownback (R-Kansas)
- S. 3260, Conservation Security Act of 2000, by Sen. Tom Harkin (D-Iowa)

(continued on following page)

Box 1 continued

Biomass
- S. 935, National Sustainable Fuels and Chemicals Act of 1999, by Sen. Richard G. Lugar (R-Indiana)
- S. 1945, Biofuels Air Quality Act, by Sen. Christopher S. Bond (R-Missouri)
- H.R. 2788, Biofuels Air Quality Act, by Rep. John M. Shimkus (R-Illinois)

Tax Incentives
- S. 1777, Climate Change Tax Amendments of 1999, by Sen. Larry E. Craig (R-Idaho)
- S. 1833, Energy Security Tax Act of 1999, by Sen. Thomas A. Daschle (D-South Dakota)
- S. 2718, Energy Efficient Buildings Incentives Act, by Sen. Robert C. Smith (R-New Hampshire)

Restrictions on Administrative Activities
- H.R. 2221, Small Business, Family Farms, and Constitutional Protection Act, by Representatives David McIntosh (R-Indiana) and Joseph Knollenberg (R-Michigan)

Sectoral Approaches: Power Plant Legislation and Fuel Economy Standards

In 1998 (the most recent year for which data are available), the U.S. electricity industry accounted for 37 percent of U.S. carbon dioxide (CO_2) emissions.[6] This fact, as well as congressional debate over electricity deregulation and EPA rulemaking and enforcement directed toward power plants, inspired legislation in the 106th Congress that would regulate CO_2 emissions from power plants. Several bills were introduced to control power plant emissions of CO_2, as well as three other power plant pollutants — nitrogen oxides (NO_X), sulfur dioxide (SO_2), and mercury. The bills were sponsored by Sen. Jim Jeffords (R-Vermont), and Representatives Henry Waxman (D-California), Thomas Allen (D-Maine), Frank Pallone (D-New Jersey), and Dennis Kucinich (D-Ohio). Most of the bills would establish performance standards governing the pollutants — a traditional regulatory approach that requires emissions to be reduced by a certain amount. However, some of the bills would also allow for trading of CO_2 emissions credits (and perhaps of other pollutants such as NO_X and SO_2) across firms.

In addition, Sen. Robert Smith (R-New Hampshire), chairman of the Senate Environment and Public Works Committee, held hearings on the power plant four-pollutant approach and may introduce legislation on the issue. This notion of a "multi-pollutant" approach has even attracted attention from ardent critics of the Kyoto Protocol. Rep. Charles H. Taylor (R-North Carolina) introduced a bill during the 106th Congress that would have applied this approach to plants owned by the Tennessee Valley Authority. Rep. Taylor and the bill's seven co-sponsors were not obvious supporters of CO_2 reduction legislation: days before introducing the bill, they all voted against the Olver amendments (mentioned above).

Another sectoral approach involves regulation of automotive fuel economy standards. In 1998, transportation was responsible for 30 percent of U.S. CO_2 emissions.[7] Since 1995, Congress has included language in the annual Transportation Department Appropriations Act effectively preventing changes to the federal automotive fuel economy standard, known as the Corporate Average Fuel Economy (CAFE) standard. Proponents of stricter CAFE standards have tried each year to have the language removed, without success. During the 106th Congress, environmental advocacy groups named the removal of the "CAFE freeze rider" as one of their top climate change priorities.

During Senate debate on the FY 2000 Transportation Department Appropriations Act in 1999, Sen. Slade Gorton (R-Washington) offered an amendment to lift the CAFE freeze provision, but it was defeated by a vote of 55–40. The next year, however, in deliberations over the FY 2001 Transportation Department Appropriations Act, proponents and opponents of the provision struck an agreement intended to advance the debate. While the FY 2001 Act still prevents changes to the CAFE standard, it also directs the National Academy of Sciences (NAS) to report on the impact that increasing CAFE standards would have on motor vehicle safety, the U.S. automotive industry, and U.S. employment in the automotive industry, and on the effect of requiring CAFE calculations for domestic and non-domestic fleets. The CAFE debate has centered on these issues, with each side apparently having a different understanding of the facts. The NAS report, due by July 2001, is intended to refine Congress's understanding of the facts in time for debate over the FY 2002 Act.

Carbon Sequestration and Biomass

Members of Congress representing farm and forestry states introduced legislation in the 106th Congress to reduce the atmospheric levels of carbon dioxide by promoting the sequestration — or storage — of carbon through forest and soil management. Bills were also introduced to advance the use of biomass as a fuel. Farm-related legislation is expected to be especially important in the 107th Congress (2001-2002), when the Farm Bill is expected to be reauthorized. (The current version of the Farm Bill is the Federal Agriculture Improvement and Reform Act of 1996, which is authorized to continue through FY 2002.)

Senators Pat Roberts and Sam Brownback (both Republicans from Kansas) each introduced legislation in the 106th Congress intended to promote soil sequestration. The Carbon Cycle and Best Practices Research Act, introduced by Sen. Roberts, would encourage carbon sequestration use and research, but also require the implications of the Kyoto Protocol on the farm economy to be analyzed under various scenarios — including with and without the use of carbon sinks and market mechanisms, and with and without the participation of developing countries.

Sen. Brownback's domestic sequestration bill would require the U.S. Department of Agriculture (USDA) to establish a carbon sequestration program. Under the program, farmers would enroll in a program intended to increase carbon storage on their land, and would receive assistance from the USDA extension service to do so. Sen. Brownback also introduced the International Carbon Sequestration Act with the aim of enhancing international conservation projects and rewarding voluntary carbon storage. The bill would amend the tax code to establish a limited carbon sequestration investment credit, and would make those owning or operating property outside the United States eligible for an extension of credit from the Export-Import Bank of the United States.

Sen. Tom Harkin (D-Iowa), the top Democrat on the Senate Agriculture Committee, introduced a bill that, among other things, would base payments to farmers on "the extent to which the [farmer's] conservation security plan incorporates practices that optimize carbon sequestration and minimize greenhouse gas emissions."

As mentioned, farm state representatives also introduced legislation to promote the use of biomass as a fuel. Sen. Richard Lugar (R-Indiana), chairman of the Senate Agriculture Committee, introduced the National Sustainable Fuels and Chemicals Act of 1999, which would provide federal grants for biomass-related activites, including "research on accurate measurement and analysis of carbon sequestration and carbon cycling in relation to bio-based industrial products and feedstocks." Sen. Christopher Bond (R-Missouri) and Rep. John Shimkus (R-Illinois) introduced bills to direct the Department of Transportation to consider the extent to which a project proposed under the congestion mitigation and air quality improvement program reduces atmospheric carbon emissions and sulfur, presumably to promote the use of biomass under the program.

In a related area, Sen. Ron Wyden (D-Oregon) introduced a bill that would amend the Energy Policy Act of 1992 to assess opportunities for increased carbon storage in national forests.

Tax Incentives

A wide variety of policy objectives, such as private home ownership and certain conservation measures, are embodied in the U.S. tax code. Sen. Larry Craig (R-Idaho), one of the most outspoken critics of the Kyoto Protocol, applied this approach to the climate change issue in introducing the Climate Change Tax Amendments Act of 1999. The bill would provide a research and development tax credit to companies that reduce greenhouse gas emissions.

Tax incentives for energy efficiency were also the subject of legislation in the 106th Congress, building on the longstanding U.S. interest in energy efficiency. Two bills merit particular attention. Sen. Thomas Daschle (D-South Dakota), minority leader of the Senate, introduced the Energy Security Tax Act of 1999, which would create tax incentives to promote the use of energy efficient technologies and renewable power generation. Sen. Robert Smith (R-New Hampshire) introduced the Energy Efficient Buildings Incentives Act to create tax incentives for energy efficient buildings.

Proponents of these measures would likely urge that they be included in any tax legislation enacted in the 107th Congress.

Petition to Regulate GHG Emissions Under the Clean Air Act

In October 1999, the International Center for Technology Assessment (ICTA) and a coalition of environmental advocacy organizations sent a petition to the EPA requesting that the agency regulate GHG emissions from new motor vehicles. The petition asked the EPA to carry out what petitioners asserted to be a mandatory duty under a specific provision of the Clean Air Act (section 202(a)(1)).

The EPA has not decided whether to grant or deny the petition, and has not begun a rulemaking process. Instead, the EPA has published a notice asking for public comment on the petition.[8] The purpose of the public comment process is to solicit opinion and information from all interested parties as to how the EPA should respond to the petition.

Emissions Trading Proposals

Two "think tanks" — Resources for the Future (RFF) and the Progressive Policy Institute (PPI) — have proposed carbon emissions trading programs modeled on the U.S. acid rain program, which was successful in reducing sulfur dioxide (SO_2) emissions.[9] These programs — not yet embodied in legislation — would allow businesses complete flexibility to choose their compliance methods and to buy and sell the right to emit carbon. While both the PPI and RFF proposals would establish an emissions trading market, they differ in several important ways.

The PPI proposal would cap emissions at year 2000 levels. This emissions cap would decline over time so that near-1990 levels would be achieved by 2012. PPI's proposal would require that credits be acquired "downstream," encompassing sources over a certain size in all sectors, including government. The PPI proposal would "grandfather" (or give away for free) 95 percent of the year 2000 emissions to existing sources, initially auctioning off 5 percent — with the auctioned amount increasing over time. The PPI proposal would also establish reporting and public notification requirements similar to those of the Securities and Exchange Commission and the EPA's Toxics Release Inventory program. It would also provide credit for early reductions undertaken between 1993 and 2000.

The RFF proposal would cap emissions at 1990 levels, but with an important escape clause — emissions credits would be capped at $25 per ton of carbon in the first year. If reduction costs exceeded $25 per ton of carbon, more credits would be issued. The cost cap would then be increased by 7 percent per year to create an incentive to reduce emissions. Under the RFF proposal, only "upstream" facilities — all domestic energy producers and importers — would be required to obtain credits. All credits would be auctioned by the government to major sources of GHG emissions. Revenues from the auctions would be returned to households to defray higher consumer prices and to states to assist with transition of affected industries, workers, and communities.

Moving Forward

Meeting the goal of the U.N. Framework Convention on Climate Change (stabilization of GHG concentrations at a level designed to prevent dangerous interference with the climate system) will ultimately require a comprehensive set of approaches, possibly including support for the development of new energy-efficient technologies, rewards for companies that act early to reduce their GHG emissions, mechanisms to allow the trading of emissions credits, and tax incentives and energy-efficiency standards to promote the use of efficient products and technologies. A domestic program should also integrate well with an international trading system, if and when such a system is established.

Formulating a politically viable and effective domestic program that will significantly reduce our production of greenhouse gases will undoubtedly be a challenge. But the sooner the United States begins to implement intelligently crafted policy aimed at creating incentives to improve technology, reduce greenhouse gas emissions, and find viable alternatives to traditional carbon-based energy sources, the more likely it will be to meet the challenge in a way that does not compromise either its economy or the environment.

Endnotes & References

1. Such actions consist largely of voluntary programs aimed at limiting greenhouse gas emissions. Such programs include: DOE's Climate Challenge program for electric utilities; EPA's Landfill Methane Outreach Program, Coalbed Methane Outreach program, Energy Star®, and the Green Lights Program, as well as the U.S. Initiative on Joint Implementation. In addition, DOE's Voluntary Reporting of Greenhouse Gas Program required by Section 1605(b) of the Energy Policy Act of 1992 records the results of voluntary measures to reduce, avoid, or sequester carbon. During 1999, a total of 201 U.S. companies and other organizations reported on 1,715 projects that achieved reductions and sequestration equivalent to 226 million metric tons of carbon dioxide, or about 3.4 percent of total 1999 U.S. greenhouse gas emissions. (*Voluntary Reporting of Greenhouse Gases 1999*, DOE/EIA – 0608(99), February 2001).

2. Article I, Section 1, U.S. Constitution.

3. Under the Montreal Protocol, adopted in 1987, production of the most damaging ozone-depleting substances was eliminated (except for a few critical uses) by 1996 in industrialized countries and will be eliminated by 2010 in developing countries.

4. These figures pertain to non-appropriations measures that specifically referred to climate change, greenhouse gases, or carbon dioxide. Other measures important to the climate change issue were introduced as well, particularly to promote energy efficiency.

5. Hillary Clinton defeated Rep. Lazio in the November 2000 election to represent New York in the U.S. Senate.

6. U.S. EPA 2000. *Inventory of U.S. Greenhouse Gas Emissions and Sinks: 1990-1998*.

7. U.S. EPA 2000.

8. 66 Fed. Reg. 7486 (Jan. 23, 2001).

9. The trading program established under the Clean Air Amendments of 1990 sets a graduated series of "caps" on total national emissions of sulfur dioxide. The government allocates credits to electric utilities corresponding to the amount of sulfur dioxide emissions permissible under law; these credits can then be traded among the utilities and others. The "cap and trade" program lowered sulfur dioxide emissions below the required levels at a cost of less than $1 billion compared to an estimated $4.5 billion price tag for a conventional regulatory program. Importantly, the sulfur dioxide program covers only a few thousand power plants — those that are responsible for 70 percent of the domestic sulfur dioxide emissions. (*Allowance Trading Offers an Opportunity to Reduce Emissions at Less Cost*, GAO/RCED-95-30, December 1994; Ellerman, A.D., R. Schmalensee, P. Joskow, J.P. Montero, and E.M. Bailey. 1997. *Sulfur Dioxide Emission Trading: Evaluation of Compliance Costs*. M.I.T. Press, Cambridge, Massachusetts).

FACTS & FIGURES

Contents

Global Climate Data

All Greenhouse Gases

Most of the sun's energy that reaches the earth is absorbed by the oceans and land masses and radiated back into the atmosphere in the form of heat or infrared radiation. Most of this infrared energy is absorbed and reradiated by atmospheric gases such as water vapor and carbon dioxide. This phenomenon, referred to as the greenhouse effect (Figure 1), serves to keep the earth some 33°C (60°F) warmer than it would otherwise be. As concentrations of gases that absorb and reradiate infrared energy (i.e., greenhouse gases — GHGs) increase, the warming effect increases.

Figure 1

The **Greenhouse Effect**

Note: Greenhouse gases are shown as a layer to simplify the drawing. In reality, they are dispersed throughout the atmosphere. Although the atmosphere consists largely of oxygen and nitrogen, neither absorbs infrared energy; thus, they do not play a role in warming the earth and are not greenhouse gases.

Source: Rekacewicz, 2000.

The data in this chapter were compiled by Naomi Peña and Geraldina Grünbaum. They and the Pew Center gratefully acknowledge Michael Gillenwater, Dale Heydlauff, Gregg Marland, Art Rypinski, Kitty Seiferlein, and Elizabeth Scheele for their review of all or portions of this section.

Table 1 shows that concentrations of some GHGs have increased since the late 18th century. Atmospheric concentrations of carbon dioxide (CO_2) have continued to climb since the 1994 values given in the table and stood at 368,000 parts per billion volume (ppbv) in 1999 (CDIAC, 2000).

The first three GHGs shown in Table 1 — CO_2, methane (CH_4), and nitrous oxide (N_2O) — are emitted to the atmosphere by both natural and anthropogenic sources, while the last four occur only as a result of industrial processes.[1]

Table 1

The Main **Greenhouse Gases**

Greenhouse gases	Chemical formula	Pre-industrial concentration (ppbv)	Concentration in 1994 (ppbv)	Atmospheric lifetime (years)*	Anthropogenic sources	Global warming potential (GWP)**
Carbon dioxide	CO_2	278,000	358,000	Variable	Fossil-fuel combustion	1
					Land-use conversion	
					Cement production	
Methane	CH_4	700	1,721	12.2+/-3	Fossil fuels	21***
					Rice paddies	
					Waste dumps	
					Livestock	
Nitrous oxide	N_2O	275	311	120	Fertilizer	310
					Industrial processes	
					Combustion	
CFC-12	CCl_2F_2	0	0.503	102	Liquid coolants	6,200-7,100****
					Foams	
HCFC-22	$CHClF_2$	0	0.105	12.1	Liquid coolants	1,300-1,400****
Perfluoro-methane	CF_4	0	0.070	50,000	Production of aluminum	6,500
Sulfur hexa-fluoride	SF_6	0	0.032	3,200	Dielectric fluid	23,900

Note: ppbv = parts per billion volume; 1 ppbv of CO_2 in the earth's atmosphere is equivalent to 2.13 million metric tons of carbon (www.cdiac.esd.ornl.gov, accessed on December 10, 2000).

* No single lifetime for CO_2 can be defined because of the different rates of uptake by different sink processes.

** GWP for 100-year time horizon.

*** Includes indirect effects of tropospheric ozone production and stratospheric water vapor production.

**** Net global warming potential (i.e., including the indirect effect due to ozone depletion).

Source: United Nations Environment Programme's Introduction to Climate Change, accessed at www.grida.no/climate/vital/intro.htm on April 17, 2001.

Figure 2

Global | **CO$_2$ Flows, Carbon Reservoirs, and Reservoir Changes** |

in Gigatons (Gt) of Carbon

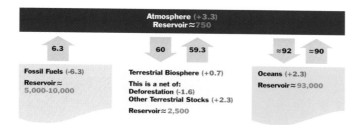

Note: Tan colored pool is decreasing in size. Blue colored pools are increasing. Intensity of blue indicates magnitude of stock change. Numbers in red indicate estimated total amount of carbon in reservoir. Numbers in green indicate average annual change in amount of carbon in reservoir.

Gigatons (Gt) = 10^9 metric tons.

Sources: Bolin et al. in IPCC, 2000a; Houghton, 1997.

Figure 2 portrays estimated sizes of the earth's carbon reservoirs, average annual changes in the amount of carbon held in the reservoirs, and recent flows of CO$_2$ between land and ocean reservoirs and the atmosphere.

As shown in Figure 2, the annual average human-induced flows of CO$_2$ — 6.3 gigatons (Gt) from fossil-fuel combustion and 1.6 Gt from deforestation in the 1990s — are a small fraction of total CO$_2$ flows. However, these flows are resulting in increased carbon in the ocean and atmospheric reservoirs.

Anthropogenic emissions of CH$_4$ and N$_2$O comprise a much larger share of total emissions of these gases than is the case for CO$_2$. Approximately 70 percent of the 550 million metric tons (MMT) of CH$_4$ emitted annually and about half of the 14 MMT of N$_2$O emitted annually are due to human activities (Bolin et al. in IPCC, 2000a).

The concept of global warming potential (GWP) was invented to allow comparisons of the total cumulative warming effects of different GHGs over a specified time period. A GWP is a measure of relative contribution to radiative forcing (see *The Science of Climate Change*). The warming effect of CO$_2$ is assigned a value of 1, and the warming effects of other gases are calculated as multiples of this value. (See Box 1.)

The GWPs shown in Table 1 are for a 100-year time horizon[2] and are from the Intergovernmental Panel on Climate Change (IPCC) Second Assessment Report (SAR). The Third Assessment Report (TAR) includes revised 100-year GWP estimates, increasing the GWP of CH$_4$ to 23 and decreasing the GWP of N$_2$O to 296 (IPCC, 2001). Virtually all GWPs used in the literature to date are from the SAR. Emissions calculations relevant to the United Nations Framework Convention on Climate Change (UNFCCC) and the Kyoto Protocol are also based on SAR GWPs.

GWP and Carbon Dioxide Equivalents (CO₂E)

GWPs are used to convert emissions of non-CO_2 gases into their CO_2 warming equivalents (CO_2Es). The CO_2E of a non-CO_2 gas is calculated by multiplying the mass of the emissions of the non-CO_2 gas by its GWP. A 100-year GWP of 21 for CH_4 (see Table 1) means that each gram of CH_4 emitted is considered to have cumulative warming effects over the next 100 years equivalent to emitting 21 grams of CO_2. Using the SAR 100-year GWP of 21 for CH_4, the CO_2E of 310 tons of CH_4 is 310 tons x 21 = 6,510 tons CO_2E. Emitting 310 tons of CH_4 would thus be considered to result in the same cumulative warming over the next 100 years as emitting 6,510 tons of CO_2.

While the 100-year time horizon GWPs shown in Table 1 are the most commonly used, GWPs based on cumulative warming over 20- and 500-year periods were also developed by the IPCC and are used in some analyses. For the HFCs and PFCs, it is common to use the GWP of the dominant gas (shown in parentheses) to calculate the CO_2E for emissions of these gases.

While most international data sources provide emissions data in terms of CO_2 or CO_2E, many U.S. sources provide emissions data in units of carbon (C) or carbon equivalents (CE). Emissions reported in CO_2 or CO_2E units are 3.67 times emissions reported in C or CE units. To convert from CO_2 to C or from CO_2E to CE, multiply by 12/44 (0.27), the ratio of the mass of C to the mass of CO_2. To convert from C to CO_2 or from CE to CO_2E, multiply by 44/12 (3.67).

Gas	SAR 20-Year GWPs	SAR 500-Year GWPs
CO_2	1	1
CH_4	56	6
N_2O	280	170
HFCs	460–9,100 (3,400)	42–9,800 (420)
PFCs	4,900–6,200 (4,400)	10,000–14,000 (10,000)
SF_6	16,300	34,900

Table 2

Table 2 shows 1990 global anthropogenic emissions of the GHGs covered under the Kyoto Protocol.

World Anthropogenic Emissions of GHGs, 1990 (MMT)

Gas	Emissions	Carbon Dioxide Equivalent (CO₂E)	Carbon Equivalent (CE)
CO_2	22,000	22,000	6,000
CH_4	310	6,510	1,775
N_2O	10.5	3,264	889
HFCs	NA	70	19
PFCs	NA	117	32
SF_6	0.006	139	38

Note: CO_2 emissions shown are from fossil-fuel combustion and other industrial processes, but do not include emissions from the conversion of forests or grazing lands to agricultural or urban land. CO_2Es shown are based on the 100-year time horizon GWPs from the SAR.

HFCs = Hydrofluorocarbons. PFCs = Perfluorocarbons.

Source: IPCC, 2000b.

Figure 3 shows the sources of emissions by economic sector in 1990 for GHGs covered under the Kyoto Protocol.

Table 3 shows the distribution of GHG emissions across the four regions used in the IPCC *Special Report on Emissions Scenarios* (SRES) (IPCC, 2000b).

Figure 3

Sources of **Anthropogenic GHG Emissions**
Worldwide, 1990, in CO_2E

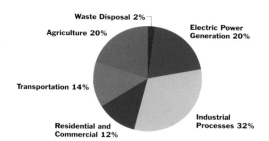

Sources: Distribution to sectors for CO_2, CH_4, and N_2O is from EDGAR, 2000. All other GHGs are assumed to be from industrial processes.

Table 3

Regional Distribution of GHG Emissions, 1990 (MMTCE)

	OECD90	ASIA	REF	ALM	World Total
CO_2	2,830	1,150	1,300	720	6,000
CH_4	418	647	270	439	1,774
N_2O	345	305	80	159	889
HGWP	60	7	15	7	89
Total	3,653	2,109	1,665	1,325	8,752

OECD90 = Australia, Austria, Belgium, Canada, Denmark, Finland, France, Germany, Greece, Iceland, Ireland, Italy, Japan, Luxembourg, the Netherlands, New Zealand, Norway, Portugal, Spain, Sweden, Switzerland, Turkey, the United Kingdom, and the United States.

ASIA = Asia except for those countries belonging to the OECD90 group (i.e., Japan, Australia, New Zealand).

REF= the Former Soviet Union and Eastern Europe.

ALM = Africa, Latin America, the Caribbean, and the Middle East except for Turkey.

HGWP = High Global Warming Potential Gases and refers to the GHGs with very long lifetimes such as the PFCs, HFCs, and SF_6.

Source: IPCC, 2000b.

The Framework Convention and the Kyoto Protocol

The United Nations Framework Convention on Climate Change (UNFCCC), which calls for "stabilization of greenhouse gas concentrations in the atmosphere at a level that would prevent dangerous anthropogenic interference with the climate system" (UNFCCC, Article 2, 1992), was adopted in 1992 and entered into force in 1994. The UNFCCC did not, however, include binding, quantitative emissions reductions for signatories. In 1997, the Kyoto Protocol was adopted to meet this need.

Under the Kyoto Protocol, 38 countries, known as Annex B countries, agreed to reduce total emissions of six GHGs — the six gases listed in Table 2 — by an average of 5.2 percent compared to 1990 emissions. The Annex B countries' 1990 emissions of the three main GHGs — CO_2, CH_4, and N_2O — and the percent reductions these countries agreed to are shown in Table 4.[3] To take effect, the Protocol requires

Table 4

Gases Reported by Parties for 1990 or Other Base Year
in Gigagrams of CO_2 Equivalent

	CO_2	CH_4	N_2O	Total	Targets for 2008–2012 as % of Total Base Year Emissions
Australia	278,669	117,122	22,620	418,411	108
Austria	62,130	11,289	2,033	75,452	92
Belgium	113,997	12,853	9,613	136,463	92
Bulgaria	103,856	28,009	25,225	157,090	92
Canada	465,755	74,376	62,793	602,924	94
Czech Republic	165,490	16,349	7,998	189,837	92
Denmark	52,894	5,848	10,825	69,567	92
Estonia	37,797	2,209	713	40,719	92
Finland	60,771	6,118	8,258	75,147	92
France	387,590	63,471	95,074	546,135	92
Germany	1,014,501	116,990	68,386	1,199,877	92
Greece	85,164	9,484	9,395	104,043	92
Hungary	83,676	13,952	4,005	101,633	94
Iceland	2,147	294	130	2,571	110
Ireland	31,575	12,836	9,086	53,497	92
Italy	432,565	39,889	45,261	517,715	92
Japan	1,124,532	32,400	18,090	1,175,022	94
Latvia	24,771	3,913	6,984	35,669	92
Liechtenstein	208	21	31	260	92
Lithuania	39,535	7,937	4,077	51,548	92
Luxembourg	12,750	502	197	13,448	92
Monaco	108	1	1	111	92
Netherlands	161,360	27,138	20,398	208,896	92
New Zealand	25,398	35,212	11,852	72,462	100
Norway	35,146	6,610	5,161	46,916	101
Poland	476,625	65,961	21,700	564,286	94
Portugal	43,132	14,494	6,232	63,858	92
Romania	194,826	49,497	20,556	264,879	92
Russian Federation	2,372,300	556,500	69,967	2,998,767	100
Slovakia	62,237	7,637	6,159	76,032	92
Slovenia	13,935	3,701	1,576	19,212	92
Spain	226,057	34,626	41,236	301,919	92
Sweden	55,443	5,966	7,990	69,399	92
Switzerland	44,409	5,080	3,516	53,005	92
Ukraine	703,792	197,448	17,980	919,220	100
United Kingdom	584,220	77,212	65,677	727,110	92
United States of America	4,914,351	652,139	396,853	5,963,344	93
Total	14,343,745	2,299,025	1,098,271		

Notes: Bulgaria and Romania used 1989 as their base year; Poland used 1988; and Hungary used an average of 1985–87.

Sources: IPCC's On-line Searchable Database of GHG Inventory Data, accessed at http://ghg.unfccc.int/ on April 17, 2001 and UNFCCC, 1998.

Climate Change: Science, Strategies, & Solutions

ratification by 55 countries that account for at least 55 percent of 1990 CO_2 emissions from Annex I countries.[4]

Global Carbon Dioxide Emissions

As can be seen from Table 2, almost 70 percent of the anthropogenic contribution to global warming in 1990 was due to CO_2. Attention has focused on CO_2 emissions both because anthropogenic CO_2 emissions far exceed other anthropogenic emissions and because of the availability of data.[5]

Anthropogenic CO_2 emissions come from two primary sources — (1) the combustion of fossil fuels, and (2) land-use change, primarily tropical deforestation.

As shown in Figure 4, emissions from fossil-fuel burning and land-use change were of very similar magnitudes from 1860 until the mid-1900s. Emissions from the use of fossil fuels have far exceeded emissions from land-use change over the last 50 years, a result of growing reliance on fossil fuels for energy.

Figure 5 shows CO_2 emissions on a regional basis. Emissions from the combustion of fossil fuels are broken out by the three major fuels — coal, oil, and gas. The triangles at the top of each bar illustrate the uncertainty associated with emissions

Figure 4

Annual Global **Carbon Emissions** from Fossil Fuels and Land-Use Change (Deforestation)

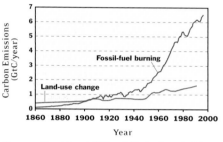

Sources: Houghton, 1996 and Marland et al., 1999.

Figure 5

1990 **CO_2 Emissions** from Fossil-Fuel Combustion and Land-Use Change

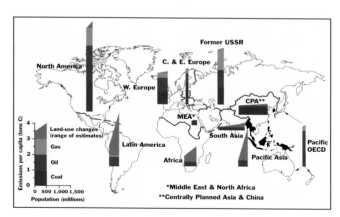

Note: The width of the bars represents population while the height represents per capita fossil-fuel emissions. Consequently, the total area of the rectangle (population times emissions per person) represents total emissions. Countries in the "Pacific Asia" group are shaded in black.

Source: IPCC, 1995.

from land-use changes. As shown in the figure, most of the CO_2 emissions in the Northern Hemisphere result from the use of fossil fuels, while land-use changes account for a significant portion of emissions in Latin America, Africa, and Pacific Asia.

Table 5 provides data on CO_2 emissions from fossil-fuel combustion, and from land use, land-use change, and forestry (LULUCF) for the seven countries with the highest CO_2 emissions in 1997.

Table 5

CO₂ Emissions from Fossil Fuels and Land-Use Change (MMTC)				
	Emissions from Fossil-fuel Combustion			Emissions from LULUCF 1997	Net Carbon Emissions 1997

	1990	1995	1997	Emissions from LULUCF 1997	Net Carbon Emissions 1997
US	1,314	1,414	1,490	-211	1,279
China	655	877	914	NA	NA
Russian Federation	506*	438	391	-229***	162
Japan	292	310	316	-26	290
India	184	248	280	NA	NA
Germany	243**	227	227	-9	218
UK	154	143	142	4	146

Note: The emissions due to land-use change shown in this table are estimates of changes in carbon stock, i.e., the result of measuring the amount of carbon in forests at two different points in time and dividing the difference by the number of years between the two measurements. Negative numbers indicate LULUCF acts as a net remover rather than source of atmospheric CO_2.

* Data are for 1992.

** Data are for 1991.

*** Data are for 1996.

Source: Information on fossil-fuel emissions is from Marland, et al., 2001. Information on land-use change emissions is from the United Nations Framework Convention on Climate Change website accessed at www.unfccc.de/resource/index.html on April 17, 2001.

The distribution of emissions from fossil fuels across economic sectors worldwide in 1998 is shown in Figure 6.

Figure 7 shows energy-related CO_2 emissions from major regions of the world from 1860-1997.

Figure 8 shows the history of CO_2 emissions from LULUCF activities for major regions of the world since 1850.

Figure 6

Worldwide Sources of CO₂ Emissions by Sector, 1998

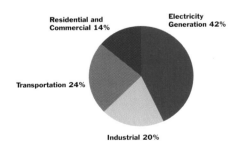

Source: IEA, 2000a.

Figure 7

Worldwide **Energy-Related CO$_2$**

Emissions, 1860-1997

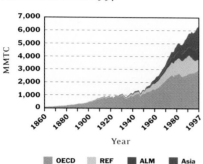

Year

■ OECD ▨ REF ■ ALM ■ Asia

Note: Emissions attributable to SRES regions were calculated by combining CDIAC regional and country data. This graph includes CO$_2$ emissions from both energy-related activities and cement production. In 1997, globally, cement production accounted for 3 percent of the total shown.

Source: Marland et al., 1999 and 2000.

Figure 8

CO$_2$ Emissions from

Land-Use Change 1850-1990

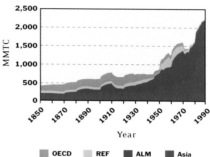

Year

■ OECD ▨ REF ■ ALM ■ Asia

Note: The emissions shown in this figure are derived by applying ecosystem-specific rates for emissions and regrowth to lands that have been deforested, logged, or are undergoing recovery from these activities.

Source: Adapted from Houghton and Hackler, 2000.

Global Emissions Drivers

Anthropogenic GHG emissions levels depend on the human population size, the level of economic activity, and the technologies in use. Increases in population and level of economic activity tend to be closely tied to increased use of energy. Insofar as fossil fuels are used as the source of this energy, increased use of energy will lead to increased CO$_2$ emissions unless sequestration, energy efficiency improvements, or other technologies (e.g., capture and storage of carbon) can balance the increased use. This section provides data on population and energy use. Gross Domestic Product (GDP) is provided for the seven countries with the highest CO$_2$ emissions as an indicator of their level of economic activity.

Population projections are "arguably the backbone of GHG emissions scenarios" (IPCC, 2000b). Examples of such projections are presented in Table 6, which shows projections used in the IPCC scenarios (see Box 2) and recent projections by the United Nations.[6]

Box 2

Future Emissions

The correlation between CO_2 emissions and population size and energy use raises the question of what CO_2 emissions may be like in the future. The IPCC examined this issue in depth, exploring a wide range of emissions and the forces underlying them. Six scenarios were then selected to illustrate the range of possible futures. The CO_2 emissions associated with the six illustrative scenarios are shown in Figure 9 below.

Scenario Descriptors

A1 family: Global population peaks around 2050 and declines thereafter. New technologies are rapidly introduced and economic disparities between regions are substantially reduced. All scenarios in the A1 group use the same basic population, technology, and economic assumptions. They differ in assumptions about how energy is supplied. Three illustrative scenarios have been used:

A1FI — fossil fuels continue to supply most of the energy.

A1T — non-fossil fuel energy sources dominate.

A1B — energy supply is balanced among fossil fuel and non-fossil energy sources.

A2 family: Global population continues to increase throughout the 21st century. Disparities in economic growth of regions persist, and technological change occurs more slowly than in any of the other illustrative scenarios.

B1 family: Global population peaks around 2050 and declines thereafter. Economies rapidly become service and information oriented, income disparities decrease and non-fossil fuel energy technologies are introduced.

B2 family: Global population continues to increase throughout the 21st century but not as rapidly as in the A2 family. Economic growth is less rapid than in the B1 family and less concentrated in the energy, service or information sectors than in either B1 or A1. Decreases in economic disparities occur primarily at local and regional levels.

Figure 9

CO_2 Emissions from the

IPCC Emissions Scenarios

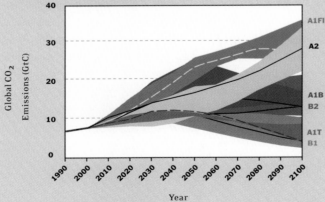

Source: IPCC, 2000b.

Table 6

| **Population Projections** by Region in 2050, in Millions |

| | | 2050 Population Projections | | | | | |
| | | IPCC Illustrative Scenarios | | | | UN | |
Region	2000 Population	A1	A2	B1	B2	Low	High
OECD90	912	1,081	1,151	1,001	976	821	1,134
ASIA	3,260	4,220	5,764	4,220	4,696	3,762	5,519
REF	412	423	519	432	406	332	479
ALM	1,496	2,980	3,862	3,055	3,289	2,428	3,541
Total World	6,080	8,704	11,296	8,708	9,367	7,343	10,674

Note: See Table 3 for key to regions. See Box 2 for brief descriptions of the IPCC scenarios.

Sources: IPCC, 2000b and Population Division of the Department of Economic and Social Affairs of the United Nations Secretariat, 1999.

Figure 10 shows the worldwide growth in the use of energy and electricity, while Figure 11 shows the fuels used to produce this energy. Type of fuel used is an important factor in emissions because fuels have significantly different ratios of CO_2 emissions per unit of energy consumed. Coal produces 21 percent more CO_2 than oil and 76 percent more CO_2 than natural gas per unit of energy consumption. Nuclear, solar, wind, hydroelectric, and biomass energy sources do not result in significant CO_2 emissions. Consequently, CO_2 emissions per unit of energy used are directly related to fuels and technologies in use, and assumptions about fuels and the rate of technological change play a significant role in emissions forecasts. (See Figure 9 in Box 2.)

Figure 10

Total World Primary

| **Energy Production and Electricity Consumption** |

1980 - 1998

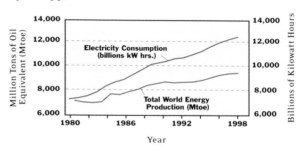

Source: USDOE, 2000c.

Figure 11

Worldwide Primary

| **Energy Supply by Fuel Type** 1998 |

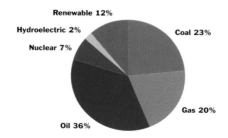

Note: Renewable includes biomass, waste, geothermal, solar, and wind.

Source: IEA, 2000b.

Table 7 provides 1998 energy and electricity use data as well as population and GDP statistics for the seven countries with the highest CO_2 emissions in 1998.

Table 7

| Population, GDP and Energy Consumption | for Seven Countries, 1998 | | | |

	Population (millions)	GDP (billion 1990 US$)	Total Energy Consumption (Mtoe)	Electricity Consumption (billion kWh)
US	269	7,044	2,182	3,603
China	1,239	805	1,031	1,080
Russia	147	334	582	716
Japan	126	3,304	510	1,013
India	980	499	476	407
Germany	82	1,884	344	532
UK	59	1,123	233	344

Source: IEA, 2000b.

U.S. Climate Data

All Greenhouse Gases

The U.S. Environmental Protection Agency (USEPA) and the U.S. Department of Energy (USDOE) are primary sources of information on GHG emissions in the United States. USEPA provides annually updated information on past and current emissions

Figure 12

U.S. GHG Emissions

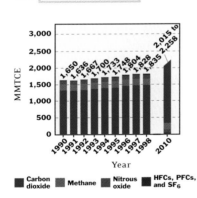

Year

Note: Excludes emissions from U.S. territories.

Sources: 1990-1998 data are from USEPA, 2000b.
2010 projections for CO_2 are from USDOE, 1999.
2010 projections for non-CO_2 gases are from USEPA, 2000a.

Figure 13

Sources of Total GHG Emissions

in CO_2E in the United States by Sector, 1998

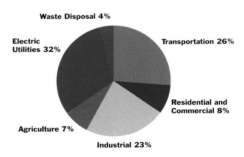

Note: Emissions from electricity produced by industries but sold to the grid is included in the "Industrial" category. Excludes emissions from U.S. territories.

Source: USEPA, 2001.

Climate Change: Science, Strategies, & Solutions

in the *Inventory of U.S. Greenhouse Gas Emissions and Sinks.* The USDOE Energy Information Administration's *Emissions of Greenhouse Gases in the United States* covers all GHGs. USDOE's *Annual Energy Outlook* is devoted to energy production and use in the United States while the *International Energy Outlook* provides data for both the United States and the rest of the world. These publications are available on the websites listed at the end of this section.

Figure 12 provides information on past and projected GHG emissions in the United States. The percent of U.S. GHG emissions attributable to major economic sectors in 1998 is shown in Figure 13. Figure 14 provides information on the sources of emissions of CO_2, N_2O, CH_4 and industrially produced GHGs in the United States in 1998.

Figure 14

Sources of GHGs in the United States, 1998

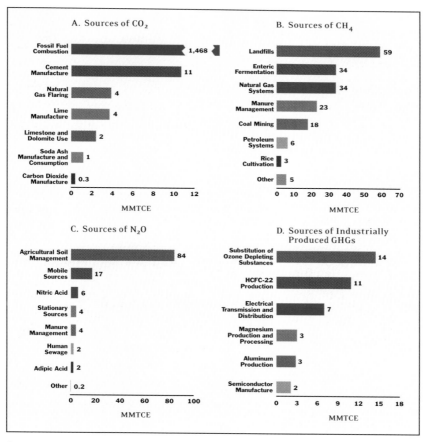

Source: USEPA, 2000b.

U.S. Carbon Dioxide Emissions

Figure 15 illustrates the rise in U.S. CO_2 emissions due to fossil-fuel combustion over the last 150 years. Figure 16 shows the sources of fossil-fuel-generated CO_2 emissions by fuel type and economic sector.

Figure 15

CO_2 **Emissions by Fuel Type**
in the United States, 1850-1996

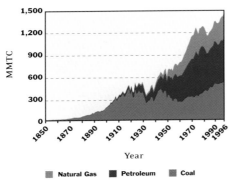

Year

■ Natural Gas ■ Petroleum ■ Coal

Note: In addition to releases due to combustion of fossil fuels, CO_2 is also released during the manufacture of cement, lime, and soda ash, flaring of natural gas, combustion of wastes, and use of CO_2. Emissions from processes other than the combustion of fossil fuels contribute approximately 2 percent of total CO_2 emissions in the United States.

Source: Marland et al., 1999.

Figure 16

CO_2 Emissions for
Major Economic Sectors
in the United States, 1998

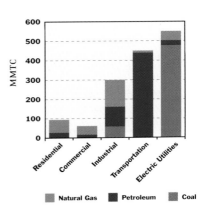

■ Natural Gas ■ Petroleum ■ Coal

Note: Excludes emissions from U.S. territories.

Source: USEPA, 2000b.

In the 1800s and early 1900s, deforestation was a source of CO_2 emissions in the United States as it is in many developing nations today. However, by the 1950s, U.S. forests had become a "sink," absorbing more CO_2 through forest regrowth than was being lost through harvesting. The average annual amounts of carbon absorbed by U.S. forests since 1950 are shown in Table 8. As depicted, although U.S. forests continue to store carbon, the rate of storage is decreasing.[7]

Table 8

Annual Carbon Stored by U.S. Forests, 1950-1998 (MMTC)

Reservoir	Average Annual for 1950–90	1990	1998
Forests (including soils)	281	274	171
Wood Products	NA	18	18
Wood in Landfills	NA	19	19
Total		316	211

Sources: 1990 and 1998 data are from USEPA, 2000b. Estimates for the 1950-1990 period are from Birdsey and Heath, 1995.

Climate Change: Science, Strategies, & Solutions

U.S. Emissions Drivers

As is true globally, population, level of economic activity, and technologies in use determine emissions levels in the United States. Population projections for the United States are shown in Figure 17.

Historically, increases in the level of economic activity have correlated strongly with increased energy use. However, in recent years there has been a "decoupling" of the connection between economic growth and consumption of energy in the United States. This decoupling is illustrated in Figure 18, which shows that GDP has grown rapidly while energy use has grown at relatively modest levels, particularly since the mid-1980s.

Figure 17

U.S. | **Population Projections** | to 2050

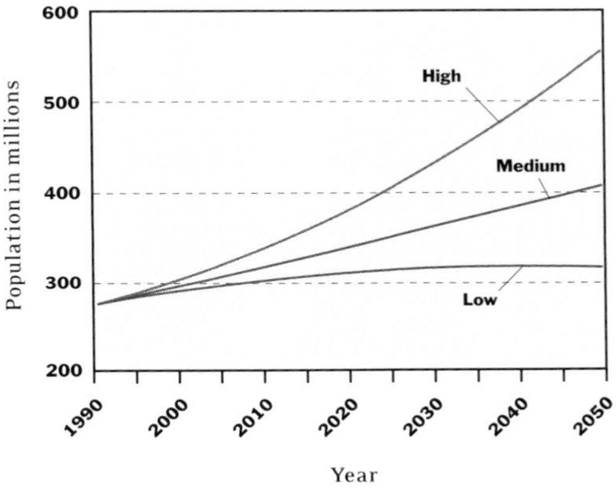

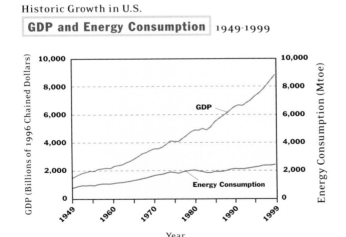

Source: U.S. Census Bureau, 2000.

Figure 18

Historic Growth in U.S.

| **GDP and Energy Consumption** | 1949-1999

Note: Chained dollars are dollars adjusted to purchasing power as of a base year, 1996 in this case.

Source: USDOE, 2000a.

The distribution of energy use across economic sectors in the United States since 1949 is shown in Figure 19. As can be seen, use of energy in the residential and commercial sectors has, since the mid-1980s, been almost as great as industrial energy use, with all three sectors sharing a common rate of increase in recent years.

Most of the energy consumed in the United States is produced by the combustion of fossil fuels — coal, natural gas, and oil — with approximately 24 percent coming from other sources, as shown in Figure 20.

Figure 19

Energy Consumption by End-Use Sector

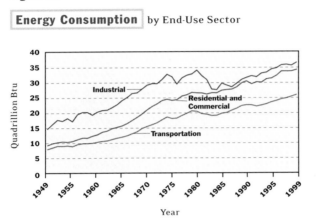

Year

Source: USDOE, 2000b.

Figure 20

Energy Production by Fuel Type and Technology in the United States, 1999

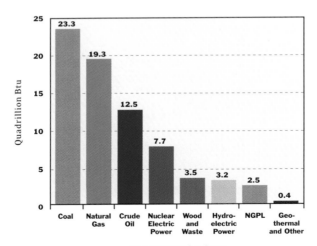

Fuel Type/Technology

Note: NGPL = natural gas plant liquids.

Source: USDOE, 2000b.

Climate Change: Science, Strategies, & Solutions

Conversion Factors & Websites

A variety of units are used to provide emissions and energy data. The following tables provide factors for converting between the units most commonly encountered in sources.

Table 9

Standard International Units

Standard International Prefix	Symbol	Multiplication Factor	Alternative Term
kilo	K	10^3	thousand
mega	M	10^6	million
giga	G	10^9	billion
tera	T	10^{12}	trillion
peta	P	10^{15}	quadrillion (United States and France)*

* In the United Kingdom and Germany the term quadrillion refers to 10^{24}.

Table 10

Factors for Converting Units

From	To	Multiply by
Emissions Reported in Mass of Non-oxygen Element*	*Total Mass of Emissions*	
CO_2-C or CE	CO_2 or CO_2 E	44/12 (3.67)
N_2O-N	N_2O	44/28 (1.57)
SO_2-S	SO_2	64/32 (2.0)
Mass in	*Mass in*	
grams	metric tons (MT)	10^{-6}
kilograms	MT	10^{-3}
gigagrams	million metric tons (MMT)	10^{-3}
teragrams	MMT	1
petagrams	MMT	10^3
gigatons	MMT	10^3
Energy in	*Energy in*	
Quadrillion British thermal units (Btu)	million tons oil equivalent (Mtoe)	25.2
Terawatt Hours **	Mtoe	0.086
Nuclear Terawatt Hours	Mtoe	0.26
Mtoe	quadrillion Btu	0.034
Mtoe	petajoules (PJ)	41.87
Barrels (bbl)	U.S. gallons	42

* Emissions of greenhouse gases are sometimes reported only in terms of the mass of the non-oxygen element of the gas, i.e., only the mass of the C, N, or S is reported rather than the entire mass of the CO_2, N_2O, or SO_2 emissions. When calculating a CO_2E, the GWP must be applied to the total mass of the emissions (see Box 1).

** Nuclear and geothermal electricity are not included.

Table 11

Websites Used in this Section

Organization	Primary Web Address	CO$_2$	Other Gases	Energy	Population	GDP
Carbon Dioxide Information Analysis Center	cdiac.esd.ornl.gov	•	•			
International Energy Agency	www.iea.org	•		•	•	•
IPCC	www.ipcc.ch	•	•	•	•	•
National Institute of Public Health and the Environment (Netherlands)	www.rivm.nl/env/int/ coredata/edgar	•	•			
United Nations Environment Programme	www.grida.no/climate/	•				
United Nations Framework Convention on Climate Change	www.unfccc.de	•	•			
U.S. Bureau of the Census	www.census.gov				•	
U.S. DOE, Energy Information Administration	www.eia.doe.gov	•		•	•	•
U.S. EPA	www.epa.gov/globalwarming	•	•	•		

Note: More in-depth information on a wide range of climate change topics and issues can be found in the reports available on the Pew Center on Global Climate Change website: www.pewclimate.org.

Endnotes

1. CFC-12 and HCFC-22 along with all other CFCs and HCFCs, are being phased out under the Montreal Protocol of 1987 due to their role in depleting stratospheric ozone. Other industrially produced gases — perfluorocarbons (PFCs), hydrofluorocarbons (HFCs), perfluoromethane (CF$_4$) and sulphur hexafluoride (SF$_6$) — absorb energy at wavelengths at which naturally occurring atmospheric gases do not. Thus, prior to the advent of the industrially produced gases, energy radiated at these wavelengths from the earth's surface was able to escape directly back into space and did not contribute to the warming effect.

2. Since gases remain in the atmosphere for different lengths of time (Table 1, fifth column), a time period over which the cumulative effect of the gas is to be measured must be specified. The Intergovernmental Panel on Climate Change (IPCC) developed three alternative sets of GWPs based on 20-, 100-, and 500-year time frames. Estimations of GWPs require considerable knowledge of atmospheric processes and are dependent on relative concentrations of gases. As knowledge increases and relative concentrations change, GWPs will be modified. At present, GWPs are considered accurate within +/- 35%.

3. Table 4 does not show emissions for the other three groups of gases covered by the Kyoto Protocol: PFCs, HFCs, and SF$_6$.

4. Annex I comprises all Annex B countries shown in Table 4 plus Turkey and Belarus.

5. Extensive records of the amount of fossil fuels used for energy are available and the correlations between fossil fuel consumption and resultant CO$_2$ emissions are well documented. This contrasts markedly from the situation for CH$_4$ and N$_2$O. Attempts to document sources and to establish correlations between activities and emissions for these gases are relatively recent. There are relatively little data, and models to correlate emissions with activities are limited. In general, emissions from controlled industrial processes are known to a much greater degree of accuracy than emissions from other sources. The majority of CH$_4$ and N$_2$O emissions are from landfills and agriculture rather than from controlled industrial processes, while the majority of CO$_2$ emissions are from industrial processes. The result is that, taken in total, CO$_2$ emissions estimates are accurate to within ± 10 percent for much of the world while emissions estimates for CH$_4$ and N$_2$O are considered to be accurate to within ± 50 percent.

6. By comparing the IPCC and UN projections, it can be seen that population projections from the IPCC A2 scenario are higher than the most recent "high" UN projections. The A2 scenario population projections are based on the International Institute for Applied Systems Analysis (IIASA) 1996 high population projections. Population projections are extremely sensitive to small changes in human fertility rates, particularly over the long term. In recent years fertility rates have declined more quickly than anticipated in a number of developing countries and countries in transition. These declines, coupled with mortality increases due to AIDS, have resulted in downward revisions of previous population projections. The population projections of the B2 scenario assume that world average fertility rates will stabilize roughly at replacement levels (approximately 2.1 children per woman) around 2050. The A1 and B1 scenarios represent fertility rates stabilizing somewhat below

replacement levels, but not at rates as low as the low UN projection that assumes an average fertility rate of approximately 1.6 children per woman.

7. U.S. forests cover 737 million acres or approximately 33 percent of U.S. land and contain some 78,000 MMTC, approximately 4 percent of the carbon stored in forests worldwide (USDA, 1992). Approximately half of the carbon stored in U.S. forests is in the soil. In addition, pastureland, rangeland, grassland, and agricultural soils also store and release carbon. Carbon stored by management of these lands is believed to amount to less than 10 percent of the carbon storage that can be stored by management of forests.

References

Birdsey, R. and L. Heath. 1995. Carbon Storage in U.S. Forests. In *Encyclopedia of Energy Technology and the Environment*. Wiley & Sons, Inc., New York, NY.

CDIAC (Carbon Dioxide Information and Analysis Center). 2000. *Atmospheric CO_2 Record from Mauna Loa*. Accessed at http://www.cdiac.esd.ornl.gov on March 19, 2001.

EDGAR. 2000. Dutch National Institute of Public Health and the Environment. *Emissions Database for Global Atmospheric Research*. Accessed at http://www.rivm.nl/env/int/coredata on April 17, 2001.

Houghton, J. 1997. *Global Warming: The Complete Briefing*. Cambridge University Press, Cambridge, UK.

Houghton, R.A. 1996. Land-Use Change and Terrestrial Carbon: the Temporal Record. In Apps, M.J., and D.T. Price, eds. *Forest Ecosystems, Forest Management and the Global Carbon Cycle*. NATO ASI Series I, Vol. 40: 117–134.

Houghton, R.A., and J.L. Hackler. 2000. Carbon Flux to the Atmosphere from Land-Use Changes. In *Trends: A Compendium of Data on Global Change*. Carbon Dioxide Information Analysis Center, Oak Ridge National Laboratory, U.S. Department of Energy, Oak Ridge, TN. Accessed at http://cdiac.esd.ornl.gov/trends/landuse/houghton on November 15, 2000.

IEA (International Energy Agency). 2000a. *CO_2 Emissions from Fossil Fuel Combustion, 1971–1997, 1999 Edition*. Organisation for Economic Co-operation and Development/IEA. Paris, France.

IEA. 2000b. *Key World Energy Statistics*. Accessed at http://www.iea.org/statist/keyworld/keystats.htm on April 17, 2001.

IPCC (Intergovernmental Panel on Climate Change). 1995. *Impacts, Adaptations and Mitigation of Climate Change: Scientific-Technical Analysis*. Contribution of Working Group II to the Second Assessment Report of the International Panel on Climate Change. Cambridge University Press, Cambridge, UK.

IPCC. 2000a. *Land Use, Land-Use Change, and Forestry*. Watson, R.T., I.R. Noble, B. Bolin, N.H. Ravindranath, D.J. Verardo, and D.J. Dokken, eds. A Special Report of the IPCC. Cambridge University Press, Cambridge, UK.

IPCC. 2000b. *Special Report on Emissions Scenarios*. Cambridge University Press, Cambridge, UK.

IPCC. 2001. *Climate Change 2001: The Scientific Basis*. Contribution of Working Group I to the Third Assessment Report of the Intergovernmental Panel on Climate Change. Houghton, J.T., Y. Ding, D.J. Griggs, M. Noguer, P. van der Linden, X. Dai, and K. Maskell, eds. Cambridge University Press, Cambridge, UK, in press.

Marland, G., T.A. Boden, and R.J. Andres. 2000. *National CO_2 Emissions from Fossil-Fuel Burning, Cement Manufacture, and Gas Flaring: 1751-1997*. Accessed at http://cdiac.esd.ornl.gov/ on December 10, 2000.

Marland, G., T.A. Boden, and R.J. Andres. 2001. *National CO_2 Emissions from Fossil-Fuel Burning, Cement Manufacture, and Gas Flaring: 1751-1997*. Accessed at http://cdiac.esd.ornl.gov/new on January 16, 2001.

Marland, G., T.A. Boden, R.J. Andres, A.L. Brenkert, and C.A. Johnson. 1999. Global, Regional and National Fossil Fuel CO_2 Emissions. In *Trends Online: A Compendium of Data on Global Change*. Accessed at http://cdiac.esd.ornl.gov/ on December 10, 2000.

Population Division of the Department of Economic and Social Affairs of the United Nations Secretariat. 1999. *World Population Prospects: The 1998 Revision*. Vol.1: Comprehensive Tables.

Rekacewicz, P. 2000. *The Vital Climate Graphic: the Impacts of Climate Change*. UNEP-Grid Arendal, Norway. Accessed at http://www.grida.no/climate/vital/intro.htm on April 17, 2001.

UNFCCC (United Nations Framework Convention on Climate Change). 1998. *The Kyoto Protocol to the Convention on Climate Change*. Climate Change Secretariat. Bonn, Germany.

U.S. Census Bureau. 2000. *National Population Projections*. Accessed at http://www.census.gov/population/www/projections/natproj.html on December 18, 2000.

USDA (U.S. Department of Agriculture). 1992. *Carbon Storage and Accumulation in United States Forest Ecosystems*. General Technical Report WO-59.

USDOE (U.S. Department of Energy). 1999. Energy Information Agency. *Annual Energy Outlook 2000.* DOE/EIA-0383. Accessed at www.eia.doe.gov/oiaf/aeo/index.html on December 18, 2000.

USDOE. 2000a. *Annual Energy Review.* Accessed at http://www.eia.doe.gov/emeu/aer/contents.html on December 18, 2000.

USDOE. 2000b. *Energy in the United States: A Brief History and Current Trends.* Accessed at http://www.eia.doe.gov/emeu/aer/eh1999/eh1999.html on April 17, 2001.

USDOE. 2000c. *International Energy Information at a Glance.* Accessed at http://www.eia.doe.gov/international on December 10, 2000.

USEPA (U.S. Environmental Protection Agency). 2000a. *Emissions and Projections of Non-CO$_2$ Greenhouse Gases for Developed Countries: 1990-2010.* Draft Report, May 2000.

USEPA. 2000b. *Inventory of U.S. Greenhouse Gas Emissions and Sinks: 1990-1998.* EPA 236-R-00-001. Accessed at http://www.epa.gov/globalwarming/publications/emissions on April 17, 2001.

USEPA. 2001. *Draft Inventory of U.S. Greenhouse Gas Emissions and Sinks: 1990-1999.* Accessed at http://www.epa.gov/globalwarming/publications/emissions/US2001/index.html on April 17, 2001.

INDEX

A

Abandonment of coastal property, 57-58
Abatement costs defined, 250
ABB, 272-73
Accommodation of sea-level change, 43, 53-55
Accountability in emissions trading, 265
Acid rain, 286-78, 370
Adaptation to change
  coastal areas, 43-44, 55, 57-59
  farm sector strategies, 25, 30, 32-38
  water resources, 75-78
Aerosols. See Sulfate aerosols
Afforestation, 39, 139-40, 194, 247
Africa, 33, 56. See also Developing nations
Aggregate economic analysis, 184-86
Agriculture, U.S., 25-42, 368-69, 384
Air pollution and air quality. See Emissions;
  Greenhouse gases
Allocation of emissions trading permits, 263
Allowances, emissions. See Emissions trading;
  International emissions trading
Alternative fuels. See Energy technologies and energy
  efficiency
American Council for an Energy Efficient Economy
  (ACEEE), 288
American Electric Power, 339, 354
American Water Works Association (AWWA), 76-77
Annex I emissions trading scenario, 198-99, 248-68
Antarctic ice sheet, 16, 44, 52
Antarctica, 7-8
Anthropogenic change, 6-11, 45, 373-88
Anthropogenic forcing, 8-10, 375-77
Appalachia, 29, 31
Applied Materials, 309
Argentina, 148-51, 159-62, 173-74
Austin, Texas, 283-84, 308-19
Austria, 88, 106-9
Automotive industry, 235, 273, 367-68, 370
Autonomous energy efficiency improvement (AEEI),
  204-5, 217, 220, 222-23
Averaging of GHG emissions, 197
Averting-behavior estimates, 179

B

Backstop technologies, 218-20
Banking of emissions permits, 248, 262
Baselines/base case assumptions of economic models,
  182, 193-97, 207-8
Baxter International, 339
Beach nourishment, 54, 58
Belgium, 89-93
Bell Atlantic, 237
Ben and Jerry's, 272
Benefit/cost analysis. See Cost/benefit analysis of
  policy options
Bequest value, 179
Berlin mandate, 363
Best management practices, 284
Biomass, 8, 39-40, 367-69
Bottom-up models, 216-21, 224, 231, 255
BP
  emissions inventory, 339, 347
  environmental leadership, 117, 273-74, 277, 284
Brave New World school, 228-29
Brazil, 27, 148-52, 159-64, 174
Bubble policy, EU, 91-92, 100, 109

Business community environmental leadership,
  284-85, 300-1
Business sector risks and opportunities, 269-79
Byrd-Hagel resolution, 362-63

C

California
  coastal resources, 45, 49
  registry of GHG emissions cuts, 282
  water resources, 67, 73-75
Canada, 27, 28-29, 66
Canadian Global Climate Model (CGCM), 65-67,
  73, 75
Canadian Recursive Trade Model (CRTM), 219
Cape Cod, 48, 67, 75
Capital stock dynamics in new economy, 237-42
Carbon cycle feedbacks, 13
Carbon dioxide (CO_2)
  cost/benefit analysis of policy options, 182-83, 188,
    196-97
  EU emissions reduction efforts, 88-115
  farm sector issues, 27-28, 30-32
  global emissions data, 374-84
  Japanese emissions forecasts, 132-47
  science of change, 7-8, 13, 15, 36-40, 373-77
  state and local policies and programs, 282-87,
    292-306
  U.S. emissions data, 384-92
  U.S. legislative initiatives, 366-67
  water resources issues, 65, 70, 78
Carbon dioxide equivalents (CO_2E), 376, 389
Carbon Emissions Trajectory Assessment (CETA)
  model, 219
Carbon emissions. See also specific types of emissions
  developing nations, 149-50
  projections, 196-97, 201-3
  technological change effects, 210, 212, 214-15
  trading. See Emissions trading; International emis-
    sions trading
Carbon management, 182
Carbon monoxide (CO), 8, 13, 286, 312
Carbon price forecasts, 194, 198-99, 201-3
Carbon reservoirs, 375
Carbon sequestration
  economic analyses, 182
  emissions inventories, 347
  farm sector, 39-40
  international emissions trading, 247
  regulatory incentives, 274, 277
  state and local policies and programs, 284
  technological change effects, 215
  U.S. legislative initiatives, 365, 368-69
Carbon sinks
  economic models, 197
  emissions inventories, 347
  international emissions trading, 259-60
  Japan, 139
  science of change, 8
Carbon tax. See Carbon price forecasts; Tax policy
  options
Carbon trading. See Emissions trading; International
  emissions trading
Causes and effects of change
  economic context, 177-92
  farm sector, 25-42
  science of change, 1-24, 373-92

United States, 123, 370
Endogenous technological change, 204-5, 217-24Energy Modeling Forum (EMF), 196, 198, 219-22, 237
Energy services management outsourcing, 235-37
Energy Star microprocessors, 285
Energy technologies and energy efficiency. *See also* Autonomous energy efficiency improvement (AEEI)
   Argentina. *See* Argentina
   Austria, 106-8
   Brazil. *See* Brazil
   China. *See* China
   corporate viewpoint, 272-74
   developing nations, 148-75
   EU policies and programs, 93
   fossil fuels. *See* Fossil fuels
   Germany, 96-99
   India. *See* India
   Japan, 132-47
   Korea. *See* Korea
   Netherlands, 103-6
   new economy scenario, 227-44
   Spain, 109-11
   state and local policies and programs, 282-86, 296-306, 308-16
   technological change, 209-11, 214-24
   United Kingdom, 100-2
   United States, 121-22
Energy use and production data, 381-84, 387-89
Enforceability of emissions trading, 265
Enhanced greenhouse effect, 7
Entergy, 348
Environmental impacts in economic models, 205-6
Environmental impacts of change
   coastal resources/sea-level change, 43-44, 48, 51-59
   corporate risk management, 270-71, 275-76
   farm sector, 28, 32, 34
   summary of, 1-5
   water resources, 63-81
Equity, international and intergenerational, 186-87
ERB model, 219
Erosion, coastal, 43, 48, 51-59
Erosion, soil, 28, 32
ETA-Macro model, 210
European Union, xi, 88-115
Evaporation and evapotranspiration, 7, 33, 63-65
Existence value, 179
Exogenous technological change, 204-5, 217-24
Extinction, 69
Extremes of temperature or precipitation
   corporate risk management, 270-71, 275-76
   farm sector effects, 26-29, 31, 34-35
   science of change, 6, 12, 16-17, 20
   water resources effects, 72-74

F

Farm sector, 25-42
Fertilization effect, CO_2, 27-28, 30, 32
Fertilizers, 25-26, 35, 39
Financial impacts of global emissions trading, 263-64
Fish populations, 69
Flood insurance, 55
Floods. *See also* Extremes of temperature or precipitation
   coastal resources impact, 50-52, 55-56
   farm sector impact, 28, 34-35
   water resources effects, 73

Florida, 48, 53, 67, 75
Food production, U.S., 25, 29-32
Food production and distribution, global, 33
Forcing, solar and anthropogenic, 8-10, 13-16, 18, 20, 375-76
Ford, 235, 273
Forecasts. *See* Computer models; Economic models; Projections
Ft. Lauderdale, 48
FOSSIL-2 model, 219
Fossil fuels
   economic analysis and modeling, 188, 195, 199-200
   EU coal-to-gas transition, 89-91
   farm sector, 39
   historical emissions data, 379-81
   science of change, 7-8
   state and local policies and programs, 285-86
   technological change, 211, 214-15
Framework Convention. *See* United Nations Framework Convention on Climate Change (UNFCCC)
France, 89-93, 211-12

G

G-Cubed model, 255-56, 259
General circulation models (GCMs)
   farm sector analysis, 31, 37, 40
   science of change, 14
   water resources, 65-66
General equilibrium, 181-82, 231, 237, 256
   Dynamic General Equilibrium Model (DGEM), 219
   General Equilibrium Environmental Model (OECD-GREEN), 219, 255-56
General Motors, 273
Georgia, 48, 286
Germany, 84, 88-99, 111-12
GHGs. *See* Greenhouse gases
Glaciers, 16, 44, 52-53, 60
Global 2100 model, 219
Global climate data, 373-84
Global-mean temperature change. *See* Temperature change
Global Trade and Environment Model (GTEM-ABARE), 255-56
Global warming potential (GWP), 375-76
Goulder and Colleagues model, 219
Great Plains, 32, 66
Green political parties, 89, 98
Greenhouse gases. *See also specific gases*
   Clean Air Act, regulation under, 122-23, 370
   emissions inventory estimates and reporting, 331-55
   emissions trading. *See* Emissions trading; International emissions trading
   EU policies, programs, and proposals, xi, 88-115
   global emissions data, 374-84
   Kyoto pact. *See* Kyoto Protocol
   overview of nations' policies and programs, xi, 83-87
   projected emissions under economic models, 193-208
   registries of emissions cuts, state, 281-83
   science of change, 6-17, 373-92
   state and local policies and programs, xi, 281-319
   U.S. emissions data, 384-92
Greenland ice sheet, 16, 44
Groundwater extraction, 45
Groundwater resources, 63-78
GTEM-ABARE model, 255-56
Gulf Coast, 43, 45-53

Maine, 49, 54
Marginal emissions abatement costs defined,
   246-47, 250
MARKAL-MACRO model, 157, 219
Markers, 12
Market-based policy instruments, 181-84, 188, 365.
   See also Emissions trading
Market penetration studies, 210-12, 220
Maryland, 48, 283
Massachusetts, 48, 67, 75, 283
Massachusetts Institute of Technology (MIT), 212
Mercury, 287, 292, 294, 367
MERGE 3 model, 255-56
MESSAGE model, 219
Methane (CH_4)
   Austin, Texas programs, 308, 312, 316-18
   economic models, 196
   EU reduction target, 91
   farm sector, 39
   global emissions data, 374-78
   ice studies, 60
   science of change, 7-8, 13
   U.S. emissions data, 384-85
Miami, 48, 52
Mid-Atlantic states, 43, 45-48
Mississippi River and delta, 34, 45, 52, 72
MIT-EPPA model, 255-59
Mitigation costs defined, 250
Mitigation of change
   coastal areas, 44, 50, 59
   coastal resources, 44, 50, 59
   cost/benefit analysis of policy options, 177-92
   economic models, 187-88, 193-208
   farm sector, 25-26, 39-40
Model for Evaluating Regional and Global Effects of
   GHG Reduction Policies (MERGE 3), 255-56
Models. *See* Computer models; Economic models;
   specific models
Monitoring of global emissions trading, 264-65
Monsanto, 117
Montreal Protocol, 8, 22, 123, 336-37
Motorola, 309

N

National Assessment, U.S., 32
Natural gas. *See* Fossil fuels
Natural greenhouse effect, 7
Negative costs of environmental regulation, 181
Netherlands, 88, 103-6, 111, 219-20
New economy, 227-44
New England, 43
New Hampshire, 281-91
New Jersey, 45, 55, 75
New Jersey Sustainability Greenhouse Gas Action Plan,
   283, 292-302
Niagara Mohawk, 339
Nitrogen oxides (NO_x)
   EU reduction target, 91
   science of change, 8, 13
   state and local policies and programs, 286-87, 292,
      294, 298, 312
   U.S. legislative initiatives, 367
Nitrous oxide (N_2O)
   economic models, 196
   farm sector, 39
   global emissions data, 374-78

science of change, 7-8
   U.S. emissions data, 384-85
No-trade emissions reduction scenario, 248-59
Northeast States for Coordinated Air Use Management
   (NESCAUM), 283
Northern Plains region, 29

O

Observed temperature change, 9-11
Ocean thermohaline circulation, 13
OECD-GREEN model, 219-20, 255-56
Offsets, CO_2, 304
Oil burning. *See* Fossil fuels
Oil extraction, 45
Opportunity costs, 180-82
Option value, 179
Oregon/Oregon Climate Trust, 283, 292, 302-7
Outsourcing of energy services management, 235-37
Ozone (O_3), stratospheric, 11, 273-74
Ozone (O_3), tropospheric, 8, 13, 28

P

Partial equilibrium costs, 182
Partnerships, public-private, 326-30
Patagonia, 272
Perfluorocarbons (PFCs), 196, 323-24, 337, 376
Permit trading systems. *See* Emissions trading;
   International emissions trading
Pesticides, 26, 34
Pests, agricultural, 28-29, 35
Petroleum extraction, 45
Planned retreat from coastal areas, 43, 53-55
Polar ice melt, 44
Policy options
   adaptation. *See* Adaptation to change
   coastal resources, 43
   command-and-control instruments, 188, 262, 278
   cost/benefit analysis. *See* Cost/benefit analysis of
      policy options
   economic analysis, 177-92
   economic models, 187-88, 193-208
   emissions trading. *See* Emissions trading;
      International emissions trading
   European Union, 89-113
   Japan, 132-47
   market-based instruments, 181-84, 188
   mitigation. *See* Mitigation of change
   new economy implications, 227-44
   overview of nations' policies and programs, 83-87
   sea-level change, 43, 53-55
   state and local policies and programs, 281-319
   taxation. *See* Tax policy options
   United States, 116-31, 215
Population growth
   coastal resources, 48-49
   economic models, 195
   projections, 40, 381-84, 387
   world hunger and, 33
Post-disaster planning, 54-55
Power plants
   developing nations electric power futures, 148-75
   greenhouse gas emissions, 367, 377, 384. *See also*
      Greenhouse gases; *specific GHGs*
   siting, Oregon law, 302-7
   technological advances. *See* Energy technologies and
      energy efficiency

Printed in Bruges, Belgium by Tiger Publishing Systems